The birth of history and philosophy of science

Kepler's *A Defence of Tycho against Ursus*
with essays on its provenance
and significance

The birth of history and philosophy of science

Kepler's *A Defence of Tycho against Ursus* with essays on its provenance and significance

N. JARDINE

READER IN THE HISTORY AND PHILOSOPHY OF SCIENCE,
UNIVERSITY OF CAMBRIDGE,
AND FELLOW OF DARWIN COLLEGE

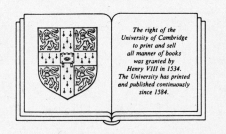

The right of the
University of Cambridge
to print and sell
all manner of books
was granted by
Henry VIII in 1534.
The University has printed
and published continuously
since 1584.

CAMBRIDGE UNIVERSITY PRESS

CAMBRIDGE

NEW YORK   NEW ROCHELLE

MELBOURNE   SYDNEY

Published by the Press Syndicate of the University of Cambridge
The Pitt Building, Trumpington Street, Cambridge CB2 1RP
32 East 57th Street, New York, NY 10022, USA
10 Stamford Road, Oakleigh, Melbourne 3166, Australia

First published 1984
First paperback edition with corrections 1988

Printed in Great Britain by the University Press, Cambridge

Library of Congress catalogue card number: 83–14186

British Library Cataloguing in Publication Data
Jardine, N.
The birth of history and philosophy of science:
Kepler's *A Defence of Tycho against Ursus* with
essays on its provenance and significance.
1. Kepler, Johannes 2. Astronomy – History –
17th century
I. Title II. Kepler, Johannes. *Apologia
pro Tychone contra Ursum*
521'.092'4   QB36.K/

ISBN 0 521 25226 1 hard covers
ISBN 0 521 34699 1 paperback

FOR DANIEL

Contents

Acknowledgements

My most outstanding debts are to Mr Gerd Buchdahl, who first excited my interest in Kepler as a philosopher of science, and to Professor Owen Gingerich, without whose constant encouragement and advice over several years this work would never have been completed. I thank Dr Penny Bulloch, Dr Elisabeth Leedham-Green and Professor Anthony Grafton for advice on points of translation, and Dr Timothy Beardsworth, who generously allowed me to consult his draft translation of a part of Kepler's *A Defence of Tycho against Ursus*. Dr Martha List and Professor Robert Westman kindly obtained for me a microfilm of the manuscript of the work from the Württembergische Landesbibliothek, Stuttgart; and the former has advised me on the state and pagination of the manuscript. The number of those to whom I am indebted for helpful advice is large. I thank especially Dr Charles Schmitt, Professor Robert Westman, Dr Geoffrey Lloyd, Dr Lisa Jardine, Dr Peter Burke, Dr Michael Hoskin, Dr Alistair Crombie, Dr Andrew Cunningham, Dr Antonio Pérez-Ramos and an anonymous referee with a sharp eye for errors. I offer my thanks for helpful responses to enquiries to the Bayerische Staatsbibliothek, Munich, the Universitätsbibliothek, Heidelberg, and the Nationalbibliothek, Vienna. I am grateful for much kind and patient help to the staffs of the British Library, the Bodleian and, above all, the Rare Books Room of Cambridge University Library. Finally, I am indebted to Ms Jenny Woodhouse for deciphering and typing several drafts of a chaotic manuscript and to Mr Jonathan Sinclair-Wilson and Ms Doreen Jones for their patience and help in seeing this book to press.

In making corrections for this second impression I have profited from the late Professor Edward Rosen's *Three Imperial Mathematicians* (New York, 1986). Rosen's book gives a fuller account of the events retailed in Chapter 1 of the present work and provides a very different perspective on them. The conclusions of Chapter 7 about scepticism in sixteenth-century astronomy are elaborated and qualified in my 'Scepticism in renaissance astronomy', in R. H. Popkin and C. B. Schmitt, eds., *Scepticism from the Renaissance to the Enlightenment* (Wolfenbüttel, 1987). In the same chapter I indicate the importance and problematic nature of renaissance views on the government of celestial motions by 'laws' (*leges*); this issue is illuminated by J. E. Ruby's fine paper 'The origins of scientific "law"', *Journal of the History of Ideas*, 47 (1986), 341–59. I am indebted for notice of errors to Edward Rosen's review in *Journal of the History of Ideas*, 46 (1985), 453–4, and to a personal communication from Volker Bialas following his review in *Journal of the History of Astronomy*, 17 (1986), 140–3. The latter review contains Dr Bialas' welcome announcement of his forthcoming edition of Kepler's *Defence of Tycho* and related documents, to be published in *Kepler, gesammelte Werke*, xx.

Abbreviations of works frequently cited

Apologia	Johannes Kepler, *Apologia pro Tychone contra Ursum*. Page references are to the translation in Part II of this book.
Epist. astron.	Tycho Brahe, *Epistolarum astronomicarum liber primus* (Uraniborg, 1596).
K.g.W.	W. von Dyck and M. Caspar, eds., *Johannes Kepler, gesammelte Werke* (Munich, 1938–).
K.o.o.	C. Frisch, ed., *Joannis Kepleri astronomi opera omnia* (Frankfurt and Erlangen, 1858–71).
M.o.o.	C. G. Bretschneider, ed., *Philippi Melanthonis opera quae supersunt omnia. Corpus Reformatorum, I–XXVIII* (Halle, 1834–52; Braunschweig, 1853–60).
T.B.o.o.	J. L. E. Dreyer, ed., *Tychonis Brahe Dani opera omnia* (Copenhagen, 1913–29).
Tractatus	Nicolaus Ursus, *De hypothesibus astronomicis tractatus* (Prague, 1597).

Introduction

Kepler's *A Defence of Tycho against Ursus* (*Apologia pro Tychone contra Ursum*), written around Christmas 1600, offers much more than the title suggests. To counter Ursus' denial of the capacity of astronomers to 'portray the form of the world' Kepler provides an elaborate defence of astronomy against a variety of cogent sceptical ploys. And to refute the history of astronomical hypotheses by which Ursus had sought to discredit Tycho Brahe's claims to originality, Kepler provides a selective history of world-systems from Zoroaster and Pythagoras to Copernicus and Brahe. The rebuttal of scepticism in astronomy is a *tour de force*, efficient and witty in its refutation of Ursus' arguments and striking in its apparent anticipations of strategies used in the present century in defence of scientific realism. Parts of his history of astronomy, notably his examination of Apollonius of Perga's models for the second anomaly and his attempts to trace a classical pedigree for geoheliocentric cosmology back from Martianus Capella to Plato, remain of interest to present-day historians of the subject. Further, the work is of considerable importance for Keplerian scholarship. Kepler is unusual amongst the major mathematical scientists of his period in the depth of his engagement in epistemological and methodological issues. All his major astronomical works are enlivened by such philosophical reflections. But the *Apologia* is his only extended treatment of these topics. As such it provides many keys to the understanding of the theory and practice of inquiry that inform both his earlier *Mysterium cosmographicum* and his mature masterpieces, *Astronomia nova*, *Harmonice mundi* and *Epitome astronomiae Copernicanae*.

The *Apologia* is a work of great originality not only in its theses, but also in its concerns. Kepler focusses throughout the work on theoretical progress in the history of astronomy and the means whereby such progress may be achieved. Concern with theoretical progress is prevalent at every level, historical and philosophical, academic and popular, of modern reflection on the natural sciences. Yet in this respect the *Apologia* appears to be without substantial precedent in sixteenth-century writing on the epistemology and history of the mathematical arts. Claims about the origins of genres and disciplines inevitably oversimplify complex processes and are vulnerable to the discovery of earlier documents. Nevertheless I conjecture that if any one work can be taken to mark the birth of history and philosophy of science as a distinctive mode of reflection on the status of natural science it is Kepler's *Apologia*.

The *Apologia* remained unpublished until 1858.[1] Since then it has received only a modicum of scholarly attention. An early reaction was that of Karl von Prantl, who praised the first chapter for its exposition of a sober inductive methodology in sharp contrast to the 'fantastic' theological and metaphysical speculations of the *Mysterium* and *Harmonice*.[2] Ernst Cassirer, a far more sensitive reader, saw in the work a plea for the grounding of astronomical hypotheses in physics and metaphysics of the kind which Kepler had attempted in his *Mysterium* and was to elaborate in his later works.[3] And Pierre Duhem in his influential essay *To Save the Phenomena* identified the clash between Ursus and Kepler as an instance of a long-standing confrontation between sound instrumentalist and unsound realist conceptions of the mathematical sciences.[4] Amongst the few more recent appreciations of the epistemology of the work, those of Ralph Blake, Jürgen Mittelstrass, Bob Westman, Eric Aiton and the author have all, following Cassirer rather than von Prantl, found in the *Apologia* a methodological stance which illuminates the practice of Kepler's major astronomical works; and they have variously emphasised the importance and originality of Kepler's views on the status of astronomy and on the means whereby theoretical disputes in astronomy may be resolved.[5] Reactions to Kepler's history of astronomy have tended to concentrate on isolated passages which cast light on issues in Copernican

[1] C. Frisch, ed., *Joannis Kepleri astronomi opera omnia*, i (Frankfurt and Erlangen, 1858–71), I, 236–76. Hereafter cited as *K.o.o.*

[2] K. von Prantl, 'Galilei und Kepler als Logiker', *Sitzungsberichte der philosophische-philologischen und historischen Klasse der k.b. Akademie der Wissenschaften zu München* (1875), 394–408.

[3] E. Cassirer, *Das Erkenntnisproblem in der Philosophie und Wissenschaft der neueren Zeit*, I (Berlin, 1906), 262–4.

[4] P. Duhem, ΣΩΖΕΙΝ ΤΑ ΦΑΙΝΟΜΕΝΑ [*To Save the Phenomena*] (Paris, 1908); transl. by E. Doland and C. Maschler (Chicago, 1969).

[5] R. M. Blake, 'Theory of hypothesis among Renaissance astronomers', in E. H. Madden, ed., *Theories of Scientific Method: the Renaissance through the Nineteenth Century* (Seattle, 1960), 22–49; J. Mittelstrass, 'Methodological aspects of Keplerian astronomy', *Studies in the History and Philosophy of Science*, 3 (1972), 213–32; R. S. Westman, 'Kepler's theory of hypothesis and the "realist dilemma"', *Studies in the History and Philosophy of Science*, 3 (1972), 233–64; E. Aiton, 'Johannes Kepler and the astronomy without hypotheses', *Japanese Studies in the History of Science*, 14 (1975), 49–71; N. Jardine, 'The forging of modern realism: Clavius and Kepler against the sceptics', *Studies in the History and Philosophy of Science*, 10 (1979), 141–73.

scholarship,[6] though recently Christine Schofield has discussed Kepler's treatment of Ursus' claim that Tycho's hypotheses had been anticipated by Apollonius of Perga, and Bruce Eastwood has offered a detailed exposition and assessment of Kepler's attempts in the last chapter of the *Apologia* to recover geoheliocentric doctrines from texts of Vitruvius, Pliny, Macrobius and Martianus Capella.[7] I know of no systematic exposition or interpretation of the work as a whole.

The lack of modern appreciation of the *Apologia* is hardly surprising. It has remained untranslated. And in addition to the usual difficulties posed by Kepler's elliptical and often contorted Latin there are many further barriers to comprehension. The work is incomplete and parts of it are evidently not fully prepared for publication. The chaotic manuscript poses many problems for the editor, and Frisch's edition is occasionally inaccurate, is in places incomplete, and so revises Kepler's punctuation as to make parts of the text very hard to follow. Kepler adopted what may appear to the modern reader as a highly artificial mode of composition, apt to obscure the main lines of argument, presenting his case as a formal Ciceronian judicial oration. And last but not least, because the work remained for so long unpublished there is lacking that type of accessibility which may accrue to works of the distant past through a continuous history of interpretation and emulation. Despite these barriers to comprehension the work is by no means irretrievable. Its central themes, theoretical progress and the resolution of theoretical dispute, remain central themes in modern history and philosophy of science. Apart from its frequent references to the contents of Ursus' published attack on Tycho, *De hypothesibus astronomicis tractatus*,[8] it is relatively self-contained; no immersion in an unfamiliar genre is needed to grasp the point of Kepler's arguments. And whilst with the decline of classical education conversancy with its literary form, the Ciceronian oration, is no longer commonplace,

---

6 In particular Kepler's attribution to Andreas Osiander in Ch. 1 of the *Apologia* of the anonymous first preface to Copernicus' *De revolutionibus* is often cited; N. R. Swerdlow, '*Pseudodoxia Copernicana*: or, enquiries into very many received tenents and commonly presumed truths, mostly concerning spheres', *Archives internationales d'histoire des sciences*, 26 (1976), 136–7, makes ingenious use of a passage from a draft of the *Apologia* (given in Ch. 3 of this work) in support of his conjectures about the role of solid orbs in the genesis of the Copernican system.

7 C. J. Schofield, 'The geoheliocentric planetary system: its development and influence in the late sixteenth and seventeenth centuries' (Ph.D. thesis, Cambridge, 1964; New York, 1981), 119–33; B. Eastwood, 'Kepler as historian of science: precursors of Copernican heliocentrism according to *De revolutionibus*, I, 10', *Proceedings of the American Philosophical Society*, 126 (1982), 367–94.

8 Prague, 1597. Hereafter cited as *Tractatus*.

the influence of that form on European literature has been so pervasive as to create a measure of tacit familiarity.

It is in the hope of rendering the *Apologia* accessible to the general reader that I offer my translation and the accompanying notes and essays on its provenance and significance. The first part of this book is concerned with the provenance of the *Apologia*. The second part consists of an edition and annotated translation. The third offers interpretative essays on the significance of Kepler's history and philosophy of astronomy. Though the highly complex background to the composition of the *Apologia* is important for a full understanding of the work, it involves much that is likely to be of interest only to specialists. The general reader may prefer to read first the *Apologia* itself and then the interpretative essays, referring back to the opening chapters for relevant information where necessary. To facilitate this way of reading the book I have given copious cross-references.

The plan of the book as a whole is as follows. The opening chapter retails the complex and, for Kepler, embarrassing circumstances which allowed Tycho to inveigle him into reluctant authorship. These circumstances turn out to be important for an understanding of the work since they help to explain Kepler's choice of themes and certain puzzling aspects of the way he handles them. The second chapter provides a survey of the contents of Ursus' *Tractatus* together with a translation of the section of the work on which Kepler's refutation is focussed. In the third chapter translations of Kepler's three earlier assessments of the *Tractatus* are given. These provide evidence of Kepler's real attitude to the controversy between Ursus and Tycho, and they give a reliable indication of the lines along which he would have completed the *Apologia*. The fourth chapter sketches the somewhat devious strategy Kepler adopts in pursuit of his primary aim in the *Apologia*, the defence of Tycho's claim to originality in the formulation of astronomical hypotheses. It offers also an analysis of the structure of the work as a Ciceronian judicial oration, and a survey of the dialectical and rhetorical devices proper to the genre which Kepler employs. After these preliminaries there follow the text and annotated translation.

In the four final essays I attempt to establish the intrinsic interest and historical significance of the epistemology and history of astronomy which Kepler offers in the *Apologia*. The first of these essays, 'Against the Sceptics', is primarily exegetical. In it I offer a detailed reconstruction of Kepler's rebuttals of the various arguments for scepticism in astronomy adduced by Ursus. In the second, 'The Status of Astronomy', I argue that the *Apologia* is not to be considered merely as a defence of an established position. When set against earlier sixteenth-century pronouncements on the nature of astronomy and its relation to other disciplines, Kepler's vision of an astronomy which both derives support from and actively contributes to natural philosophy is seen to be profoundly original, being opposed not only to the pragmatism of his avowed opponents Osiander and Ursus, but also to the essentially conservative stances of Copernicus, Brahe and his own teacher Maestlin. Both the epistemology of astronomy of the first chapter of the *Apologia* and the history of astronomy of the subsequent chapters can, I argue, be seen as serving, besides the overt purposes of discrediting Ursus' scepticism and exposing the inadequacies of his historical scholarship, the deeper purpose of validating Kepler's vision of a new role for astronomy. In the third essay, 'Historiography and Validation', I contrast Kepler's history of astronomy with earlier histories of astronomy and the mathematical arts, and show that his treatment shows a quite remarkable advance in historiographical sophistication. The historical sense that Kepler displays in his interpretation of the sources and his concern with theoretical change and the means whereby it has come about mark the *Apologia* as an unprecedented venture into the genre of history and philosophy of science. The essay ends with some speculations on the historical conditions that made it possible to compose such a work. In the final essay I single out certain themes of the *Apologia* that are seminal both with respect to Kepler's mature works and with respect to later philosophy and historiography of science, and I reflect tentatively on the implications of the historical context in which they emerged.

No pretence at an exhaustive interpretation and assessment of the *Apologia* is made in this book. For example, though passages from Kepler's later works are often used to cast light on passages from the *Apologia*, I make no systematic attempt to substantiate my claim that the *Apologia* is crucial for an understanding of the methodology of Kepler's mature works. And though an assessment of certain general historiographic themes is offered, no attempt is made to supplement the existing specialised studies of Schofield and Eastwood so as to provide a full scholarly assessment of Kepler as historian of astronomy. Had the *Apologia* been published in Kepler's lifetime it would now, I conjecture, be a classic on a par with such seminal reflections on the nature of human inquiry as the *Novum organum* and the *Discourse on Method*. These selective and partial essays will have succeeded if they stimulate others to offer more comprehensive and more penetrating philosophical and historical assessments of the work. Thus it may become a classic.

I

The provenance of the *Apologia*

1

The circumstances of composition

'From June to October I was travelling and took my family. From October 1600 to August 1601 I was gripped by quartan fever. Meanwhile to Tycho's delight I wrote against Ursus and helped him with my efforts in other studies at his will. To Tycho's annoyance I investigated Venus, Mercury and the moon...Moreover, travelling from April to the end of August I went to Styria, leaving my wife in Prague. From September to July 1602 I devoted my effort to children and made a most beautiful little daughter. From September, I say, I began to seek most laboriously the extent of the second eccentricity of the sun; in the course of this labour Tycho died'.[1] This account given to Christen Longberg (Longomontanus) in 1605 and confirmed by circumstantial evidence indicates that the *Apologia* was composed between October 1600, when Kepler and his family first took up residence in Prague, and April 1601, when he returned to Graz to wind up his wife's financial affairs.[2] It was a wretched period. He and his wife were in ill-health; he was in financial straits, resentfully dependent on Tycho, and without definite prospects of salaried employment; and Tycho was proving unexpectedly reluctant to share his coveted astronomical observations. Further, the composition of the *Apologia* was a thoroughly unwelcome duty. He had, as we shall see, long before urged Tycho to consider Ursus' attack unworthy of reply, and to his patron Herwart von Hohenburg he had protested at the unseemliness of the kind of concern with priority which had led Tycho and Ursus into public controversy and mutual libel.[3] In the preface to the work

[1] W. von Dyck and M. Caspar, eds., *Johannes Kepler, gesammelte Werke*, (Munich, 1938–), xv, 139. Hereafter cited as *K.g.W.*

[2] Further evidence is provided by the references in the *Apologia* to his meeting with Ursus in January 1600 and to Gilbert's *De magnete* of 1600, and by the account of the work given in his letter of February 1601 to Maestlin, cited below. The question of the date of composition is considered by E. Rosen, 'Kepler's defense of Tycho against Ursus', *Popular Astronomy*, 54 (1946), 408–11.

[3] In a letter of March 1598 to Herwart von Hohenburg Kepler attributes geoheliocentric systems to Ursus, Tycho and (wrongly) Magini. Such systems are, he says, merely mutations of the system of Aristarchus of Samos and Copernicus. He adds: 'I am, indeed, of the opinion that since astronomers are priests of almighty God

itself the way in which he exonerates himself from the charge of impropriety
in attacking one recently deceased, and the care he takes to associate Tycho
with the plan for such an attack, suggest that he had in fact grave doubts
about the propriety of the enterprise.[4] That he found the work an unwelcome
distraction from astronomy is intimated both by the terms in which he
describes it to Longomontanus and by the account he offers in a bitterly
resentful letter of February 1601 to his teacher, Michael Maestlin.[5] Here he
complains of Tycho's miserliness over his observations and notes that the only
writing he has undertaken during his illness has been a refutation of Ursus
which has turned out to be 'philological rather than mathematical'. That
he found the work onerous is indicated both by the state of the manuscript,
with its massive and chaotic deletions, revisions and insertions, and by a barely
legible marginal jotting: '...the same perhaps below with hard work and
good luck. But it is not easy.'[6] How did Kepler come to have this unwelcome
task foisted on him?

The story starts with a letter. In November 1595 Kepler wrote at the
instigation of the courtier Sigismund Waganer to the Imperial Mathematician,
Nicolai Reymers Baer (Nicolaus Raimarus Ursus).[7] In the letter Kepler
sketches the construction which he was to elaborate in his *Mysterium
cosmographicum*, an embedding of Copernican planetary circuits in the nested
Platonic solids, and humbly seeks Ursus' opinion of it. The opening courtesies
are fulsome, understandably, as Kepler was later to explain, given Ursus' high
office and his own youth and ambition. They include the fatal words: 'I
admire your hypotheses.' The hypotheses to which Kepler refers are to be
found in Ursus' single major work, *Fundamentum astronomicum*, published at
Strasbourg in 1588. This book, on the strength of which Kepler describes
Ursus as his teacher, is primarily devoted to geometrical and trigonometrical
methods of use in practical astronomy. It is highly competent, and is of great
interest to historians of mathematics for its accounts, largely derived from
Jost Bürgi and Paul Wittich, of prosthaphaeresis – that is, methods of
obtaining products and dividends of trigonometrical functions by addition
(*prosthesis*) and subtraction (*aphaeresis*).[8] In addition to its technical mathematical contents the work includes in the last chapter an account of a 'new'
world-system, differing from the world-system Tycho had presented in his
De mundi aetherei recentioribus phaenomenis of 1588 only in its admission of

with respect to the Book of Nature, we should concern ourselves not with praise
of our cleverness but with God's glory' (*K.g.W.*, XIII, 193).

[4] *Apologia*, p. 136.   [5] *K.g.W.*, XIV, 165.   [6] See Ch. 5, fn. 16.
[7] The letter, of which Kepler kept no copy, is known only from Ursus' *Tractatus*,
sig. Di, r–v, translated in Ch. 2.   [8] See Ch. 2, pp. 31–2.

diurnal rotation of the earth, in its placement of the orb of Mars outside that of the sun rather than intersecting it, and in its suggestion of variable distances for the fixed stars. Ursus provides the design of a planetarium representing his new system, and offers in support of his hypotheses twenty 'physical theses', including a denial of the triple motion of the earth on the Aristotelian ground that a simple body can have but a single natural motion and a denial of the existence of solid spheres. The system is illustrated by an enormous fold-out diagram interpolated between the dedicatory letter and the first chapter. It was on the strength of his trigonometry that Kepler hailed Ursus as his teacher, and it was of Ursus' allegedly new world-system that Kepler expressed his admiration. As we shall see, Tycho maintained that both the world-system and a crucial portion of the trigonometry had been plagiarised from him by Ursus.

In 1596 Tycho published a selection of his correspondence. Ursus found himself publicly denounced by Christoph Rothmann and Tycho as a dirty scoundrel and charged not only with plagiarism of astronomical hypotheses and theft of documents from Tycho, but also with plagiarism of his mathematics from Rothmann and other German mathematicians.[9] And Ursus may have got wind of a more specific charge that Tycho had levelled at him. Tycho had claimed in letters to a number of correspondents, including the Imperial Physician, Tadeáš Hájek (Hagecius), Ursus' colleague at court, that the methods of prosthaphaeresis retailed in the *Fundamentum astronomicum* were derived *via* Paul Wittich and Jost Bürgi from discoveries Tycho had made in collaboration with Wittich in 1580.[10] In 1597 Ursus struck back. The title page of his *Tractatus* announces, in the Greek of the Septuagint, 'I will meet them as a bear' – that is, to complete the verse, 'as a bear that is deprived of her whelps'. The work is, as we shall see, savage and scurrilous even by the ferocious standards of sixteenth-century polemic. Unfortunately for Kepler it includes his adulatory letter, evidently introduced, as Kepler surmises in the preface to the *Apologia*, to show that 'there really are men who acknowledge the worthiness of his case'.

The upshot of this publication of Kepler's letter can be reconstructed in extraordinary detail from subsequent communications between Kepler, Tycho, Maestlin and Kepler's patron, the Bavarian Chancellor Herwart von

---

9 *Epistolarum astronomicarum liber primus* (Uraniborg, 1596), 33 (Rothmann to Tycho, September 1586) and 149–50 (Tycho to Rothmann, February 1589); J. L. E. Dreyer, ed., *Tychonis Brahe Dani opera omnia*, (Copenhagen, 1913–29), VI, 61–2 and 179–80. Hereafter cited as *Epist. astron.* and *T.B.o.o.*

10 *T.B.o.o.*, VII, 281 (letter of March 1592 to Gellius Sascerides) and 323 (letter of March 1592 to Hagecius).

Hohenburg. In December 1597 Kepler sent a copy of the *Mysterium* to Tycho at Wandesbek, humbly requesting his opinion in terms similar to those in which he had addressed Ursus.[11] At this stage, as the letter makes clear, Kepler did not possess any of Tycho's works, knowing of his world-system and treasury of observations only indirectly through Maestlin and others. Though Kepler already knew from Maestlin of Tycho's dispute with Ursus he did not, as he later testified to Tycho with incomparable tactlessness, take it seriously since he could not understand a fuss about priority over a mere 'alteration' of the Copernican hypotheses. Further, he had gathered from an inquiry about Tycho to Georg Limnäus that Ursus was amongst those who respected Tycho.[12] So the news which Kepler, who had still had no reply from Tycho, received from Maestlin in August 1598 must have come as rather a shock.[13]

Finally, I have had a letter from the most noble Lord Tycho Brahe in which he steadfastly defends his opinion...In it there were many noteworthy things. But though they impressed me not a little, I am much more impressed by your relation of the heavenly bodies to the five regular solids. He included your letter to him and his reply to you, in which likewise there are many noteworthy things. Further, I cannot forbear to mention what he wrote about that man Ursus. I understand that Ursus has published a certain book in which the most offensive jibes are directed at Tycho, to which he prefixed a letter of yours in which you honour him in the most glowing terms. I, indeed, have not seen the book, nor can I believe that you wrote such a letter. For you know my opinion of that man. The things he published in his book are not his own, nor did he understand them properly, so that what is good in the book he expresses in the wrong words. He derives many things from Tycho and retails them as his own, things which I have shown you in Tycho's book.[14] His doctrine of triangles contains nothing memorable that is not to be found elsewhere better expressed. And the things in it that are noteworthy he did not understand, and so expressed them falsely and ineptly, etc. Hence it would seem to me a marvel were he thus praised by you to the skies. So it would be proper for you to vindicate yourself against him for having so greatly abused your goodwill, for I truly do not believe that you honoured him with praises of this kind.

Kepler, who did not yet know which of his several letters to Ursus had been published, replied at once uneasily explaining his dealings with Ursus.[15]

[11] *K.g.W.*, XIII, 154–5. [12] *K.g.W.*, XIII, 207–8.
[13] *K.g.W.*, XIII, 236–7.
[14] The books to which Maestlin here refers are evidently Ursus' *Fundamentum astronomicum* and Tycho's *De mundi aetherei recentioribus phaenomenis*.
[15] *K.g.W.*, XIII, 261–2.

I wrote to Ursus in the summer of 1595 before I came to Tübingen.[16] One of the courtiers, Waganer, who had become acquainted with the man and made much of him since he frequently went to Prague, urged me to do so. My reasoning was almost exactly the same as in the case of my letter to Tycho – I would seek his criticism as a student from a senior about the business of the five bodies, since he was a widely-known man from whose writings I would learn (as far as I could tell on slight reflection). This business I sketched using very few words and figures. If there is anything in the way of adulation here it is the fault of the splendour of the Imperial Court and its member, my protagonist Waganer. After a year and a half he replied in frank terms asking for copies of my book and sent his chronology.[17] I wrote again sending copies. If I added anything in the way of praise (I do not know if I did so, for I had only just written to you), I think I did so because of the earlier letter. I mentioned certain chronological matters; I said that it pleased me that he did not follow those authors who are unreliable; I said that I marvelled at the extent of the reading of one who had been able to compose such a chronology. I begged him in every way to send me his criticism of my book, whose publication was undertaken so that I might discuss such matters with learned men. I begged him to send a copy to Tycho and if he had any work of Tycho's to send it to me as my payment. Later, within a year as far as I remember, I wrote again for he had not replied, but he had praised me in the presence of our treasurer, who was at Prague. I reproached him for his absurd restraint: he who is in the habit of giving back to the Emperor the money he receives from him (so they say here), could help those who participate in the same disciplines – me for example. I wrote this jocularly. But he did not reply. If I spoke to him face to face [I would say this]. He who perverts to his own nature and manners the tribute I frankly paid does wrong. It was none of my business to criticise what he had filched from the labours of others; I acted only on my own behalf. But he sees fit to exhibit in public as a judge between himself and Tycho one who had declared himself his disciple. I rejoice in the honour conferred on me, which I shall in future use openly. He had heard a disciple – what is that to gloat about? If he should appeal to me as a judge perhaps he would hear something different.[18] If the letters I pour out are made so much of by those to whom for the sake of honour I address myself as a disciple, very well then – I too shall value my letters; I shall write with caution; I shall keep copies! He behaves like Frischlin;[19]

---

16 Kepler wrote in November 1595 from Graz. He left Graz on leave in January 1596, and was at Tübingen in March when he consulted Maestlin about the publication of the *Mysterium*.

17 *K.g.W.*, XIII, 124–5 (Ursus to Kepler, May 1597). The work of which Ursus sent a copy is the exceedingly rare *Chronotheatrum* (Prague, 1597).

18 According to the preface of the *Apologia* Kepler did eventually address Ursus along these lines in January 1600 (*Apologia*, p. 135).

19 Nicodemus Frischlin (1547–90), polymath, classical scholar, poet and Lutheran apologist. The reference is presumably to his *Oratio de laudibus vitae rusticae*, delivered in 1577, in which he scurrilously lampoons colleagues at Tübingen and

he wrongs good men; to give vent to his rage he abuses one's goodwill in writing flatteringly rather than seriously. He sins much more grievously in trying to dishonour with my words a most excellent man of whom my admiration is evident. This crime is, I submit, a matter for lawyers to redress. But what need is there for me to write to him? For I see that he is one who puts his time to ill-use, and it is reasonably to be feared that he may commit further offences. He seems to have done this as an insult to me no less than to Tycho. For at first he praised me, but afterwards angered when he found out that I admired Tycho he made a fool of me in this way with his calumny and satire. If only I could see this monster who carries that letter of mine on his horns! For the rest, I shall avoid the man altogether and when the occasion arises I shall offer an apology to Tycho, which I beg you also to do. Let there be sent to me a copy of my published letter. For even though I cannot after such a long time remember any more of it, I shall nevertheless easily be able to tell whether it is what I wrote or whether he has enlarged on it, perhaps attributing to me words which are not my own.

In February 1599 Kepler received first a copy of Tycho's letter from Maestlin, then the original. Maestlin again urged him not only to apologise to Tycho, but to protest to Ursus about his behaviour.[20] Tycho's reception of the *Mysterium* was cordial, though critical on points of detail.[21] And he issued the invitation for which Kepler, knowing of his treasury of observations, had hoped: an invitation to join him, once his affairs were settled, in discussion of astronomical matters. The sting was in the characteristically arrogant and petulant postscript.

By goodness knows what accident it happened that the same messenger who brought your letter to me at Helmstedt brought to me there at the same time bound up with your letter the notorious and abominable writing of a certain Ursus, more a wild beast than a man. When I read through its impudent calumnies and the unspeakable lies and insults in which it everywhere abounds altogether shamelessly and beyond measure, I found there as well a certain letter of yours with which he attempts to adorn himself and hide his shame. When I read it I was amazed that you made so much of him. But I excused your mistake, for you would not have done so but for hearsay and that book which he wrote which he calls *Fundamentum astronomicum*. In it he provides almost nothing that is his own. Rather everything is snatched and stolen from others as is his custom. And not only the crass error which he commits in depicting the

courtiers to the Duke of Württemberg. The oration was published at Tübingen with his paraphrases of Virgil in 1580. The Duke eventually imprisoned him, and he died as a result of a fall sustained whilst attempting to escape. On Frischlin's life and works see D. F. Strauss, *Leben und Schriften des Dichters und Philologen Nicodemus Frischlin* (Frankfurt, 1856).

[20] *K.g.W.*, XIII, 276. [21] *K.g.W.*, XIII, 197–201.

orb of Mars,[22] but also many other things which I shall reveal in due course, prove incontestably the plagiarism of my hypotheses of which I accuse him. I still have with me certain notes of his which he secretly copied out in '84 when he was servant to a certain nobleman related to me and was visiting Uraniborg with him. In them he deals with not unrelated matters. I, indeed, had already long before thought up this basis for hypotheses; nor would I have introduced it into the book about the comet of '77 had I not feared such plagiarists to whom this had become known in my house.[23] So I am indeed amazed at your calling the hypothesis his, when both in your book and in this letter you attribute it (not without justice) to me. Moreover, at the same time you extol him with a great, if not hyperbolic, encomium. In addressing him you call him 'most noble mathematician'. (I shall not say what sort of mathematician he is; as for his nobility, if you look at his pedigree it is of peasants from Dithmarschen, and even they disdain him.)[24] Then you say that what brought him to your attention was his most illustrious reputation according to which he alone outshines the mathematicians of this age as the orb of Phoebus outshines little stars. You would have added more if time allowed. But how cleverly you excuse yourself from saying more on the grounds that garrulity is not proper for mathematicians. I wonder whether when your wit inclined you to say this you then tacitly acknowledged something of the sort. Besides saying that all learned men make much of him, you call him your teacher and attribute my hypotheses to him. And finally you conclude by wishing that he may flourish amongst the stars (of which he has very little knowledge) and in science (of which, if he has any, he procured and filched it from here) as an ornament to Germany. You seem to have written such things rashly about a man not known to you, but I shall dwell on them no more. Indeed, I should prefer it to be true that there existed someone of this sort to whom such great encomia in mathematics and astronomy would be fitting. I would not be the last to celebrate his renown. But I would not proclaim him to be superior to Ptolemy and Copernicus and to other excellent persons, of whom Maestlin in Germany is at present one, unless I held him to be clearly worthy enough. But be these things as they may, it is no laughing matter that your letter is thus abused in that accursed[25] book against me, in order, in effect, to provide an excuse for himself and, given that he would like to show (God willing) how high an opinion one ought to have of him, in order as far as possible to praise himself to the skies, to flatter himself and by implication to denigrate others. But you wrote this in your youth through ignorance, nor perhaps did you think that he would ever publish that letter, let alone that he would ever abuse it to disparage and libel others. That is why I bear this quite equably. I should, nevertheless, like you to let me know in person at the very first opportunity whether

---

[22] In his *Fundamentum astronomicum* Ursus had represented the orb of Mars as lying outside that of the sun, whereas in the Tychonic system their paths intersect.

[23] *De mundi aetherei recentioribus phaenomenis* (Uraniborg, 1588).

[24] J. Moller, *Cimbria literata* (Copenhagen, 1744), quoted in *K.o.o.*, 1, 217–18, confirms Ursus' peasant stock.   [25] Reading *maledicto* for *maledico*.

what he did was welcome to you and what you, and likewise any other learned men into whose hands it has come, think of this virulent work. And I have no doubt that what was brought here was indeed printed at Prague, but with the name of the printer suppressed as in notorious libels. I shall say nothing more of these matters, for they are unworthy to be recalled at length. You should interpret these incidental addenda with an open mind, and consign this note to the flames.

Kepler at once sent a long and frank formal apology to Tycho. The letter is known only from Kepler's draft and an incomplete copy sent to Maestlin.[26]

When I gathered from the letter Maestlin sent to me last August that you had replied to my letter and that he had a copy of that reply, I urged him to send me that copy at the next opportunity since your letter had not been delivered to me. Thus by chance and your diligence it came about that, wherever your letter may have got to, after ten months I knew what you wished me to know. At the outset I give you my assurance – for by Christ how very justly you request it in this dispute of yours – that at the very first opportunity I shall reveal my opinion to you lest greater suspicion of my guilt should arise in you because of this ten months' delay. As for my letter to Ursus, from the words which you excerpt I acknowledge it to be for the most part mine. . . . Thus it is that though I shudder at the thought of some of them about which I am in doubt, I cannot deny any of them. But I am in grave doubt whether Ursus published the entire letter, a doubt which would easily be dispelled if I had a copy of the publication. For in the same letter I expressed the wish to read your works and I sought Ursus' help in the matter. But, by immortal God, how greatly and on how many scores that savage man wrongs me. In the letter I sent I solicited his friendship. If he granted my request, I find his loyalty lacking; if he spurned it, he is still uncivil. The talk of friends, or rather their letters, should be private; but he published mine; and he did not consult me, nor did he tell me before or after doing so, nor did he send me a copy. Fair play protects them and one would not dare to do such a thing even to an enemy. He should, as the ancients say, have renounced me[27] or destroyed the letter if I was not acceptable to him as a friend. But if his letter eventually written to me a year later and his judgement of me [in it] are to be trusted, then the proposal was indeed agreeable to him and he had begun to cultivate our friendship. But these are trifles. The serious wrong, worthy of retribution if there is any hope or pleasure to be derived from public affairs, is this. What I wanted to say to him as a geometer, he, as you inform me, attached to his scurrilities. Most unpleasantly of all, with my letter he dishonoured as far as he could the man whom he gathered from it to be most highly regarded by me. Thus, to be sure, he was pleased to take revenge on me for my praise of his enemy. But as I air the matter of my behaviour in this exaggerated praise, I could not be more annoyed at the behaviour of the man. Why? Does he really value my tribute

---

[26] *K.g.W.*, XIII, 286–9.

[27] *Remisisset mihi, quod veteres ajebant, nuncium. Remittere nuncium* meant literally 'to send formal notice of divorce or annulment'.

so much? What kind of man is he? If he was a gentleman, he would have condemned it; if prudent, he would not have broadcast it. As an unknown person I sought one who would honourably praise my newborn discovery. I begged alms from him: behold, he extorted alms from the beggar. I called him teacher, and truly he gained little honour from the adulation of a disciple. Let him seek the praise of those who regard themselves as his teachers. But if he offers me the role of judge (and he could not have done so more openly than by thus revealing my letter), let him know that manners other than those of a disciple will befit me. Adulation of any sort is allowable in a disciple if it arises from a good opinion of the teacher and from esteem and is directed to the goal of learning. In a judge adulation is a disgrace. Though what I was thus led to write shows lack of caution, there is little that is very harmful to be angry with me about. Moreover, if there is no other way to preserve your honour, you may have as a remedy my public judgement of the published calumnies; but let me do it when finally [the work] has been sent to me. However, the question is not how badly he has behaved, but how well I have behaved. There are indeed in that letter of mine many things that can be defended, some that should be cut out and some that should be absolved of the gravest suspicion.

Let us start with the most serious matter, that which you impute to me as a crime, namely that I said to him things which I did not believe. In dealing with this you speak so as to seem to attribute to me an altogether conclusive judgement of others. If I can remove this suspicion from myself, which is quite easy, I shall think that a great part of my absolution has been achieved. There is now, surely, no dispute about Ursus' nature; undoubtedly at the time I wrote the letter I was more foolish in my judgement, and I would not still pay the same tribute to him. To my joy there came a certain learned man just back from Italy who showed me the book. For three days on end I read it avidly. In it I found diagrams dedicated to the principal mathematicians of our age; I found in the preface a magnificent challenge to attempt something similar, to construct astronomy from the foundations; I found certain splendid tables of figures which I remembered Maestlin having praised at Tübingen; I found the theory of angles and the theory of triangles. These are commonplaces indeed, but were mostly new to me – let him realise how he acquired his reputation! For I have since found in Euclid and Regiomontanus most of what I then attributed to him. Do not fear that what I now say [about my ignorance] could well apply to my discovery! Let Maestlin himself testify on my behalf with what meagre elements of learning I set out at the beginning; and undoubtedly when he sent me off on the second occasion he would never have bestowed on me such fine commendation (of which as long as I live I shall be unworthy) had I been as insignificant as I was when he first sent me off in '94.[28] Thus roused to a most excellent opinion of Ursus by this semblance of learning, I afterwards met certain of the Styrian nobility, admirers of mathematical studies, to whom Ursus was known because they were bound on

---

[28] Kepler left Tübingen for Graz in March 1594. In February 1596 he left Tübingen for Stuttgart and Maestlin sent a fulsome testimonial to Friedrich, Duke of Württemberg (*K.g.W.*, XIII, 67–9).

occasion to stay at the Imperial Court. They declared that his cleverness and his discoveries gave me grounds for attempting to gain his acquaintance and boded well for his help. But some of them described him to me as a sort of Diogenes from whom nothing could be extorted unless it was agreeable to him or was not by dishonourable means. For these reasons I was moved to write to him shortly afterwards about my discovery, since there was no mathematician nearer to hand to whom I could write more conveniently. My mind was loosed by delight in my recent discovery, and confident of the outcome I adapted the form partly to what was said to be the nature of the man, partly to the letters of telling brevity that I happened by chance to be reading. It is a matter of youthful rashness if in this way, thanks to those who praised him and my esteem for him, I let slip a little more than I really thought of him. I shall offer no further excuse, if only I have removed the suspicion of crime. From what I have said you will see in the first place that it was with a clear conscience that I attributed to him the glory with which it appeared that he reigned in the Imperial Court and amongst the mathematicians he addresses in his treatise. But I cannot recall the reason why I called [his glory] 'most illustrious', unless it was that our noblemen [*Illustres*] praised him, and the Emperor [*Illustrissimus*] seemed to praise him since he has him at court, and because a pompous mind seeing a fly easily makes an elephant of it. Moreover, since it was derived from others, I ascribe to the same [cause] my having said (if I really did say) that he was highly esteemed by all learned men. For I certainly held Maestlin and Tycho to be learned men who I knew did not think very highly of him. But I do not remember having promised that I would say more in this vein if time permitted. Indeed that protestation about garrulity being foreign to mathematicians was not made to excuse yet more in this vein, but rather to compensate for what went before. But if each and every one of these words – that he alone outshines the mathematicians of this age as the sun outshines minute stars – really is mine, then, by Christ, I did a great injustice to many excellent men and, moreover, something repugnant to my own conscience. In very truth I could never willingly and deliberately, seriously or in jest, in public or in private, out loud or in my heart have said that Ursus alone was to be valued above Regiomontanus, Copernicus, Rheticus, Reinhold, Tycho, Maestlin and others. I never thought so; I never in my right mind said so; and I never approved of such shameless adulation. But if my actual words say something of the kind – of which I am unsure – then chance, my haste and the fact that I did not read through what I had written are to blame. All this is, as you will see, poetical, derived from the poets and said in a poetic spirit. For I do not now wish, and I did not wish when I wrote, to be taken to say anything or to express any view other than that he excelled the mass – the mass, I declare – of mathematicians, which is something I believed and which can be true without outstanding excellence. I would not testify with such confidence against [my] evident words (if they really are of this sort), but for the fact that there is to be found in that letter abundant testimony to my innocence of which I have already spoken. Now though I do not know in what words I referred to Tycho, I know that I did so. And the reference was of a kind, and sprang from an opinion, that would surely have mortified

Ursus, if these remarks about him so pleasantly excited him. But as I now realise I must in future write with caution. Likewise it warns one never to pay one's respects to anyone by whispering in the ear of a bear. The other points you recount are trifles. I called him 'teacher' because I derived some useful things from his book, and 'most noble' because erudition and an office at court are accounted noble in these provinces of Austria. I ascribed to him hypotheses in which he made some alteration. For if I remember rightly he both makes the earth turn and has a different opinion, or as you say an erroneous one, about the circle of Mars. For though I had some time ago heard from Maestlin of this controversy between you and Ursus, nevertheless since both Magini in Italy[29] and Roeslin in Alsace had each derived for himself something similar from Copernicus' hypotheses,[30] I thought the same could easily have happened as a result of Ursus' cogitation as well. Finally, with regard to your last point nothing remarkable is really meant by, nor is there any impropriety in, the wish that 'he may flourish in knowledge of the stars'. For he is indeed a student of it, if one sets aside the malice and vain ostentation. 'Ornament of Germany' it is proper for me to attribute freely as I did to him, for mathematical studies are an ornament of Germany alone. I have written at such length both to express my mortification at your reasonable complaint and to make it clear that I take what you think of me to be of great importance for my reputation.

A note appended to the partial copy sent to Maestlin indicates how Kepler ended his apology: 'Finally, I added that it did not seem worthy of Tycho's stature to be so violently upset by this disparagement. He should rather allow their works to speak for both of them, confident of what the learned would think.' The covering letter to Maestlin reveals that Kepler planned to travel to Prague and thence to Wittenberg, where Tycho had taken up temporary residence, in order to 'clear up this matter of Tycho and Ursus'.[31]

Kepler had cogent reasons for wishing to be on good terms with Tycho, and his consequent embarrassment over Ursus' trickery is amply revealed in this letter. He already knew that Tycho was likely in due course to come to Prague under imperial patronage and he was anxious to gain access to Tycho's observations. Further, he was bound to take very seriously the possible effects of the incident on his prospects for employment. Following Archduke Ferdinand's return from Italy to Styria in June 1598 his situation

---

[29] The attribution to Magini is, as Tycho later informed Kepler, mistaken.

[30] Helisaeus Roeslin's system, set out in his *De opere Dei creationis* (Frankfurt, 1597), and alleged by the author to have been thought up before he had read Tycho's *De mundi aetherei recentioribus phaenomenis*, differs from the Tychonic system only in its acceptance of solid orbs and in its dimensions, according to which the orb of Mars does not intersect that of the sun and the orb of Saturn reaches up to the *primum mobile*.   [31] *K.g.W.*, XIII, 292.

at Graz had become increasingly insecure.[32] In September all Protestant teachers had been expelled and Kepler alone had been allowed to return to his post, perhaps through the intervention of his powerful Catholic patron Herwart von Hohenburg. But the terms of his personal exemption by the Archduke – 'He should in every way show proper discretion and avoid offence so that His Highness will not have cause to withdraw such mercy again' – indicate that he remained strictly on sufferance. Kepler was in a dilemma: failure to placate Tycho could destroy his prospects of employment; but so too could denunciation of the Imperial Mathematician in a letter that Tycho might well use in pursuit of his quarrel with Ursus. Further, as Kepler explained to Maestlin, too open a condemnation of Ursus, whom he had once praised so fulsomely, would make a fool of him and perhaps provoke Ursus to publish yet more of the praises Kepler had bestowed on him.[33] No wonder he hoped to settle the issue by taking Tycho's side in his dispute in private conversation rather than in writing. But for all his hesitation Kepler does commit himself to eventual publication of a judgement of Ursus' book should Tycho insist. It is perhaps to this offer that Kepler refers in the preface to the *Apologia* when he says that he had voluntarily undertaken to compose such a work.

It is hardly surprising that Kepler's formal and somewhat sanctimonious apology failed to placate Tycho. It is clear from Tycho's letter that what had infuriated him was not the apparent insincerity of the adulation, which caused Kepler such agonies of conscience, but the attribution to Ursus of what Tycho regarded as his own hypotheses. This Kepler blandly dismissed as a trifle. And the grounds on which he so dismissed it gave further cause for irritation. Ignorant as he was of the contents of Ursus' book, Kepler characterised Ursus' system, and thus by implication Tycho's too, as a mere emendation of the Copernican hypotheses.[34] On top of these failures of tact Kepler presumed to question Tycho's dignity in reacting so violently to Ursus' attack. By the time Kepler's apology reached Tycho it was anyway too late to persuade him to drop the affair. Throughout the year since Tycho had first heard of the work he had been busily soliciting testimony against Ursus with a view to taking up legal proceedings against him.[35] Already in April 1598, had

---

[32] Here, as elsewhere, details of Kepler's life are derived from M. Caspar, *Kepler*, transl. and ed. by C. D. Hellman (London, 1959); and C. Frisch, *Vita Joannis Kepleri* (*K.o.o.*, VIII, ii).   [33] *K.g.W.*, XIII, 290.

[34] An attitude evidently shared by Maestlin, who had dismissed such systems as 'patching up an old and worn-out garment with new cloth' (preface to his edition of Rheticus' *Narratio prima* appended to Kepler's *Mysterium* [*K.g.W.*, I, 84]).

[35] *T.B.o.o.*, VIII, 24 (letter of February 1598 to Rosencrantz), 34 (letter of March 1598 to Longomontanus), 61 (letter of May 1598 to Hagecius), 94 (letter of August 1598 to Craig).

Kepler but known it, Tycho was on the lookout for some third party to publish a defence of himself against Ursus.[36] And by January 1599 he was sufficiently confident of the outcome of his counteroffensive to offer to secure for his former student Longomontanus the post Ursus was about to vacate.[37]

On 25 May Kepler at last obtained a copy of the *Tractatus* from Herwart, who requested his opinion of the work.[38] Kepler's response is given in Chapter 3. It anticipates parts of the *Apologia* and provides a valuable indication of the way in which he might have completed the work. In it he does circumspectly incline to the view that Ursus was guilty of plagiarism, but his position on the crucial question of Tycho's originality remains ambivalent. Though he refutes Ursus' claim that the Tychonic hypotheses are actually set out in *De revolutionibus*, he continues to maintain that they are easily derived from the Copernican hypotheses and is prepared only to concede that Tycho could well have arrived at them independently. In August Kepler reported briefly to Maestlin on the contents of the book. This report, likewise given in Chapter 3, is similar to his report to Herwart, but is of some independent interest since it undoubtedly represents Kepler's frank opinion, whereas the report to Herwart may well be coloured by Kepler's knowledge that Herwart was in communication with Tycho and might well reveal its contents to him – Herwart did, in fact, eventually do so. And it was probably at this time that Kepler, in pursuance of Tycho's request for him to find out what others thought of the book, wrote to his friend Helisaeus Roeslin, another target of Ursus' attack, seeking his opinion of the *Tractatus*.[39]

In the course of the summer of 1599 Kepler's situation in Graz became ever more hopeless as the persecution of the Styrian Protestants gathered momentum. There is sheer desperation in his reports to Maestlin on the persecution and in his fruitless attempts through Maestlin to solicit the offer of a post at Tübingen. In September Herwart informed him that Tycho had entered the Emperor's employment at a salary said to be enormous. He added significantly: 'I wish you had such a position, and who knows what fate has in store for you.'[40] In December Kepler had still not heard from Tycho, but evidently in response to an inquiry about the prospects for a formal association with Tycho at Prague he received an offer from the Privy

---

[36] *T.B.o.o.*, VIII, 56–9.   [37] *T.B.o.o.*, VIII, 136.

[38] *K.g.W.*, XIII, 332.

[39] As Roeslin in his *Historischer, Politischer und Astronomischer natürlicher Discurs von heutiger Zeit Beschaffenheit*... (Strasbourg, 1609) proudly reported (passage quoted in *K.o.o.*, I, 229); Kepler admits this with evident irritation in his published reply to Roeslin's book (*K.o.o.*, I, 504–5).   [40] *K.g.W.*, XIV, 59.

Councillor Baron Hoffmann to act as mediator between him and Tycho.[41]
In January 1600, still having had no reply from Tycho, he set out for Prague
in Hoffmann's retinue to explore for himself the prospects of employment.
Ironically, Tycho had in fact already written inviting Kepler in the most
cordial terms to join him in Prague. But if he imagined that his apology to
Tycho would end his involvement in the dispute with Ursus the letter would
have disappointed him.[42]

The matters you recall at length at the beginning in order to absolve yourself in
the matter of that idle scoundrel Ursus did not really merit so many words and such
a detailed declaration, since for other reasons I hold you to be sufficiently excused.
And I do not blame you for his having inserted your letter, unknown to you and
against your wishes, mutilated and distorted, moreover, as I now gather from you,
into his scurrilous and accursed book for the purpose of decking himself with honours
of which, maybe, he stood in need. As for my hypotheses snatched away by him,
it would not be a question of my honour and reputation but for the fact that in
that shameless book, swarming with fabrications and calumnies, he is not ashamed
to attack with his furious and scurrilous pen my country, my family and my most
honoured home with the most impudent and evil lies, as far as he can to dishonour
them, and to spread such things abroad in print. He does so despite the fact that
he himself declares in his preface that I deserved well of him;[43] and I possess a certain
poem of his, inept to be sure (as is usual with him), an outpouring of his spirit in
his own hand, in which he thanks me not only for having kept him in food and
drink for quite a few days once when he was my guest at Uraniborg, but also for
having given him some money when he left. Look how he pays me back for these
and other things! He took as his pretext for these insults the fact that there is to
be found in the first volume of my *Letters of the Astronomers* a letter of Christoph
Rothmann, Mathematician to the Landgrave, in which he calls Ursus 'that dirty
rascal'.[44] For he had already learned what kind of man he was from his own
experience at the court of his prince as well as from the reports of others from the
places where he had previously lived. I, indeed, in another letter to the said Rothmann
said the same merely in imitation, and at the same time I indicated by what trickery
he had secretly stolen some things of mine, some of which I recovered and still have
with me written down in his own hand (which he cannot deny).[45] But certainly,
had there been time to spare from my more serious studies, when these letters were
being printed at my press I would not, on seeing their particular contents, have
allowed that book published by me to be sullied with the name of that unspeakable

---

[41] *K.g.W.*, XIV, 98–9.

[42] *K.g.W.*, XIV, 89–98. The letter was sent *via* Vienna, where Kepler finally saw it
on his return journey (*K.g.W.*, XIV, 128).

[43] The mention of Tycho's hospitality in the prefatory letter merely retails insults
Ursus had received at Tycho's table (*Tractatus*, sig. Fi, r).

[44] *T.B.o.o.*, VI, 61–2.   [45] *T.B.o.o.*, VI, 179.

and foul Ursus – not, indeed, because his name could be thought undeserving of any epithet of the sort applied to it or indeed of any of the other generally agreed condemnations of him that men retail (for he did commit the crime, as not only Rothmann but also many others who know him well enough hold to be evident), but rather because in my writings I would not willingly touch on anyone's life or manners whatever they might be like. That is why in the letter of the late noble Lord Jacob Curtius later appended by me at Wandesbek to my book on mechanics I omitted his name, although the said Lord Jacob Curtius openly mentioned him and also called him my plagiarist.[46] But it happened that when I entrusted to certain of my students the task of arranging the printing of the book of letters and handed the originals over to them as copy, they inferred that nothing was to be changed and did not dare to disturb me, occupied as I was with more serious matters. And so they allowed everything to be printed just as it was in the originals. Thus it came about through negligence that his name with the epithets applied to it was retained. I shall briefly consider whether this is an adequate and sufficient ground for raging with such insane and boundless malice against me and my family and my country as well. He could, indeed, have complained about the injustice, if any was done him, and sought to redress it by legitimate means, without raging wildly with such unbridled lust for revenge and fighting back with far worse things. Nor is it true that I feel more strongly about this silly man than my status allows, as you declare (but frankly enough). What makes you think so? For I do not think he deserves the Emperor's wrath, as someone has, indeed, suggested to me in writing.[47] But because against all justice and equity and against the most praiseworthy laws of the Empire he has shamelessly tried to dishonour me beyond all truth and desert with his more than notorious writing, I shall make sure that he is prosecuted and curbed and punished in accordance with the laws. He took himself off from Prague to Silesia shortly after my arrival here, but recently as I have ascertained he came back secretly. Perhaps he did so because he had understood that I was no longer there. But he will realise how long my arm is, even when I am not present, when he is arrested and then brought to trial.[48] Moreover, because I gather from your letter that you have not read these filthy and scurrilous pages, I send you a copy that you may the sooner see your letter there and judge the nature of the rest. You will certainly see what an insane, accursed, insulting and stinking compilation it is and how its author stinks to high heaven. Never, perhaps, will you read more impudent libels. The ears of all good and honest men will recoil from them. This is more than enough about that unspeakable Ursus of Dithmarschen, nor is he worthy of all the words you and

---

46 *Astronomiae instauratae mechanica* (Wandesbek, 1598); *T.B.o.o.*, v, 121. Ursus had evidently seen this letter of 1590, for he reproduces the part in which he is called Tycho's plagiarist in his *Tractatus* (sig. I3, v). It was, according to Hagecius, Curtius who had secured Ursus' appointment as Imperial Mathematician in 1591 (*T.B.o.o.*, vii, 305).

47 *Caesaris ira*: Tycho appears to mean the death penalty.

48 Tycho's arm was not all that long. He was at Benatky, about 20 miles from Prague.

I waste on him. And if you knew him inside-out, as the saying goes (not to mention the syphilis that he spreads everywhere because of his filthy way of life), you would protest along with me; indeed, you have already protested some time ago. I will add just one point as a reason for asserting that the hypotheses are mine, though it is a very slight one and I have to hand other reasons far more cogent. Knowing that he was never in the habit of making any of the celestial observations from which new hypotheses have to be constructed, who could believe that the hypotheses first thought up by me in the year 1584, that he is unafraid to arrogate to himself, really were discovered by him?[49] Indeed, he was unable using this new supposition to demonstrate in full detail the apparent position of a single planet, let alone one of the superior planets; and he could scarcely have done so. Who, I ask, realising this, would be persuaded by him that it was his own discovery?

Tycho was evidently stung by Kepler's assertion that others, including Roeslin and Magini, had arrived independently at similar 'alterations' of Copernicus' system. He goes on to charge Roeslin and Duncan Liddel with plagiarism, and, on firmer ground, to deny that Magini published anything of the sort. In a postscript he reverts to the question of Ursus' plagiarism. He encloses two letters from a doctor, Christian Hansen (Johannis), and a testimonial affidavited before a notary at Cassel by Michael Walther, secretary to Erik Lange, the nobleman in whose retinue Ursus had visited Uraniborg in 1584. These, he claims, confirm the charges of theft and plagiarism that he had made in his letter to Rothmann.[50] The testimonial tells how a student warned Tycho of Ursus' suspicious behaviour and how a certain Andreas then searched Ursus when he was asleep. In one of his trouser pockets he found some papers, which he removed; the other he could not search for fear of waking him. On discovering his loss Ursus became wildly disturbed and calmed down only when he was assured of the return of all papers which did not have to do with Tycho's affairs. Walther goes on to relate how Ursus became increasingly deranged after the visit to Hveen, until Lange had to sack him.

Unaware either of Tycho's cordial invitation or of the counterattack on Ursus retailed in this paranoid letter, Kepler arrived in Prague in mid-January. He stayed at Baron Hoffmann's house and, as he relates in the *Apologia*, he

49 On the vexed question of the date at which Tycho arrived at his world-system see W. Norlind, 'Tycho Brahes Världssystem', *Cassiopeia*, 6 (1944), 55–75; and C. J. Schofield, 'The geoheliocentric planetary system: its development and influence in the late sixteenth and seventeenth centuries' (Ph.D. thesis, Cambridge, 1964; New York, 1981), 50–8.

50 In *K.g.W.*, XIV, 469, these documents are said to remain unpublished; however, the testimonial and one of the letters are in *K.o.o.*, I, 230–1. A letter of September 1599 to Andreas Velleius indicates that Tycho regarded the testimony of Lange as decisive (*T.B.o.o.*, VIII, 180–1).

met Ursus and told him that he would take Tycho's side in the dispute.[51]
Three weeks later he joined Tycho at Benatky. During his stay there he
composed an adjudication of the dispute between Tycho and Ursus for
Tengnagel von Camp, Tycho's familiar and future son-in-law. In this, which
is given in Chapter 3, he comes out openly in support of Tycho both on
the question of Ursus' plagiarism and on the question of Tycho's originality.
But the visit ended with a quarrel, apparently occasioned by Kepler's failure
to secure terms of association with Tycho which would guarantee him access
to Tycho's observations and a measure of independence in his research. Kepler
returned in a huff to Prague and dashed off a violent letter of complaint to
Tycho. Tycho reacted, for once, with moderation, merely forwarding the
letter to his colleague Jessenius asking him to find out if Kepler had taken
up with Ursus.[52] Tycho's suspicion had some foundation, for Kepler did
communicate with Ursus at this time as he unwisely admitted to Herwart,
who duly reproached him.[53] However, Hoffmann and Jessenius managed
to bring about a reconciliation, and Kepler returned to Benatky for a further
month before returning to Graz armed with a fulsome testimonial from
Tycho.

On his return to Graz Kepler found his position there untenable. The
councillors, who had little use for an astronomer, wanted him to go to Italy
to qualify as a doctor. And there was rumour of yet more ruthless action
against Protestants. But for whatever reasons Kepler was now reluctant to
reassociate himself with Tycho. He again petitioned Maestlin about the
possibility of a post at Tübingen and he even made an approach to the
Archduke Ferdinand, prime mover in the persecution, apparently in the hope
of becoming Archducal Mathematician. Nothing came of these overtures.
In July the Archduke decreed an examination of faith for all inhabitants of
Graz and on 12 August, having refused to convert, Kepler was duly ordered
to leave Graz within 45 days.

In these straits he again appealed to Tycho. In September, in the nick of
time, he received Tycho's reply.[54] In it Tycho, who had taken up residence
at Prague after serious setbacks in his building programme at Benatky, reports
that at an audience the Emperor has nodded his approval to the suggestion
that Kepler should be attached to his observatory for a year or two, and that
he has requested the Vice-Chancellor to ask the Styrian authorities to grant

---

[51] *Apologia*, p. 135. [52] *K.g.W.*, XIV, 114.

[53] Writing to Herwart in July 1600 Kepler remarks that Ursus had told him that
Ptolemy's *Harmonics* had been published (*K.g.W.*, XIV, 131).

[54] *K.g.W.*, XIV, 145–9. Tycho refers to letters from Kepler, not now extant, dated
25 June and 10 August.

Kepler two years' salaried leave. In a postscript he reverts to his remorseless onslaught on Ursus.

You added, moreover, that you would refute the mathematical matters to be found in that notorious Ursine production insofar as they are incorrect and I trust that you really will do so. But above all I request you not to regard it as an imposition to refute more fully and clearly than before what he fraudulently and deceitfully alleged about our discovery of new hypotheses, namely that it was derived from Copernicus or from Apollonius of Perga.[55] And I request you to say, as is proper, that it is mine, as you have already said on plausible grounds. I instituted legal proceedings against him after I came here to Prague, understanding that he had stayed here for some time and that there was risk in delay, for he was seriously ill. And I have petitioned the Emperor (despite certain men who caused delay for as long as they could) to depute four commissioners, two Barons and two Doctors of Law, to deliver sentence in the matter. For, indeed, when questioned by my agents he was unwilling to retract what he had written, but rather appealed to law. It happened, however, that in the very hour that the writ was to be served on him he died.[56] But nevertheless I am persisting in my uncompleted quest for justice. For the suit has already in effect been embarked on and the action clearly concerns not a person but a thing, because of the notorious book that will forever speak on behalf of its author. I do this to deliver myself and my family legitimately from the calumnies in which the book abounds. And already the most illustrious Lord Archbishop has been commissioned by the Emperor to seek out all the copies that are to be found here and have them burned, and to punish the printer.[57] And Lord Corraduc has promised a decree to be promulgated by the Emperor, so that the book will be annulled throughout the whole Empire and declared to be a notorious and scurrilous nonentity. Afterwards I have, moreover, decided to publish the entire proceedings of the trial together with the judgement of the commissioners and the decree of His Majesty the Emperor as a special book, which will assume quite considerable proportions. And with the most generous consent of the Emperor and the approval of Lord Corraduc already assured me, I am, God willing, about to do so. And in the other part of the book I am about to reply to the matters which are mathematical and concern hypotheses. I would like to have your views on these matters in due course; but if (as I hope) you come here as soon as possible, it could await your arrival.

The account Tycho offers of his attempts to bring Ursus to trial has significant

---

[55] The earlier refutation to which Tycho refers is presumably Kepler's judgement on the dispute offered to Tengnagel at Benatky.

[56] On 15 August 1600.

[57] That this was done is confirmed by a decision of the Hofkammer dated 20 October 1600 awarding 300 florins compensation to Ursus' widow for the confiscated books – a human conclusion to a grim tale: see M. List and V. Bialas, 'Die Coss von Jost Bürgi in der Redaktion von Johannes Kepler', *Abhandlungen der Bayerische Akademie der Wissenschaften: math.-nat. Klasse*, n.F., 154 (1973), 108–9.

omissions. The delay, it seems, was occasioned partly by the removal of the court in the autumn of 1599 from Prague to Pilsen to avoid the plague and partly by Ursus' presentation of a petition to the Chancellor at Pilsen complaining about Tycho's malicious behaviour towards him.[58] From this postscript it appears that despite Ursus' death, and whatever may have been Kepler's understanding about possible terms of collaboration with Tycho,[59] composition of a defence of Tycho against Ursus had become a condition of employment. Tycho was apparently still suspicious about Kepler's real attitude to the dispute. On the same day he wrote to Herwart asking for a copy or summary of the judgement of the *Tractatus* Kepler had sent him. Herwart complied.[60]

At the end of September, despite the authorities' refusal to grant him salaried leave, Kepler left with his family for Prague. In mid-October, sick with intermittent fever, they were lodged again at Baron Hoffmann's house. Shortly before his arrival he wrote to Tycho urgently requesting a clarification of his financial prospects now that the Styrian salary had fallen through, and talking in more confident terms than the prospects revealed by his fruitless petitions to Maestlin and Herwart warranted of possibilities of employment elsewhere.[61] His future remained unsettled throughout the next five months, the period during which he composed the *Apologia*.

At the beginning of February 1601 Kepler reported to Maestlin on the work as follows.[62]

I write against Ursus and nothing else. In it I deal with nothing except what pertains to science. The principal concern is with antiquity and with the explication of opinions of the ancients. So the treatise is to be hardly mathematical, but rather philological.

There is no further mention of the work in Kepler's correspondence until after Tycho's death in October 1601 and his own appointment as Tycho's successor to the post of Imperial Mathematician. The letter to Longomontanus, cited earlier, is slightly ambiguous, but appears to indicate that he abandoned the work on leaving Prague for Graz in April 1601. In September 1602 David Fabricius, who had visited Tycho whilst Kepler was away on that final visit to Graz, wrote: 'I hear that you have refuted Ursus' absurd

---

[58] As Tycho had reported in his letters of March 1600 to Jessenius and of August 1600 to Daniel Cramer (*T.B.o.o.*, VIII, 281 and 334).

[59] The terms of collaboration drawn up by Kepler which had contributed to the quarrel at Benatky included the proviso 'that he should not burden me except with astronomical matters insofar as they are necessary for his publications' (*K.g.W.*, XIX, 40).

[60] *K.g.W.*, XIV, 149–50 and 152.

[61] *K.g.W.*, XIV, 152–5.

[62] *K.g.W.*, XIV, 165.

hypotheses. I ask you to publish at the first opportunity as you promised Tycho.'[63] Kepler replied: 'I have written against Ursus, but it does not satisfy me; I must first read Proclus and Averroes on the history of hypotheses. I will publish sometime when it can be done with less odium than now. For he was my predecessor in office.'[64]

---

[63] *K.g.W.*, XIV, 281.

[64] The reference to Proclus is to his *Hypotyposis astronomicarum positionum*. The reference to Averroes' writing on the 'history of hypotheses' is puzzling. Nothing of the sort is to be found either in his *De substantia orbis* or in his commentaries on *De caelo* and *Metaphysics*. What evidence there is suggests that Kepler had in mind Averroes' *Paraphrase of Ptolemy's Syntaxis*, which he knew of from Copernicus' reference to it in *De revolutionibus*, I, 10. It was, in fact, available only in Hebrew translation and contains no relevant account of astronomical hypotheses. Kepler's later correspondence reveals attempts to locate a copy of the work (*K.g.W.*, XV, 418 and 462). For further details of Kepler's confusion about Copernicus' reference to Averroes see Rosen's note on the passage in J. Dobrzycki, ed., *Nicholas Copernicus on the Revolutions*, translation and commentary by E. Rosen (Warsaw and London, 1978), 356–7.

2

Ursus' *Tractatus*

The tone of the *Tractatus* is set by its full and truculent title page.

An astronomical and cosmographical treatise concerning astronomical hypotheses, that is, the world-system, by Nicolaus Raimarus Ursus of Dithmarschen, mathematician to His Most Holy Majesty the Roman Emperor, a work both edifying and most profitable ro read. Also: a vindication and defence of the astronomical hypotheses discovered, presented and published by him against certain men who have importunately, or rather with criminal audacity, arrogated them to themselves; a demonstration of them from the Sacred Scriptures; and an account of their application, in which application the whole hidden genuine astronomy, and the very astronomical foundation, is seen, is revealed and is made manifest. Together with: certain new and most subtle tables and devices, in a clearly new doctrine of angles and triangles, which is now presented again here in a completely new way; as well as some most edifying mathematical exercises propounded both so as to vanquish everyone, and in particular his critics and detractors, on the question of the palm and mastership in mathematics, and for the sake of the study of mathematics; and, finally, practical instructions for the whole process of astronomical observation, that is, the business of observing the phenomena.

Hosea Chapter 13: 'I will meet them as a bear'.[1]

Prague, at the house of the author, without any licence, in the year 1597.

The critics and detractors Ursus is out to vanquish are Christoph Rothmann, Helisaeus Roeslin and Tycho. Against Rothmann and Roeslin Ursus had modest grounds for resentment. Rothmann had described him in a letter published by Tycho as a 'dirty scoundrel',[2] and Roeslin had in print described Ursus' hypotheses as physically absurd and contrary to the scriptures.[3] But against Tycho Ursus' grounds for resentment were ample. For it was in Tycho's publication that he had been variously defamed as a scoundrel, as a former servant, as having stolen Tycho's astronomical hypotheses and as having derived most of the mathematics of the one major

---

[1] '...that is deprived of her whelps': a play on his name, 'Ursus' = 'bear', his 'whelps' being his astronomical hypotheses.   [2] *T.B.o.o.*, VI, 61–2.

[3] *De opere Dei creationis* (Frankfurt, 1597), 47.

work on which his reputation rested from Rothmann and other German mathematicians.[4] The defence of his modest claims to originality in the elaboration of trigonometrical methods with which he concludes the *Tractatus* and the mathematical challenge to Tycho which he issues in the opening poem strongly suggest that he had got wind of Tycho's more specific charge, repeatedly made in unpublished correspondence, that in trigonometry it was from Tycho and Wittich that he had plagiarised.[5]

Ursus's indignation over the charge of plagiarism may well have been righteous. On the score of plagiarism of his astronomical hypotheses Tycho had, after all, levelled charges against others that were definitely ill-founded.[6] The testimony Tycho regarded as crucial, that of Erik Lange, is inconclusive. For it fails to substantiate the specific charge of copying of a diagram of Tycho's system, indicating only that Ursus had made drawings of Tycho's buildings. In the *Tractatus* Ursus confirms that the night before he left Hveen his papers were secretly filched from him, he admits that they may have included drawings of Tycho's buildings and he challenges Tycho to produce from the filched papers the diagram of the hypotheses.[7] Far from suggesting Ursus' guilt this suggests his innocence, for he is unlikely to have known of the contents of Lange's testimony. The supposed confessions of guilt in the *Tractatus* itself, to which Kepler appeals in his judgement addressed to Tengnagel, are not in fact confessions. Thus Ursus reproduces the letter in which Tycho had charged him with theft of the diagram and plagiarism with his own marginal comments and notes, and one of the marginal comments, which Kepler takes as a confession, reads: 'Let it be a theft, but a philosophical one. It will teach you to look after your things more carefully in future.' However, elsewhere Ursus denies the theft of hypotheses: '[The theft] is a philosophical one, which I firmly deny having committed.'[8] In context it

---

[4] *T.B.o.o.*, VI, 176–80.   [5] See Ch. 1, fn. 10.

[6] Thus Tycho had voiced the suspicion that Paul Wittich had claimed for himself Tycho's improvements in astronomical instruments (*Epist. astron.*, 7 [*T.B.o.o.*, VI, 36]). And in his letter of December 1599 to Kepler, the letter with which he sent testimonials allegedly proving Ursus' plagiarism, he charged Duncan Liddel with having represented the Tychonic system as his own in lectures, and he sent Kepler copies of correspondence between himself and Liddel and of a testimonial by Daniel Cramer to that effect (*K.o.o.*, I, 226–8). Tycho later withdrew the charge against Wittich (*Epist. astron.*, 113 [*T.B.o.o.*, VI, 141–2]). The evidence for the charge against Liddel has been reviewed in detail by C. J. Schofield, 'The geoheliocentric planetary system: its development and influence in the late sixteenth and seventeenth centuries' (Ph.D. thesis, Cambridge, 1964; New York, 1981), 146–60; she concludes that Tycho did not have a good case against Liddel.

[7] *Tractatus*, sig. Fii, r; for Lange's testimony see *K.o.o.*, I, 230–1.

[8] *Tractatus*, sig. Fi, r, and Div, r.

is clear that Ursus is merely using, somewhat ineptly, the standard defensive
tactic of *concessio*, argument of the form 'even if your claims are conceded,
your case is still not proven'. This is a strategy he employs elsewhere in the
Tractatus to equally confusing effect. Tycho's letter also reports that Georg
Rollenhagen had warned him that a fugitive servant of Tycho's had shown
him the Tychonic hypotheses at his home in Magdeburg in 1586. Ursus
admits to having told Rollenhagen about Apollonius of Perga's hypotheses
and his new emendation of them.[9] Kepler takes this to imply Ursus' guilt
on the grounds that Tycho's hypotheses are not to be found in Apollonius
as reported by Ptolemy. But though Kepler is surely right in suggesting that
Ursus misunderstood Apollonius' models for the second anomaly, it is quite
possible that this misunderstanding was indeed Ursus' source of inspiration.
Ursus' innocence is further suggested by the fact that his system, as presented
in the *Fundamentum astronomicum*, differs from Tycho's in several respects:
in its placement of Mars' orb; in its attribution of diurnal rotation to the
earth; and in its suggestion that the various magnitudes of the fixed stars
indicate their various distances. In the published letter to Rothmann Tycho
tries to preempt this obvious line of defence by claiming that Ursus had copied
a certain diagram with a defective representation of Mars' orb which he still
possesses. This renders Tycho's case highly conjectural. Even if both Lange's
testimony and Tycho's are entirely correct, the charge of theft rests on the
suppositions that Ursus saw the defective diagram and that his copy eluded
the search which Tycho had instigated. Further, it leaves the other
discrepancies between Ursus' and Tycho's hypotheses unexplained. Roeslin's
claim that Ursus introduced these modifications to conceal his debt to Tycho
seems forced, given that in the *Fundamentum astronomicum* Ursus presents
reasonably coherent physical arguments in support of these peculiar features
of his system.[10]

Tycho's case for plagiarism in mathematics appears to be yet weaker. At
issue are the formulae for prosthaphaeresis retailed on f. 16v and f. 17r of
the *Fundamentum astronomicum*, formulae of great practical interest for the
savings in computation they entail.[11] Tycho claimed that he had discovered

9 *Tractatus*, sig. Fii, r.
10 *K.o.o.*, 1, 229 (Roeslin to Maestlin, December 1588). Both J. L. E. Dreyer, *Tycho
    Brahe* (Edinburgh, 1890), 185, and Schofield, 'The geoheliocentric planetary
    system', 133–5, hold Tycho's evidence against Ursus to be inconclusive, though
    the latter inclines to accept Ursus' guilt. It should be noted, however, that Dreyer
    did not have access to Ursus' *Tractatus*.
11 *Tractatus*, sig. I3, r. Details of the sixteenth-century development of prosthaphaeretic
    methods are to be found in A. von Braunmühl, *Vorlesungen über Geschichte der
    Trigonometrie*, 1 (Stuttgart, 1900), Kap. 8, and 'Zur Geschichte der prosthaphaer-

these formulae in collaboration with Paul Wittich in 1580. But Dreyer has provided a strong circumstantial case for assigning the credit to Wittich alone.[12] In the *Fundamentum* Ursus does not explicitly indicate his sources for the two formulae and the proofs of their validity which he retails. In the dedicatory letter he fulsomely acknowledges Jost Bürgi as his guide and mentor in the 'doctrine of triangles';[13] and his dedication to Paul Wittich of the diagram which illustrates the proof of validity of the first of the prosthaphaeretic formulae may be intended as a hint of his debt to Wittich.[14] In the *Tractatus*, however, Ursus is perfectly explicit. The first formula was, he says, brought to Cassel by Wittich; Bürgi devised its proof; and he himself derived the second formula and its proof as easy corollaries of Bürgi's proof.[15] Given that these are in fact easy corollaries of Bürgi's elegant and sophisticated proof, and that Ursus was, as Kepler was prepared to concede, a competent mathematician, there is little reason to doubt Ursus' rather belated acknowledgement of his sources.[16] It seems that Tycho's charge of plagiarism is without foundation. And his fury over the publication of the *Tractatus* may be in part attributable to the fact that Ursus' account of the sources of the prosthaphaeretic methods of his *Fundamentum* threatened to reveal as fraudulent his own claim to have had an important part in their invention.

Before we turn to the strategies Ursus deploys in the *Tractatus* to vindicate

etischen Methode in der Trigonometrie', *Abhandlungen zur Geschichte der Mathematik*, 9 (1899), 17–29. J. L. E. Dreyer, 'On Tycho Brahe's manual of trigonometry', *Observatory*, 39 (1916), 127–31, is informative on the role of these methods in practical astronomy.

[12] Dreyer, 'On Tycho Brahe's manual of trigonometry'. Kepler himself more than once speaks of *prosthaphaeresis Wittichiana* (see, e.g., *K.g.W.*, XVI, 189) and in *Die Coss von Bürgi*, his introduction to and edition of Bürgi's *Canon*, Wittich is credited with having brought the prosthaphaeretic method to Cassel: see M. List and V. Bialas, 'Die Coss von Jost Bürgi in der Redaktion von Johannes Kepler', *Abhandlungen der Bayerische Akademie der Wissenschaften: math.-nat. Klasse, n.F.*, 154 (1973), 7. Tycho's claims to originality in this field are scarcely consistent with his request of March 1600 to Daniel Cramer for help with the problems set as a challenge to him in the *Tractatus* (*T.B.o.o.*, VIII, 292). The prosthaphaeretic formulae had in fact been discovered prior to 1580 both by Viète, who published them in his *Canon mathematicus* of 1597, and much earlier by Johann Werner, whose *De triangulis sphaericis* Rheticus had planned to publish. But Wittich may, as List and Bialas suggest ('Die Coss von Jost Bürgi', 104), have been the first to apply the method as a computational device. I am indebted to Professor O. Gingerich for drawing my attention to relevant secondary literature on the issue of priority.

[13] f. [*4r].

[14] f. 16v: *Diagramma casus prioris. Paulo Vuitichio Vratislavensis dedicatum.*

[15] Sig. I3, r.

[16] For Kepler's preparedness to concede Ursus' skill as a mathematician see the assessment of the *Tractatus* offered to Herwart (Ch. 3, p. 63).

his own originality and to discredit Tycho's claims to originality, it is worth considering briefly some general points about the content and significance of Tycho's and Ursus' charges and countercharges of theft and plagiarism. Nowadays both in mathematics and in the natural sciences we have relatively clearcut criteria for priority and plagiarism, the issues which gave rise to the controversy between Tycho and Ursus. Thus we take priority in printed publication to be the crucial form of priority. We recognise mathematical and scientific theories as distinctive, indeed typical, subjects of claims of priority and charges of plagiarism. And we regard priority in the rediscovery of theories, whether from the sources or independently, as a form of priority that reflects little credit compared with the construction of new theories. Concern with priority of the kind which exercises Tycho, Ursus and Kepler presupposes the concept of a public domain of scientific practice and knowledge, a domain to which contributions are made by publication. But this should not lead us to assume that sixteenth-century authors attached the same weight to priority in printed publication as we do. Whilst both Tycho and Ursus evidently regard such publication as an important issue, great importance is also attached to other forms of publication – for example, Ursus' alleged communication of Tycho's hypotheses to Rollenhagen at Magdeburg and Bürgi's construction of a mechanical model of the Ursine system for Wilhelm IV, Landgrave of Hesse.[17] The difference is hardly surprising given the extent to which, before the development of a periodical press, teaching, private communication and the annotation of printed works served as vehicles for the dissemination of scientific ideas. Likewise, though Ursus is for other reasons at pains to belittle the credit to be assigned for the construction of hypotheses, he evidently does not regard the admission that his hypotheses were anticipated by Apollonius of Perga as vitiating his claim to originality. Indeed, he presents his enterprise as a restoration of ancient astronomy and evidently held that it is not the anticipation of Tycho's and Roeslin's hypotheses itself that disgraces them, but their refusal to admit it. For those like Ursus who held that ancient mathematics had far surpassed modern achievements, anticipation by the ancients added lustre and plausibility to hypotheses. And even Kepler, who vigorously asserted the superiority of the moderns to the ancients in astronomy as in other fields, clearly sees in the alleged anticipation of the Copernican hypotheses by the Pythagoreans something which redounds to Copernicus' credit.

Equally important for an understanding of the terms in which Tycho and

[17] See *Epist. astron.*, 149–50 (*T.B.o.o.*, VI, 179), for Tycho's allegation about the communication to Rollenhagen; and *Tractatus*, sig. Fii, r, for Ursus' rebuttal. On Bürgi's model of Ursus' system see *Tractatus*, sig. Aii, v.

Ursus conduct their dispute is the inapplicability to the controversy of modern assumptions about the role of theories in science. In Chapter 8 I argue that the modern conception of a theory as a systematic body of hypotheses with an associated domain of relevant evidence is not to be found in sixteenth-century writings. In this connection it is significant that the terms in which Ursus and Tycho charge each other with plagiarism tend to assimilate what we would regard as theories to what we would regard as applications or byproducts of theories. Thus Tycho charges Ursus with having copied a diagram of his world-system and with having derived trigonometrical formulae from him, and Ursus takes it as an adequate defence to show that he did not copy the diagram and that long before he published his hypotheses he had presented his world-system to Wilhelm IV, Landgrave of Hesse in the form of a planetarium. This, to us strange, conflation of intellectual products with artefacts is reflected throughout the dispute in the terminology used. Ursus and Tycho charge each other with theft of their 'constructions' and 'inventions', and Ursus characterises astronomical hypotheses as being constituted by 'circles and their mechanical delineation'. The absence of a modern conception of theory is evident in another way as well. Tycho complains of Ursus' plagiarism of formulae, and Ursus defends the originality of his contributions to the 'constructions' and tables associated with one of the formulae. The issue is seen in terms of isolated items, not in terms of a mathematical theory in the modern sense, in which formulae, 'constructions' and derived tables are seen to form a conceptual unit. Likewise where Tycho and Ursus charge each other with the theft of hypotheses, the issue does not concern an astronomical theory in the modern sense, a fully articulated world-system with associated planetary models and derived tables, but rather a bare specification of answers to a limited number of basic questions: Does the earth rotate or not? What is the order of the celestial orbs? etc. Only in the *Apologia*, where Keper insists that Tycho's claim to originality rests on his articulation of a detailed system of astronomical hypotheses answerable to careful observation, is something approaching the modern notion of a theory invoked.

However legitimate Ursus' grounds for seeking revenge, the *Tractatus* is a work of quite extraordinary scurrility and virulence. Stylistically the work combines invective with Menippean satire, a type of satire characterised by a jocular semicomical tone and the mixture of prose with verse. Nicodemus Frischlin's satire on his colleagues at Tübingen, to which Kepler compared it on the strength of Tycho's description of its contents, pales by comparison. A closer approximation to Ursus' virulence is to be found in the *Spongia* of 1578 and 1579 with which Ursus' colleague Hájek, the Imperial Physician, countered Annibale Raimondo's attacks on his works on the new star of 1572 and the comet of 1577.[18] But though close to the *Tractatus* in style and actually surpassing it in Rabelaisian fruitiness, they fall short in personal malice. To find full precedents in virulence and scurrility we must rather turn to vernacular political and religious satires of the period – such works as the notorious *Satyre Ménippée*, directed by the defenders of the liberties of the Gallican church, Jacques Gillot, Pierre Pithou and others, against the Catholic League, or the Calvinist Théodore de Bèze's anti-Papist *Epître de Benoit Passavant*.[19] Tycho did not exaggerate when he doubted if Kepler would ever read a more scurrilous libel. His rage can only have been increased by Ursus' shrewd choice of targets: his truncated nose, the illegitimacy of his children, and the arrogance of the monuments to himself erected on the island of Hveen. Our primary concern is with the relatively sedate section of the work to which Kepler replied in detail. But a couple of extracts from the meatier sections may help to explain the obsession with Ursus that Tycho so amply reveals in his correspondence. The first is from the invective poem which follows the dedicatory letter.

What you make much of, discerning double stars through the triple holes in your nose, is nothing to me. You carry on with great boasts about such things as handling sextants and placing quadrants, things even sailors know how to do. You don't know anything else, anything involving skill, that is. Indeed you know nothing if you don't know anything others don't know. Let me tell you a few things you don't know. Critic of others, answer these questions... 1. Critic, work out the proportions of a triangle, given only the angles, whatever its least angle may be. Don't do it by hand, and don't perform any multiplication or division...He who can untie these knots will be an astronomer. He'll be the astronomer, the Prince of Astronomers, as Tycho dares to boast himself. But he's a mere mechanic, for he doesn't understand these mathematical matters.[20]

The second excerpt is drawn from Ursus' reply to the published letter in which Tycho had accused him of *plagium*, literally 'abduction', of his hypotheses.

The word *plagium* applies to persons, and strictly to a wife or a daughter. But Tycho never married or had a wife. And his daughter, though the most nobly-born of girls,

---

[18] *Spongia contra rimosas & fatuas cucurbitulas Hannibalis Raymundi...* (Prague, 1578); *Ad secundas insanas cucurbitulas Hannibalis Raymundi...spongia secunda* (Prague, 1579). I have not seen Raimondo's attacks on Hájek, but from Hájek's quotations it seems that they may rival Ursus' *Tractatus* in nastiness.

[19] On the *Satyre Ménippée* see P. Demay's introduction to his edition (London, 1911); on de Bèze's pamphlet see P.-F. Geisendorf, *Théodore de Bèze* (Geneva, 1949), 47–51.

[20] *Tractatus*, sig. Biii, v–Biv, r.

was not yet nubile at the time I was there and so not of much use to me for the usual purpose. But I don't know whether or not the merry crew of friends who were with me had dealings with Tycho's concubine or his kitchen-maid.[21]

On the surface the *Tractatus*, with its repetitious digressions, anecdotes and interpolated letters, poems and tags, appears chaotic. But underneath it is, I think, systematic and closely thought out both in its attempt to discredit Tycho and in its promotion of a sceptical attitude to astronomy.

The work opens with two poems in execrable verse, one announcing the savagery with which he intends to pounce upon his ignorant detractors, Tycho, Rothmann (here, as elsewhere, written as *Rotzmann*, that is, 'sniveller' or 'snotface') and Roeslin, the other an encomium to his own skill as a mathematician. There follows the prefatory address to Moritz, Landgrave of Hesse. Here Ursus claims that he thought up his new world-system in October 1584. Jost Bürgi, clockmaker to the Landgraves of Hesse, can confirm that in 1586 he made a model of Ursus' system for Wilhelm IV, Moritz's father. He suggests that Rothmann saw this model and told Tycho about it. But instead of pressing this countercharge of plagiarism, Ursus' counterattack now takes an unexpected turn. There is here, he maintains, no issue of priority. For the Tychonic hypotheses were anticipated both by Apollonius of Perga and by Copernicus, though Tycho either is, or pretends to be, ignorant of this. Further, in claiming that he has derived his hypotheses from observations Tycho shows his inanity. There follows a diatribe against Tycho's arrogance which includes an account of the insults Ursus had received at Tycho's table during his visit to Hveen.[22] Ursus concludes by citing the public attack on his reputation by Rothmann and Tycho as the justification for his work and appeals to the Landgrave's goodwill as an impartial arbiter in the dispute. Next comes the invective poem, excerpted above, in which Ursus offers his mathematical challenge to Tycho in the form of five problems for solution.

The body of the work falls into seven fairly well-marked sections: 1) an account of the nature and purpose of astronomical hypotheses; 2) a history of hypotheses from antiquity; 3) a 'demonstration' of Ursus' own hypotheses from the scriptures; 4) a point by point rebuttal of the charges made in the published letters of Tycho and Rothmann; 5) an attack on Tycho's published claims to originality in the design of astronomical instruments; 6) an attack on Roeslin's pretensions as an astronomer and natural philosopher; 7) a

---

[21] *Tractatus*, sig. Fii, r. It was perhaps this passage which inspired Tycho, writing to Andreas Velleius in September 1599, to retail scurrilities about the morals of Ursus's wife (*T.B.o.o.*, VIII, 181).

[22] On these insults see Dreyer, *Tycho Brahe*, 274.

summary of his own and others' contributions to trigonometry as retailed in his *Fundamentum astronomicum*, a summary which provides the means for solving the problems posed in his initial challenge to Tycho.

Ursus' overall strategy is clear enough. The work is to discredit Tycho by refuting his claims to originality in the invention of his world-system, by revealing the dishonesty of his charges against Ursus and by showing that unlike Ursus he is not a true mathematician but a mere 'mechanic' concerned with instruments and observations. It is to discredit Roeslin by showing that his charge of blasphemy against Ursus' hypotheses is ill-founded and by mocking his pretensions to competence in natural philosophy. Certain of the points Ursus makes in his account of the nature and history of hypotheses clearly contribute to this general strategy. Thus by showing that rival hypotheses may provide equally effective means of saving the phenomena, Ursus exposes the fallacy in Tycho's claim that he had derived his hypotheses from observations. And the history is clearly designed to show that Tycho's claim to originality is false. But it is harder to discern the content or polemical purpose of certain others of Ursus' claims. Some of the difficulties of interpretation are as follows.

1) In his account of the nature of astronomical hypotheses Ursus variously asserts that it is impossible to arrive at true hypotheses, that all hypotheses are by definition false since *hypothesis* means 'false postulate', and that it does not matter whether astronomical hypotheses are true or false.[23] What view of the status of astronomy and astronomical hypotheses underlies these somewhat ill-assorted pronouncements?

2) As Kepler is again quick to point out,[24] there are gross inconsistencies between the claims Ursus makes about astronomical hypotheses in general and the claims he later makes about his own hypotheses. Of astronomical hypotheses in general he claims that all are probably false and that it does not matter if they conflict with the scriptures and with principles of natural philosophy since they are postulated only as predictive devices.[25] Yet later he claims that his own hypotheses are demonstrable from the scriptures and that he has, in addition, natural philosophical grounds for them.[26]

3) Only a part of Ursus' history of hypotheses is concerned with the question of Tycho's priority. What, apart from a display of erudition, is the purpose of the rest of his rambling history with its oddly facetious tone?

The first of these questions is examined in Chapters 6 and 7. There it is argued that Ursus is an exponent of a sceptical position widely adopted by astronomers of the period and that Kepler's response is to be considered not

23 *Tractatus*, sig. Biv, v.   24 *Apologia*, p. 148.
25 *Tractatus*, sig. Biv, v.   26 *Tractatus*, sig. Div, v.

as a defence of realism in general, but rather as a defence of astronomy against cogent sceptical attack. In the polemical strategy of the work Ursus' sceptical stance plays a double role. Given that the true nature of the world is inaccessible to astronomers the sole criterion of acceptability for astronomical hypotheses is their capacity to save the phenomena. It is therefore absurd for Tycho to make such a fuss about priority, as if a discovery about the nature of the world were at issue. And it is equally absurd for Roeslin to charge Ursus with impiety in proposing hypotheses which conflict with the scriptures, for his hypotheses are not proposed as novel truths about the world, but rather as predictive devices.

The second question is extremely awkward. The salient explanations of the apparent inconsistency between his general assessment of astronomical hypotheses as all probably false and of his own hypotheses as demonstrable on scriptural and other grounds are as follows: *either* one of the assessments is not to be taken at face value; *or* Ursus is prepared to adopt inconsistent premises for his various polemical purposes; *or* Ursus is greatly confused.

As a first step let us consider the arguments by which Ursus purports to 'demonstrate' the truth of his own hypotheses.[27] He considers half a dozen scriptural passages which have been alleged to show the immobility of the earth. Some of these can, he maintains, when properly interpreted, be seen to predicate immobility not of the earth but of the sphere of the fixed stars. The rest, when properly interpreted, can be seen to deny only the locomotion of the earth, not its rotation. Thus his hypotheses, the only hypotheses which keep the sphere of the fixed stars at rest whilst attributing only diurnal rotation to the earth, are demonstrated. He concludes: 'To him [Roeslin] I say that my hypotheses do not contradict the Sacred Scriptures and their authority, nor are they false as he falsely and dishonestly insists on several occasions.'[28] In addition Ursus argues that his own hypotheses are to be preferred to the Copernican hypotheses on the Aristotelian ground that a simple natural body can have but a single natural motion.[29]

If we elect to take these pronouncements at face value, how are we to understand the opening sceptical pronouncements? As a start we may note that Ursus uses the term 'hypothesis' in two distinct senses. In the opening section he treats astronomical hypotheses as planetary models designed to save the phenomena. Indefinitely many such models adequate to this task can, he implies, be thought up. But in the section of the work in which he defends his own hypothesis the term assumes a quite different connotation, that of a bare specification of basic principles. Thus his own hypothesis is specified

---

[27] *Tractatus*, sig. Div, v–Ei, v.   [28] *Tractatus*, sig. Eii, r.   [29] *Ibid.*

as being that the earth rotates, the moon revolves around it, the sun revolves around the earth, the other planets revolve round the sun, and the fixed stars are at rest. In this sense of the term there are just four such hypotheses, the Ptolemaic, the Apollonian (= Tychonic), the Aristarchan (= Copernican) and his own.[30] His demonstration of the truth of his own amounts simply to an elimination of the other three on scriptural and physical grounds. This suggests a reading which would go some way towards restoring consistency to the text. On this reading Ursus' demonstration of his own hypothesis is to be taken at face value; his opening sceptical comments are, however, to be taken not as a denial that it is possible to know which world-system is correct, but merely as a denial that we can know which out of the various planetary models consistent with the preferred world-system and adequate to save the phenomena is in fact the correct one. But whilst this reading saves Ursus from a flat contradiction, it is extremely forced. The opening section strongly implies, though it does not explicitly state, that it is not merely the detailed disposition and motions of the planets but the entire fabric of the world that is inscrutable. And it is this strongly sceptical stance that is needed if Ursus is to dismiss the priority dispute over the geoheliocentric system as much ado about a merely predictive contrivance. A reading according to which Ursus is prepared to adopt inconsistent premises for varied polemical purposes seems implausible. For what is involved here is not a cunning shift of stance, but a blatant apparent contradiction, hardly conducive to the task of persuading the reader of the justice of his case. And a reading according to which Ursus is grossly confused is scarcely plausible: idiosyncratic and uncontrolled though the work may be, it does not evince gross confusion elsewhere.

There remains only the possibility that it is the sceptical opening, not the demonstration of his own hypotheses, that is to be taken at face value. Here, I think, we are on firmer ground. Ursus' demonstration of his own hypotheses forms part of his reply to Roeslin. And this raises the possibility that his demonstration may be an argument *ex concessione*, that is, an argument from premises he denies but which his opponent accepts. His reply to Roeslin would then assume the following form. As a matter of fact the charge that my hypotheses are inconsistent with scripture and physics is irrelevant. For astronomical hypotheses are proposed only as predictive devices, not as portrayals of the fabric of the world. So it does not matter if they contradict the scriptures and the axioms of physics. But even if the

---

[30] *Tractatus*, sig. Diii, r, where the three rejected world-systems are listed and the physical arguments for and against them tabulated.

charge were relevant, it would be ineffective, for I can show that my hypotheses are consistent with the scriptures and with the axioms of physics. This reading, alas, faces certain difficulties. Ursus' *Fundamentum astronomicum* gives it little support. Though it does contain a Job-inspired outburst on man's moral and intellectual feebleness,[31] and expresses doubt about our capacity to resolve certain questions about the form of the universe – its finitude, for example – [32] it also offers twenty physical theses in support of the new hypotheses, which he asserts to be 'true and natural', to be 'not in the least at odds with the *regulae* of nature' and to provide the basis for a 'perfect astronomy'.[33] Further, if Ursus is arguing *ex concessione*, he surely ought not to have concluded as he does that Roeslin's charge of falsity against his hypotheses is false and dishonest, but merely that it is ill-founded.[34] Despite these difficulties there are solid grounds for accepting this reading. Ursus uses argument *ex concessione* elsewhere in the *Tractatus*, and to equally confusing effect, when he claims that he did not steal Tycho's hypotheses, but that had he done so it would only have been a 'philosophical' theft. (As was noted above, Kepler treated this unfortunately phrased defensive argument as a confession of guilt.) Further, his interpretations of scripture are so offhand as to make it hard to believe that he intended his demonstration of his hypotheses from Holy Writ to be taken seriously. And in his attack on Roeslin's pretensions as a natural philosopher he pours scorn on 'physics' and its practitioners – a puzzling stance if he seriously intended to vindicate his own hypotheses on physical grounds. If seen as offered *in propria persona* Ursus' 'demonstration' of his hypotheses is a bizarre performance; but as argument *ex concessione* to Roeslin, who took the role of scripture and physics in astronomy seriously, Ursus' demonstration assumes the form of effective parody and mockery. By contrast it seems hard to interpret the opening sceptical stance, backed as it is by cogent if somewhat incoherently presented arguments, as anything but a serious profession of belief. I conclude, as evidently Kepler concluded, that the sceptical arguments with which Ursus opens his *Tractatus*, arguments which are crucial for his attack on Tycho, are to be taken at face value.

The final question, that of the motives behind the parts of Ursus' rambling history of hypotheses that do not contribute directly to the refutation of Tycho's claim of originality, is more straightforward. Kepler may well be right in claiming that Ursus sets out to provide a convincing historical narrative in order to foist on the reader as historical facts certain 'lies'.[35] For

[31] *Fundamentum astronomicum*, f. 40r.
[32] *Fundamentum astronomicum*, f. 38v.
[33] *Fundamentum astronomicum*, f. 37r–39r.
[34] *Tractatus*, sig. Eii, r.
[35] *Apologia*, p. 158.

though Ursus' claim that Tycho was anticipated by Apollonius of Perga is not, as Kepler concedes, entirely unreasonable, the interpretation of passages from *De revolutionibus* on which he bases his claim that Copernicus had described the Tychonic system does appear to involve deliberate misrepresentation. But this is not the only motive. Ursus has already claimed, in support of his sceptical stance, that all hypotheses to date contain crass absurdities, and that the contrivance of predictively adequate but absurd hypotheses is an easy business.[36] His mocking presentation of the historical sequence of astronomical hypotheses as a series of silly fabrications is surely designed to substantiate this claim.

The extract from the *Tractatus* which follows includes both Ursus' account of the nature of astronomical hypotheses and his history of astronomical hypotheses, the parts of the work to which Kepler addresses his refutation. In translating I have taken considerable licence; too literal a rendering of Ursus' often barely grammatical invective would be unreadable.

ON ASTRONOMICAL HYPOTHESES, OR THE WORLD-SYSTEM

A hypothesis or fictitious supposition is a portrayal contrived out of certain imaginary circles of an imaginary form of the world-system, designed to keep track of[37] the celestial motions, and thought up, adopted and introduced for the purpose of keeping track of and saving the motions of the heavenly bodies and forming a method for calculating them. I say a contrived portrayal of an imaginary form of the world-system (not a true and genuine one, for that we cannot know), not the system itself, but a form of it of the kind which we think up by imagining and proclaim as a conception of the mind. These contrived hypotheses are nothing but certain fabrications which we imagine and use to portray the world-system. So it is not in the least necessary, nor is it necessarily required of the creators of hypotheses, that those hypotheses correspond altogether, that is to say in all respects or in all ways, to the world-system itself (I believe it to be scarcely possible to establish such hypotheses; for instance, there remain certain most crass absurdities everywhere and in whatsoever form of established hypotheses, of which forms people have indeed been able to think up and offer very many kinds), provided only that they agree with and correspond to a method of calculation of the celestial motions, even if not to the motions themselves, and that the calculation of the celestial motions can be preserved and maintained by or through them. For otherwise they would not be hypotheses, or

Margin notes:
Biv, v
What a hypothesis is

A hypothesis is a fiction

True hypotheses cannot be established

---

36 *Tractatus*, sig. Biv, v.
37 *Observare* clearly does not here mean 'to observe': cf. sig. Ciii, r, where Ursus cites Pliny's remark that the path of Mars is *inobservabilis*, that is, 'impossible to keep track of'. Kepler, perhaps deliberately, mistakes Ursus' meaning here: *Apologia*, p. 144.

(which is the same thing) fictitious suppositions, but true (not contrived) images of a true (not imaginary) form of the world-system. So hypotheses do not err in the least if they contradict the commonly held principles of other arts and disciplines, or, indeed, even if they contradict the infallible and certain authority of the Sacred Scriptures. For it is the distinguishing characteristic of hypotheses to inquire into, hunt for and elicit the truth sought from feigned or false suppositions. And so it is permitted and granted to astronomers, as a thing required[38] in astronomy, that they should fabricate hypotheses, whether true or false and feigned, of such a kind as may yield the phenomena and appearances of the celestial motions and correctly produce a method for calculating them, and thus achieve the intended purpose and goal of this art. Just this is customarily done in many other branches of learning, in which very often things which are not true, nor indeed even plausible, are cleverly assumed, and are, nevertheless, wisely proposed because things which are of the greatest usefulness follow from them; as, for example, in those hypothetical arithmetical rules of algebra, when a fictitious unity has been posited, and in the rule of the false, when a false and feigned number has been posited, we customarily find the true number sought.[39] We do not, however, err thereby, nor do we do anything contrary to the Sacred Scriptures or the principles of other branches of learning, even though in themselves these fictitious numbers, being false and feigned, would generally deceive. And so what we do could not be said with any justification to be the perpetration of an error, // still less to be absurd, even though it may appear to be so or be regarded as such. Therefore let them remember above all – those men who scoff at and hate me, those men who are, I declare, pettily spiteful, envious, trifling, ignorant dabblers and scoffers, men who with their childish little scribblings and tracts, recently published out of mere vanity and so as to be regarded as astronomers and as that which they are not in the least, spitefully hurl and vomit up at Copernicus and me also this charge: namely that we have sinned (in hypothetical matters, that is) against the authority of the Sacred Scriptures – let them remember, I say, those envious, etc., and wretched little men (what is, indeed, known to everybody) that hypotheses are hypotheses, that is, fictitious suppositions, and not at all things true or, in the common parlance, substantial. Behold how the wretched natural philosophers and petty academics, who like to appear to be and to be regarded as amongst the greatest natural philosophers, are nevertheless ignorant of this utter commonplace. But they carry on anyway, those men with their black bile, their envy and their most crass ignorance. Oh why, even though they wouldn't have understood them (for they don't understand even an iota in them, otherwise

---

[38] *Aetema* from Greek *aitema*, literally 'thing required' and hence 'thing a speaker asks an audience to concede' (cf. Aristotle, *Rhetorica*, XIX, 1433b, 17–18). In *Posterior Analytics*, I, 10, 72b, 23–4, Aristotle defines *aitema* as 'an assumption capable of proof but assumed without proof' and says that such an assumption is, relative to the learner, a *hypothesis*. Elsewhere he uses the term for 'a false or illegitimate assumption', and it is this pejorative sense that Ursus attaches to the term *hypothesis*.

[39] On these examples see Ch. 6, pp. 212–14.

they wouldn't have carried on as they did at all), didn't they at least read Copernicus' books, or even take a glance at the first little preface to Copernicus' books about the revolutions? Indeed, given that it clearly expresses and retails the same opinion as ours (in slightly different words), and given that some readers of these words of ours may perhaps not have it to hand, it is appropriate and fitting, and does not come at all amiss, to retail and present that preface with, however, a few passages scarcely relevant to the matter in hand deliberately omitted.

The distinguishing characteristics of hypotheses, from the first preface to Copernicus, by an author clearly extremely learned, but unknown.

It is the task of an astronomer to compose a history of the celestial motions through careful and skilful observation. Then, since he cannot by any means apprehend the true causes, he must conceive and devise causes or hypotheses of such a kind that when assumed they enable those motions to be calculated correctly from the principles of geometry, for the future as well as for the past. [...] Nor is it necessary for those hypotheses to be true or even probable; provided that they yield a reckoning consistent with the observations, that alone is sufficient. [...] For it is quite evident that the causes of the apparent motions are completely and absolutely unknown to this art. And if using his imagination he thinks up any causes, and he will certainly think up AS MANY AS POSSIBLE, he on no account // does so to persuade anyone that that is how things are, but merely to establish a correct method of calculation. Now when on occasion different hypotheses are available for one and the same motion (as, for example, eccentricity and an epicycle in the case of the sun's motion), the astronomer will adopt above all the one which is the easiest to understand. The philosopher will perhaps seek rather the semblance of truth. But neither will understand or state anything certain, unless it has been divinely revealed to him. [...] Nor should anyone expect anything certain from astronomy, so far as hypotheses are concerned, since it cannot provide anything of the sort, lest he should accept as the truth matters devised for another purpose and leave this discipline a greater fool than he entered it.[40]

The testimony of others

Ci, v

These are his words.

Now it is evident that without the assistance and help of such fictitious suppositions, or hypotheses, there is no way in which the motions of the heavenly bodies can be kept track of, the appearances of the heavenly motions saved, or, finally, the calculations of the appearances carried out; even though certain men ignorantly

Hypotheses are necessary

[40] Translation based on E. Rosen, *Three Copernican Treatises*, 3rd edn, revised (New York, 1971), 24–5.

affirm and state the contrary, dreaming and pretending about I know not what antique astronomy of the ancient Chaldaeans and Egyptians.[41] So the madness and insanity of certain men is to be hissed off the stage and laughed out of court – men who rashly deny the need for hypotheses and affirm that astronomy can be altogether established, completed and constructed by logical means alone (why, I ask, don't they do it, if they are able to?), as if logic alone would suffice for understanding everything and there were no need for the individual arts, arts for whose handing on logic ought to be considered only as the catholic organon, that is, the general and universal tool, since a distinction ought to be maintained between a tool and what it produces, that is, between the form and matter of philosophy, nor ought everything to be compounded and mixed together and every art stuck into one jar labelled 'logic'.[42] Besides, you would not deny that there was once practised by the ancients a more perfect, more complete and much more expedient astronomy than

Comparison with other arts

our present-day astronomy, just as there was once a Thessalian medicine more perfect than Galenic medicine, which was lost sight of owing to Galen; and the more sensible of the modern medical writers suppose and conclude that Galen maintained, propagated and left behind the husks in place of the kernels, the chaff in place of

[41] The reference is, as Kepler notes in the preface to the *Apologia*, to Ramus' claim that the Babylonians and Egyptians had a perfect astronomy without hypotheses. Ramus presents this claim briefly in his *Prooemium mathematicum* (Paris, 1567), 214–15, quoted in Ch. 8, pp. 266–8; he had developed it in more detail in a letter of September 1563 to Rheticus, published in J. Freigius, *P. Rami Professio Regia...* (Basel, 1576), sig. ☞ 1r–2v, in which he urges Rheticus to attempt a restoration of the pristine astronomy without hypotheses. As H. Hugonnard-Roche *et al.* suggest in '*Georgii Joachimi Rhetici Narratio Prima*: édition critique, traduction française et commentaire', *Studia Copernicana*, xx (1982), 238, Ramus may have been encouraged by Rheticus' enthusiastic remarks about Egyptian astronomy in the preface to his edition of Johann Werner's *De triangulis sphaericis* and *De meteoroscopiis* (Cracow, 1557).

[42] Again the primary target is Ramus, who had attacked the use of hypotheses in astronomy and had suggested that a restored astronomy could be derived by logic alone from tables of observations. Another possible target is Rheticus. Nowhere in his *Narratio prima* does he imply that hypotheses can be dispensed with in astronomy, but in the course of his reply to Ramus' letter urging him to construct a hypothesis-free astronomy, Rheticus had announced his commitment to 'the task which you too have in mind, to free the astronomical art of hypotheses, being satisfied with observations alone': see K. H. Burmeister, *Georg Joachim Rhetikus, 1514–1574: eine Bio-Bibliographie*, iii (Wiesbaden, 1968), 190. This letter was published in Josias Simler, *Bibliotheca instituta et collata primum a Conrado Gesnero* (Zurich, 1574), 288. On the Ramus–Rheticus correspondence see R. Hooykaas, *Humanisme, science et réforme: Pierre de la Ramée (1517–1752)* (Leiden, 1958), 68–9, and E. Rosen, 'The Ramus–Rheticus correspondence', *Journal of the History of Ideas*, I (1940), 363–8. The remark may even be intended to apply to Tycho, whose claims to have derived a restored astronomy from observations Ursus mocks in the preface and elsewhere. Tycho had met Ramus at Augsburg in 1570, and it may be that, as Dreyer suggests, he was influenced by Ramus' plea for a restoration of astronomy founded on observations: Dreyer, *Tycho Brahe*, 34.

the grain and, in fine, the corpse instead of the living body.[43] Likewise it is agreed that there was once practised amongst the most ancient Greeks a far sweeter and more delightful music than that customary with us today; and it was so to such an extent that when there was a mere change in the mode of singing and harmony, as from Dorian to Phrygian or Lydian, or *vice versa*, suddenly and immediately the most diverse of men's affections would follow along with the diverse modes of incantation. So, for example, when men who had once heard one mode and were fighting most fiercely amongst themselves, immediately on hearing another different mode, behold, they would embrace one another lovingly, something which present-day singers are not able to effect with the music customary today,[44] just as no present-day orator could move men by an // encomium of death, and so drive them to madness and coax them on that they killed themselves. But would to God that that ancient medicine and that pristine music could be brought to light and restored by present-day physicians and musicians to the same extent as, with the help of almighty God, that antique astronomy of the ancients will be restored by us in a few words, as we are about to attempt the foundation of this art at the end of our little work and produce a sample of it. But, to return to the point, I declare that no argument could lead me to believe that without hypotheses that old astronomy of the ancients was preserved and handed on, and the calculation of the heavenly motions duly carried out. For, I ask you, what sort of astronomy would there be if that Platonic pair of wings, I mean arithmetic and geometry, were taken away, cut off or rejected?[45] Surely it would be lame, mutilated and altogether imperfect. For in the calculation of the heavenly motions one has to know the periodic time and hence the periods themselves. But these very periods are most suitably portrayed and represented by imaginary circles. And, to conclude, those very circles and their mechanical delineation comprise and constitute the hypotheses themselves. Behold, the utter necessity of hypotheses is now clear and evident. So hypotheses are necessary to such an extent that without them nothing certain or true can be established and attained in making calculations and observations of the

Cii, r (margin, right)

Astronomy was never without hypotheses (margin, right)

---

43 By Thessalian medicine Ursus presumably means the medicine of Thessalus of Tralles. Paracelsus may be one of the authors Ursus has in mind. He castigated Galen, admired Hippocrates' *Aphorisms* and wished to restore the *prisca medicina*. A more circumspect believer in the superiority to Galenic medicine of a *prisca medicina* was Vesalius: see W. Pagel and P. Rattansi, 'Vesalius and Paracelsus', *Medical History*, 8 (1964), 309–28.

44 There are numerous classical allusions to the specific affective powers of the modes, and stories of this sort are commonplace in the period. A typical example, and a possible inspiration for Ursus' remarks, is provided by Agrippa von Nettesheim, *De occulta philosophia* (Cologne, 1533), who characterises the Dorian mode as soothing and the Lydian and Phrygian as inflammatory (p. CLVII) and retails, along with the stories of David and Saul and of Alexander and Timotheus, the tale from Saxo Grammaticus' *Historia Daniae* of an ancient Danish king first soothed and then roused to murder by a musician.

45 On this image see Ch. 5, fn. 168.

heavenly bodies. It is therefore clearly reasonable to suppose that the method of hypotheses was known even to the most ancient astronomers, and is much older

History of the invention of hypotheses

than is commonly supposed. Indeed, one Eudoxus (of Cnidos or Gnidos), a contemporary of Plato, first established the concentric hypotheses. And later the Pythagoreans, having spurned the inadequate concentrics, adopted eccentrics. The warrant for this view is to be found in the authority of the writers who maintain it.[46] For I know that in Aristotle's *Metaphysics* mention is made of certain hypotheses peculiar to Eudoxus and Calippus, contemporaries of Plato.[47] But as to their nature and whether they were the first invented hypotheses of all, the first point is dealt with only obscurely in the text and the second not at all. Now it is generally agreed that during the reign of Kings Nebuchadnezzar the Great of Babylon, Astyages of Media and Halyattes of Lydia, about the time of the 49th Olympiad and the 170th year of the city of Rome, that is, in 580 BC, a little after the time of the destruction of Jerusalem by the Chaldaeans and a little before the Medo-Lydian war waged between Astyages and Halyattes, Thales of Miletus predicted a solar eclipse.[48] And Thales lived about two hundred years before Eudoxus and Calippus. But I simply do not see by what means he could have foretold it without a method of calculation or by what reasoning he could have employed or set up a method of calculation without the help of periods or circles and hypotheses. Hence it turns out to be credible, probable and in accord with reason that either Thales himself, the father

The ancient hypotheses are unknown
Cii, v

of Ionian philosophy, or his near contemporary Pythagoras, the founder of Italian philosophy, first introduced hypotheses, or else they were known even before the time of these men. But the nature of these original hypotheses of the ancients before Aristotle and Plato is clearly unknown and altogether uncertain. For in the time // of Plato and Aristotle hypotheses of the sort that are nowadays employed and published in various forms were as yet unknown and not made public. This is made clear and obvious by the following considerations. Both Plato in *Timaeus* and Aristotle in his books about the universe, *Meteorologica* and *De caelo*, place and locate the lights or luminaries, the sun and moon, beneath the five wandering planets, contrary to the laws, postulates and principles of the customary hypotheses,[49] not to mention the opinion of Alpetragius, cited by Copernicus in Book 1, Chapter 10, according to which Venus is above the sun and Mercury below it. So to me it would seem probable that it was a little after the time of Aristotle and Plato that hypotheses or types of hypotheses of the sort that are nowadays commonly contrived and proclaimed as 'physical hypotheses' by practitioners of physics were introduced and became

The physical hypotheses

prevalent. In these the earthen globe, together with the water and the air which

---

[46] Ursus' source for these claims is probably Ramus, *Prooemium mathematicum*.

[47] *Metaphysics*, XII, 8.

[48] The sources include Diogenes Laërtius, *De clarorum philosophorum vitis*, I, 23–4; Herodotus, *Historiae*, I, 74; Pliny, *Naturalis historia*, II, 53; and Clement of Alexandria, *Stromata*, I, 65. Of these, Laërtius gives no date, Herodotus the sixth year of the Medo-Lydian war, Pliny the 48th Olympiad and Clement the 50th Olympiad.

[49] *Timaeus*, 38C–39D; *Meteorologica*, I, 4; *De caelo*, II, 12.

everywhere and on all sides surrounds it, with the moon riding on the air (just like a little bird or boat swimming on water), occupies the middle place of the heavens and is like a midpoint or centre of the outermost firmament of all. Between the earth and the heaven the remaining planets (starting with the earth) are placed in the order Cynthia, Mercury, Venus and the sun, Mars, Jupiter, Saturn – the reverse of the order in which they are customarily listed and enumerated. Thus arranged all of them, like the mobile heaven itself, go round the earth, their immobile centre, concentric to it and hence homocentric or (what comes to the same thing) parallel with each other. And these hypotheses, being very simple and in the closest agreement with natural phenomena and the system of the universe itself, were certainly in use before the others that are now employed, and they remained in use right up to the time of Ptolemy Philadelphus (or, according to others, Ptolemy Philometor), King of Egypt, during whose reign one Aristarchus, by nationality either a Samian or a Samothracian, is said to have flourished. He is said to have been the author and inventor of those hypotheses which, in the memory of our grandfathers, now almost a hundred years ago, were again adopted and restored to light by Nicolaus Copernicus the Prussian.[50] In these hypotheses he transposed and converted the places *The hypotheses* of the sun and earth, leaving all the other bodies in their places; and by an act of *of Aristarchus of* imagination he, so to speak, transferred and relocated the sun to the place of the *Samos* earth, and thus in the middle of the heaven or firmament; and, conversely, he transferred and relocated the earth, together with the air surrounding it and the moon that rides upon the air, to the place of the sun. And he most skilfully and cleverly postulated and imagined the heaven and sun, that were previously mobile bodies, as being immobile bodies; and, conversely, he postulated and imagined the earth, that was previously still, as being unstable or in motion, being moved by a two or threefold motion (that is, by diurnal rotation – *kulisis* – and annual revolution – *dinesis* – and, lastly, by inclination or nutation, pitching to and fro like a ship), and as revolving amongst the planets in the aether and being like a planet. And, moreover, *Why the* he did not do this gratuitously, but in order to account for an anomaly or irregularity *hypotheses were* of the apparent motions, which was quite impossible to account for on the physical *changed* hypotheses I have mentioned. For had the earth been in the midst of all and truly placed in the middle with all the bodies going round it concentrically, and if in going round the earth each had traversed equal arcs of its circle in equal times, then of necessity and on optical grounds no irregularity or anomaly whatsoever in the apparent motions would have appeared or been observed; // rather, simple analogy *Ciii, r* and regularity would have followed and appeared; but since in fact the contrary happened and manifested itself, it was necessary to assume and introduce something contrary [to the physical hypotheses]. Now these hypotheses of Aristarchus were afterwards changed into another and inverted form by a man called Apollonius of Perga, who is said to have flourished a little after the birth and passion of Christ our Lord. And in that form they were lately revived by Tycho Brahe, a Danish

---

[50] Ursus takes at face value Copernicus' claim in the preface that his work had lain among his papers 'by now for the fourth period of nine years'.

47

nobleman, but without the same justification as Copernicus had in reviving those of Aristarchus. And he published them as his own and impudently tries to pass them off as his own work either through ignorance or knavery. He does so despite the fact that Copernicus clearly gives an account of them in Book 3, Chapter 25, and in Book 5, Chapters 3 and 25, derived from Martianus Capella, the ancient encyclopaedia writer, whose very work he cites after his own to provide a more evident and solid demonstration of the matter.[51] And moreover this year a certain Helisaeus Roeslin has imitated Tycho in the little book he published on the work of the creation, like an ape aping an ape, and has published as his own the same hypotheses of Apollonius with no less disgrace and shamelessness than Tycho – whether he plagiarised Tycho, or, like Tycho, he plagiarised Apollonius. And he did not blush at all to assert most impudently that these hypotheses – I do not waste time worrying about his other hypotheses – were, by divine grace (if the gods approve), eventually discovered by him. And he wickedly added and appended this: 'When I had these thoughts about the world-system and had embodied my opinion in the theses set out here, the treatise of Tycho Brahe, the Danish nobleman and most excellent astronomer, about the comet of the year 1577 was unknown to me. In that book he did indeed derive a new system of the world almost the same as mine.'[52] Why not, indeed, 'exactly', since they are precisely the same. That is what he says, as if he had not earlier seen that book of Tycho's published in 1588, I think, ten whole years before.[53] That he had never read Copernicus, still less understood him, I can easily believe and readily concede to him; for it becomes very evident from the many crass absurdities in his hypotheses, illegitimate, spurious and already known, rashly and ignorantly perpetrated by him. Especially so, in that as a man clearly altogether unskilled in the business and the observations of astronomy (thus he cackles, like a goose among swans)[54] he arranges and disposes the circuit of the sun and the three superior planets at distances from each other as great as that of the sun's circuit from the surface of the earth. This is against all reason, authority and testimony of observation. For Georg Joachim Rheticus, at once the most excellent astronomer and mathematician, disciple and emulator of the preeminent astronomer Copernicus, says in his first account to Johannes Schöner of the books on the revolutions by his master Copernicus: 'Although he (that is, Pliny, Book 2) states that the path of Mars is impossible to follow, and the difficulties in correction of the account of Mars' motion are greater than for the others, there is no doubt that it sometimes shows greater diversity of aspect (which they call parallax) than

*An error in
Roeslin's
hypotheses*

*The testimony of
others*

---

[51] Both here and below Ursus completely misrepresents *De revolutionibus*, III, 25, which has no connection whatsoever with the views of Apollonius or Martianus Capella. The references to Apollonius' lemmas in *De revolutionibus*, v, are derived from Ptolemy, *Almagest*, XII, I, and not from Martianus Capella, who never mentions Apollonius.

[52] Roeslin, *De opere Dei creationis*, 49–50 [52].

[53] I.e., Tycho's *De mundi aetherei recentioribus phaenomenis* (Uraniborg, 1588).

[54] Cf. Virgil, *Eclogae*, 9, 36.

the sun itself.'[55] This is what Rheticus says. But if Mars sometimes shows greater parallax than the sun itself (and that it does so when it is setting and in opposition to the sun is surely established by the corroborating and confirming testimony of observation),[56] // it is also necessary for optical reasons that it sometimes draws nearer *Ciii, v* to the earth than the sun itself, and so on occasion descends perceptibly within the solar circuit and from time to time, almost once a year, crosses and passes through that solar circuit and descends far below it. This crossing is commonly called the interpenetration of orbs, and Roeslin himself has the cheek to count it among the absurd factual impossibilities contrary to physical principles. Behold the astronomer resisting the evidence of his own eyes and heedlessly raving on despite the evident testimony of the parallaxes. Truly, after originating and perpetrating this absurdity he attempts to establish, confirm and found too large a gap and intervening distance between the orbs of the planets in his (as he would have it) hypotheses, though the establishment and foundation are clearly at odds with experience itself. And except for this they differ in no respect whatsoever from the hypotheses of Tycho or Apollonius. But not content with this he puts forward as the principal purpose and *A more serious* end of his gappy hypotheses something else besides, namely, the discovery of the *error in* altitude of the heavens,[57] which he would like to discover and know at will, as if *Roeslin's* he agreed precisely with that Brentian[58] attitude to the journey and to the distance *hypotheses* to be covered or traversed in ascending from earth up to heaven. He says this in thesis 55 on page 23. On this basis the distance from the spheres of the superior planets to the fixed stars will be known. Why not? Read, I ask you, the whole of this most absurd thesis, which contradicts and errs against experience and the Sacred Scriptures themselves – Syracides, Chapter 1, 18 (or 17) and 43 – [59] for it wearies and shames me to describe it all. He eventually concludes as follows: 'Since God's work of creation is closed in and bounded by the outermost heaven of the fixed stars, *I judge'* (these are his words)[60] 'that God the Creator so distributed the spheres or places of the planets out towards the fixed stars, and gave each of the planets such an amount of space that a definite boundary could be established. But by no other distribution of spheres except MY system of the world' (to abolish MINE and YOURS would

[55] Rheticus, *Narratio prima* (Danzig, 1540); transl. by Rosen, *Three Copernican Treatises*, 136.

[56] Ursus' source is probably *Epist. astron.*, in which Tycho claimed that he had observed the supposed phenomenon in 1582 (*T.B.o.o.*, VI, 70 and 179); later he refers explicitly to Tycho's claim (sig. Fii, v).

[57] Roeslin, *De opere Dei creationis*, 23.

[58] Brentius = Andreas Althamer of Brenz, a Lutheran theologian (not to be confused with Luther's celebrated colleague Johann Brentz), composed a popular work reconciling apparent inconsistencies in the Bible, *Diallage S. conciliatio locorum scripturae qui prima facie inter se pugnare videntur* (Nuremberg, 1528).

[59] Syracides = Jesus ben Sirach, author of *Ecclesiasticus*. The verses are: 'Who hath measured the height of heaven?' and 'Seek not the things that are too high for thee'.

[60] αὐτὸς ἔφα [αὐτὸσ ἤφα].

be to abolish all evil, says Seneca)[61] 'can a constant boundary be established on definite grounds.' This is what he says. What a noble and excellent discovery or rather fabrication, by which, praise God, we can find out the height of heaven quite exactly to a hairsbreadth! Behold the discovery of this learned person (that plebeian)[62] and how pleasant it is for him to be pointed out and have it said: 'Here he is, the gappy Doctor.'[63] But, you inept and bungling, not to say stupid, man – 'Who hath measured the height of heaven, etc.?' How does that famous preacher appeal to you? Surely you yourself, my dear Roeslin, have done just what you blurt out as an accusation against me and Copernicus, saying that we have posited something contrary to the Sacred Scriptures. You do it quite obviously, nor can you deny it without shame and with a clear conscience, nor is there any way of escape for you – all are blocked. But, in contrast, I shall show incontestably in my demonstration of my hypotheses from the Sacred Scriptures that I did no such thing. But this is in passing and by way of enlarging on the theft of Apollonius of Perga's hypotheses. Now follows a description of them and the reason and motive for producing them by modification of the Aristarchan hypotheses. Apollonius, indeed, changed two things in them. Firstly, he made the heaven or firmament and orb // of the fixed stars concentric with the earth not the sun; and (before the next innovation, which followed soon after) he made it, like the moon, rise up and fall down along with the earth and always be the same distance from the earth. For he observed that if the earth does not stay in the middle of the fixed stars, then necessarily for optical reasons those fixed stars should appear larger when brought nearer the earth (the motion by which it travels round necessarily bringing this about) and, on the contrary, smaller when further from it and separated by a greater distance. But since this does not happen or appear he could not agree with these Aristarchan eccentricities relating the orb of the fixed stars to the earth, except by conceding and admitting a vast and immense gulf of space between the circuits of the planets and the orb of the fixed stars. And when a certain Christopher Sniveller,[64] bitterly and tenaciously defending these Copernican hypotheses (as he thought them, but in fact they are Aristarchan), was not able to escape this detestable and laughable consequence, he fell so far into idiocy as to say and write with his most impudent mouth and pen that a single fixed star even of the third magnitude is equal in size to the whole of the annual orb,[65] that is, to the solar circuit, in diameter and

Syracides,
Chapter 1

The hypotheses
of Apollonius of
Perga
Civ, r

Sniveller's new
monstrosity

Tycho's letters,
pages 186, 191
and 192

61 *Tolle MEUM atque TUUM, tollitur omne malum*: this is not in any work now attributed to Seneca. *Quieta vita iis qui tollunt, Meum, Tuum* occurs in the *sententiae* attributed to Seneca in the period, and a more verbose version occurs in the *Liber de quattuor virtutibus* (= *Formula vitae honestae*) of Martinus of Bracara, formerly attributed to Seneca: see I. C. Ortelli, ed., *Publii Syri mimi et aliorum sententiae* (Leipzig, 1822), 251.   62 ἀστὸς ἐστι ἐκεῖνος.
63 *Doctor syncopatus*: presumably a play on the 'gappiness' of Roeslin's hypotheses, mentioned above.
64 Here, as elsewhere, Ursus has *Rotzmanus* for *Rothmannus*.
65 Ursus' reference to *Epist. astron.*, 186, is to a letter of April 1590 to Tycho in which Rothmann makes this claim (*T.B.o.o.*, VI, 214–17). The other two references are to relevant parts of the reply to Rothmann's defences of the Copernican hypotheses

circumference. But it follows that the orb or sphere, of which the solar circuit is the circumference or great circle, corresponds to that magnitude. Behold, the little star of Sniveller! But if he had either known or understood the Apollonian means of escape just indicated, the wretch would not have fallen into such idiocy, and there would have been no need for him to posit so vast a star and such an immense aberration of nature. Truly (as the poet says) 'while there are fools around, they'll get into all sorts of trouble'.[66] But enough of Sniveller's little star and the first of Apollonius' alterations. Secondly then, as if taking a compass in his hand, one foot of the compass placed in the centre of the earth at rest, and the other in the centre of the sun and there delineating and describing a circle, he returns as if in imagination each body, both the earth and sun, to its original and proper place from which it had earlier been removed by Aristarchus; he quite rightly gives back to each its place; and he, so to speak, recalls them to their previous positions, the sun to the place of the earth and the earth to the place of the sun, and leads each of them back home to its own dwelling. But he leaves the other five planets in that place into which they had been relocated from the physical hypotheses and maintained by Aristarchus. So he agrees that in the case of the five planets nothing should be changed except their centre. Further, he makes the earth immobile, but on the other hand he again holds and resolves that the heaven is moved by one proper motion, and that the sun is moved by one proper motion and besides by one which happens and is brought about by the impulsion of the heaven. And this is the second alteration due to Apollonius, made and introduced for the purpose of escaping the absurdity of the revolution round the great orb or period that the earth had accomplished and completed [according to Aristarchus], and for the sake of avoiding its triple motion. And this deals with these three hypotheses and with all the concentrics, whether, as in the physical hypotheses, centred on the earth or, // as in those of Aristarchus Civ, v and Apollonius, centred on the sun. And, indeed, since this syncentricity or concentricity did not yet fully suffice for the motion of anomaly and the more particular aspects of its inequality, another form and disposition was contrived and introduced. It was for this reason that Ptolemy of Pelusium,[67] or otherwise of Alexandria, about four hundred years after Apollonius, in 140 AD, under the Highest, Greatest and Most Excellent Roman Emperor Antoninus Pius, set out and expounded different hypotheses, whether thought up by himself or held by others and handed on. And it is these, which are in his great and magnificent published work which he dared to call *Kat'exochen megale suntaxis*,[68] or *Great Compilation*, that are still

which Tycho interpolated into his letter book immediately after this letter (*T.B.o.o*, VI, 218–22).

[66] *Dum vitant vitium stulti in contraria currunt* is a misquotation of Horace, *Sermones*, I, 2, 24: *vitium* should be *vitia*.

[67] A misunderstanding of *Phelud(i)ensis* attached to Ptolemy's name in mediaeval texts, itself a corruption of the Arabic *qualudi*, in turn a misunderstanding of the Greek *klaudios*: see G. J. Toomer, 'Ptolemy', in C. C. Gillispie, ed., *Dictionary of Scientific Biography*, XI (New York, 1975), 186–206.

[68] 'Supremely Great Compilation'.

extant and to be seen today and have always held first place in the academies and still hold it today. And later Proclus Diadochus of Lycia explicitly dealt with and wrote about these hypotheses that Ptolemy, so to speak, intimated through examples.[69] In these the hypotheses of Aristarchus and Apollonius are reduced almost to the original form of the physical hypotheses; for epicycles are posited, and the epicycles of the three superior planets go round the earth as their common centre on eccentric deferents in the place and with the magnitude of the period of the sun; and in place of the periods of Venus and Mercury, which go round the sun in the other [hypotheses], are the epicycles which are in conformity with their greatest egressions from the mean motion, that is, their anomaly. Moreover, the ratio between the distances of the eccentric circles bearing epicycles from the centre of the earth and the distance from the earth to the sun is made evident by the sines of maximum egression on either side of the mean motion (that is, the anomalies) of the planets set in and carried by their epicycles. For by Chapter 4 of Book 6 of Euclid on the principle of comparison of sines of angles and sides, in each case the ratio of the sine of the anomaly or maximum egression from the mean motion to the radius of the epicycle (or, as comes to the same thing in the three superior planets, the radius of the period of the sun or distance of the sun from the earth) is equal to that of the whole or right sine to the radius of the eccentric bearing the epicycle.[70]

The hypotheses of Ptolemy

A method for finding out the amounts of the differences or distances of the planets from the earth

Another demonstration of the error in Roeslin's hypotheses

From this reasoning and its antithesis or converse the manifest falsity of Roeslin's spurious hypotheses comes to light and is made clear. For if one posits a period of Mars with a diameter twice as great as the period of the sun, as, indeed, Roeslin supposed, the maximum egression or anomaly of Mars in its epicycle on either side of its mean motion will be precisely and exactly 30 degrees; and whether that is true I leave to practical men and especially to those who observe the stars, indeed to Tycho himself, Tycho to whom Roeslin so suavely appeals at the end of the fanatical and aberrant little work, whose judgement he wants to hear and to whose criticism he wants to submit his theses – why not his hypotheses as well?[71] So,

[69] Ursus' point is that although in the *Almagest* Ptolemy presented only individual planetary models, Proclus in his *Hypotyposis* attributed to him a complete world-system. In fact Proclus had in mind Ptolemy's *Planetary Hypotheses*, a work not known to Ursus and his contemporaries.

[70] The 'sines' here are lengths of chords subtended by angles. The proposition from Euclid is: 'In equiangular triangles the sides about the equal angles are proportional, and those are corresponding sides which subtend the equal angles.' Its relevance may not be immediately obvious: as pointed out to me by Dr Michael Hoskin, the proportion is invoked to justify a trigonometrical formula which applies to all members of a class of similar triangles. The method of calculation of the ratios of the radii of the planetary orbs to the earth–sun distance which Ursus retails here is clearly explained by T. S. Kuhn, *The Copernican Revolution* (Cambridge, Mass., 1957), 174–5. As Kepler remarks in the preface to the *Apologia*, this was a commonplace method in the period.

[71] The body of the text of Roeslin's *De opere Dei creationis* consists of 124 physical theses, a hotch-potch of Aristotelian, Paracelsian and neo-Platonic cosmological dicta rather than a systematic body of natural philosophy. His 'new' world-system

Tycho, you shall be our most impartial arbiter. // And by the same method one   Di, r
could find out whether the opinion of M. Johannes Repler, Mathematician to the
Noble Provinces of Styria, about the mutual distances of the periods or orbs of the
planets, is in all respects true – though I am sure it is truer than Roeslin's hypotheses.
On this opinion he has announced in the latest catalogue a book by the title of
Prodromus dissertationum cosmographicarum, etc.[72] And he wrote me, now about two
years ago, this letter worthy of publication about his arguments.

> To Nicolaus Raimarus Ursus of Dithmarschen, most noble Mathematician to the
> Emperor, in Prague.
>
> Unknown men who write letters to those they do not know in distant parts
> are to be wondered at. Your remarkable fame, according to which you
> outshine the mathematicians of this age as the orb of Phoebus outshines minute
> stars, first made you known to me. But this is not the right occasion to say
> more, nor is garrulity proper to mathematicians. Accept just this: I make only
> as much of you as do all learned men, whose judgement it is arrogance to
> ignore but befits the modesty of youth to applaud. So, since it was with you
> as teacher, that is, your books, that I had acquired what modicum of
> knowledge I have in mathematics, I was led to consult you in a matter which
> is difficult, but not, it seems to me, unworthy. Should you approve what I
> affirm, I shall count myself blessed. And I shall account it the next degree of
> happiness should I be corrected by you, so high is my opinion of you. I admire
> your hypotheses; but there is no end to my admiration of Copernicus, whose
> hypotheses contain that which I have woven into these verses.
>
> What the nature of the universe is; what were God's purpose and plan in creating
> it; whence God derived the numbers; what rule governs so great an edifice; why
> the circuits are six in number; for what reason spaces lie between all the orbs;
> why Jupiter and Mars, whose orbs are scarcely the first, are separated by so great
> a gulf: behold, it is revealed by the five figures of Pythagoras.[73]
>
> For between Saturn and Jupiter is a cube, so placed that the inner period of
> Saturn is a circumscribed orb and the outer period of Jupiter an inscribed orb.
> Between Jupiter and // Mars is a tetrahedron; between Mars and the earth   Di, v
> a dodecahedron; between the earth and Venus an icosahedron; between Venus
> and Mercury an octahedron. Neither metaphysicians nor mathematicians have

and his account of rival world-systems constitute an 'Appendix of Hypotheses'.
At the end of the work he asks for Tycho's judgement on his theses.

[72] Kepler's work was advertised in the Frankfurt Book-Fair catalogue for 1597 in the
form *Prodromus discertationum Cosmographicarum a M. Joan. Repleo Wurtembergico
Mathematico Tubingae Georgius Gruppenbach in 4* (*K.g.W.*, VIII, 479): hence Ursus'
misspelling of Kepler's name. In the *Erratula* to the *Tractatus* Ursus corrects the
spelling (sig. K1, r).

[73] An expanded version of this poem appears as a greeting to the reader at the
beginning of the *Mysterium* (*K.g.W.*, I, 4). In translating it I have been helped by
A. M. Duncan's translation in *Johannes Kepler: the Secret of the Universe* (New York,
1981), 49.

ever with good reason changed the order of these bodies. Now the mean motions too are in agreement. For they are doubly dependent on the distances: firstly, in that (supposing the impulsion of their parts to be the same in all of them) the greater the amplitudes the more slowly they are moved round; and then there is the weakening [of impulsion] in the outer ones of the kind which happens by extenuation of rays of light. There is little discrepancy from Copernicus in either direction; more, however, if the distances are derived from the mean motions rather than from the bodies. For from the correction of the distances by [considering] the bodies there follow differences from the prosthaphaereses of the apogees no greater than 12′ in the case of Saturn, 25′ in the case of Jupiter, 1° 45′ in the case of Mars, 1° in the case of Venus and 4′ in the case of Mercury.[74] I shall say no more, awaiting your judgement. And you will not be imposing on the goodwill of the noble youth Master Sigismund Waganer, at whose instigation I write, if you send it [by him] on this or the next occasion. Flourish ornament of Germany for the sake of the stars and our science. 15th November in the year of Grace 1595.

From your disciple, Master Johannes Repler, Mathematician of the local Nobility of Styria.

But so far I have dealt with the Ptolemaic hypotheses, or, at least, those first disseminated by Ptolemy, for it is uncertain whether he was their author and inventor. For he indicated that Hipparchus (who flourished 266 years before Ptolemy) rejected the tenets of Eudoxus and Calippus, and was the first to introduce eccentric circles.[75] These Ptolemaic hypotheses alone remained in use in later generations, and they were adopted by the Arabs and by the Alphonsines, and the Alphonsine Tables were based and composed on their model. And they prevailed until Nicolaus Copernicus and the year of our Lord 1500. And about that time, or a little later, that same Copernicus, as we said earlier, revived and readopted the hypotheses of Aristarchus and emended them by means of the Ptolemaic hypotheses; and other things that were imperfect and insufficient to the appearances of the motions he rendered and set out in a perfect and sufficient form. And he did so by an application and adaptation of the eccentricities of the Ptolemaic eccentric deferent

Dii, r circles // to Aristarchus' periods of the planets described about the sun; and he provided an epicycle in the periphery of the periods almost as Ptolemy did in the circumference of the eccentric, maintaining the period of the sun (which he wishes to declare to be an annual great circle) in place of the epicycles of the three superior planets. On this basis, moreover, he attempted like Ptolemy to escape from and remove the other particular diversities (as they call them) in the inequality of anomaly

---

[74] These figures differ from those given in Ch. 15 of the *Mysterium*. Kepler was led to revise his original estimates by Maestlin's criticisms; for a detailed account of Maestlin's role in this and other emendations of early drafts of the work see A. Grafton, 'Michael Maestlin's account of Copernican planetary theory', *Proceedings of the American Philosophical Society*, 117 (1973), 523–50.

[75] *Almagest*, III, 4.

which remained in the hypotheses of Aristarchus because of the concentricity; and the attempt was carried out most felicitously. So too in our age did Tycho Brahe begin to attempt to do for the hypotheses of Apollonius what Copernicus did for the hypotheses of Aristarchus; and he resolved to base his repaired astronomy on them (Would that there were no need for repair!) and, so he says, to recall and *Letters, pages* restore the whole art and emendation of the celestial motions from reliable *104 and 117* observations, as if from first elements.[76] And after that he applauds himself with his rattling stork's bill,[77] arrogating to himself with shameless effrontery his *page 246* QUID SI SIC and vainly putting forth as his own handiwork his new discovery of hypotheses.[78] And so far we have considered the Ptolemaic hypotheses and, as well as the physical hypotheses, the Aristarchan and the Apollonian, both of which are to be found and are explicitly described in Copernicus' books of the revolutions in these words in Book 3, Chapter 25.

> In this way the reckoning of the apparent sun is obtained through the motion *Book 3, Chapter* of the earth in agreement with ancient and modern methods, so that in addition *25* the future motion has presumably already been foreseen. Nevertheless, I am also not unaware that if anybody believed the centre of the annual revolution (of the earth)[79] to be stationary as the centre of the universe, but the sun to be moved with two motions equal to those which we have described in connection with the centre of the eccentric, all the phenomena would appear as before – the same figures and the same proof. Nothing else would be changed in them, and in particular nothing pertaining to the sun, except the position. For then the motion of the earth's centre around the centre of the universe would be perfect and simple, the two remaining (motions) being ascribed to the sun. For this reason there will still remain a doubt about which of these two positions is occupied by the centre of the universe, as I said ambiguously at the beginning that the centre of the universe is in the sun or near it. I shall discuss this question further, however, in my treatment of the five planets.

[76] *Epist. astron.*, 104 (*T.B.o.o.*, VI, 133), Tycho to Rothmann, autumn of 1588: 'I have tried besides this to provide an emendation of the celestial motions and thus to revive and restore the whole art of astronomy from certain observations, as if from first elements, if only the Holy Spirit will breathe favourably on our enterprises and provide the strength and opportunity for undertaking so great an endeavour.'

[77] Cf. Ovid, *Metamorphoses*, VI, 97.

[78] Ursus mocks Tycho's description of the foyer of his observatory at Stjerneborg: 'Higher up, below the arch where the hypocaust runs, there is portrayed TYCHO's new invention of celestial hypotheses, as large as the vault can take. Immediately under this, on the wall below on the north, where his image is represented with one hand gesturing aloft, he holds between two fingers a sheet of paper on which is this inscription – QUID SI SIC? – as if he were to say to the ancients and astronomers depicted around him: What do you think of this invention?'
Epist. astron., 246 (*T.B.o.o.*, VI, 276).

[79] The round-bracketed words in these quotations are Ursus' interpolations.

And these are the inverted hypotheses of Copernicus. Moreover it becomes evident and is clearly to be elicited from the following words that they were the hypotheses of Apollonius of Perga (though his were concentric).

Book 5, Chapter 3
Dii, v

[...] to the planet's course according to the matters demonstrated by Apollonius of Perga, as will be mentioned hereafter. (And a little later:) // Therefore, if there were no other irregularity in the planet's motion, as Apollonius thought, these would be adequate. (And afterwards, at the end of the chapter:) Contrary to the view of Apollonius and the ancients, however, the planet's motions are not found to be uniform, this being revealed by the earth's irregular revolution with respect to the planet. Consequently the planets are not carried in concentrics, but in another way, which I shall describe next.

Book 5, Chapter 35

It seems that knowledge of the stations, regressions and retrogradations and their places, times and extents is relevant to the reason for the motion in longitude. Moreover these matters were discussed not a little by mathematicians, especially Apollonius of Perga. But they discussed them as if the planets moved with only one inequality, that which occurs with respect to the (mobile) sun, due to the motion of the earth's great circle, which we have called 'commutation', etc. (And a little later:) For the purpose of demonstrating these things, Apollonius adduces a certain lemma, [based on] the hypothesis of the immobility of the earth (and so of the mobility of the sun). Nevertheless it is compatible also with our principles based on the mobility of the earth, so we too shall use it.

This is what he says. From these explicit words of Copernicus I think it can definitely be agreed how disgracefully my opponents Tycho Brahe, Sniveller and Roeslin (ye gods, what barbarians with yet more barbarous names!) either failed to read

Letters, page 149

Copernicus themselves or reading him failed to understand him. Thus Tycho denies, whether cunningly and mendaciously or ignorantly and carelessly, that these hypotheses of his (for so it pleased him to call those of Apollonius) are the inverted

Letters, page 128

Aristarchan or (as he calls them) Copernican hypotheses.[80] Thus Sniveller is disgracefully ignorant on the question of the derivation of the new hypotheses of Tycho (Apollonius) by inversion from the Copernican (Aristarchan) ones.[81] Thus

---

[80] *Epist. astron.*, 149 (*T.B.o.o.*, VI, 178–9), Tycho to Rothmann, February 1589: 'I did not derive the basis for constructing these hypotheses from the inverted Copernican hypotheses, and if you ever thought any such thing, that never occurred to me, as well you know, nor can anything of the sort be gathered from Rheticus or Reinhold.'

[81] *Epist. astron.*, 128 (*T.B.o.o.*, VI, 157), Rothmann to Tycho, October 1588: 'Whether indeed your hypotheses conform to and are the same as the [inverted Copernican] I am not able to tell for sure. For though I see clearly enough that the general hypothesis and the order of the spheres and the planetary circles, whether concentrics or eccentrics, are the same, nevertheless since you call yours new and in your last letter say that they satisfy the particular appearances, something which neither [the Copernican nor the inverted Copernican] do, it seems they are different from the inverted Copernican hypotheses.'

Roeslin impudently asserts that Copernicus held his (or rather the Aristarchan)   De opere
hypotheses to be true.[82] Behold, behold, look at these vain babblers, either false and   Creationis, *page*
lying or ignorant and careless – mathematicians indeed! Behold Tycho, Prince of   *44*
Astronomers! Behold the astronomer Sniveller, subject of so great a prince! Behold
Roeslin, the restorer of hypotheses and author of a new physics!

> [82] *De opere Dei creationis*, 44: 'He [Copernicus] introduced other hypotheses different
> from and almost the opposite of the Ptolemaic hypotheses; and since he had
> obtained from them truer consequences, he held the hypotheses also to be true.'

3

Kepler's initial reactions to the *Tractatus*

Three extant assessments by Kepler of Ursus' *Tractatus* antedate the *Apologia*. The earliest records his immediate reaction to the book. It is in a letter of May 1599 to Herwart von Hohenburg, who had sent him a copy requesting his opinion. The next is from a letter to Maestlin of August 1599. The third and fullest is dated March 1600. It was composed during Kepler's first visit to Tycho at Benatky and is addressed to Tycho's familiar and later son-in-law, Franz Gansneb Tengnagel von Camp. This is almost certainly the document to which Tycho referred in the postscript to his letter of 28 August 1600, a postscript which made it clear to Kepler that composition of a fuller refutation of Ursus was, in effect, a condition of employment in association with Tycho.[1]

In these early assessments Kepler concentrates on two of Ursus' historical claims: that the Tychonic hypotheses had been entirely anticipated by Apollonius of Perga, and that they had been explicitly spelled out by Copernicus. His refutations of the latter claim are of particular interest, for the last chapter of the *Apologia* raises but does not answer Ursus' allegation that Tycho's hypotheses are in Copernicus, breaking off with the rhetorical question: 'Where precisely in Copernicus were they described?'

There are a number of further points of interest. In the *Apologia* Kepler refrains entirely from comment on the issues of Ursus' plagiarism and defamation of Tycho. As we have seen, he had good reasons for doing so. The matter of defamation was *sub judice*. He regarded Tycho's obsession with the issue of priority as undignified. And it may even be that he doubted Ursus' plagiarism; for the evidence with which Tycho had supplied him was inconclusive, and as his original somewhat tactless apology to Tycho indicates, he had originally regarded inversion of the Copernican system as a trivial emendation at which several authors could well have arrived independently.[2] However, in these assessments Kepler inclines to the view that Ursus is guilty of plagiarism. But he does so in a very circumspect way,

[1] See Ch. 1, p. 26.   [2] See Ch. 1, p. 19.

merely indicating certain passages in the *Tractatus* which might be construed as admissions of guilt. The assessment sent to Herwart is revealingly lukewarm in its support of Tycho's case in two other ways. Kepler continues to treat Tycho's system as a mere emendation of the Copernican hypotheses rather than as a fully-fledged world-system in its own right. And despite the charges levelled against Ursus by Tycho and Maestlin in their letters to him,[3] he hesitates to deny Ursus originality as a mathematician, pleading inadequate conversancy with the relevant literature.

EXTRACT FROM KEPLER'S LETTER TO HERWART VON
HOHENBURG, 30 MAY 1599[4]

As for Ursus's little book on which you seek my judgement,[5] although I think it a serious matter to express an opinion about the writings of others, nevertheless I have a special and, I think, respectable reason for doing so quite readily about [those] of this author in particular and, moreover, about this little book; and I can well believe that the same was not the least of your reasons for asking this of me. For on f. Di he published a letter written by me. Though I praise him to the skies in that letter, I also make much of Tycho, his adversary. You hoped, perhaps, that it would be fair for me to become judge in this dispute. But I am unable to conceal from your Lordship the fact that I am already prejudiced by a certain bias in favour of the other party. For, disgraceful to relate, if the truth be told, when I wrote that letter to Ursus nearly four years ago when I was scarcely grown up and had just read the little book which he entitled *Fundamentum astronomicum*, there fell upon my ears the pronouncements of Sigismund Waganer (the man I refer to at the end of my letter) about Ursus's reputation in Italy and Germany and about his admirable expertise. And seized by I know not what poetic spirit I immediately poured out that letter (for there was little time to write) and with my pen, as I still remember, went beyond the bounds of reason and conscience. For my wanton declaration that I had acquired my modicum of learning in mathematics from him, that is, from his books, was certainly excessive if not worse. I had derived something most noteworthy from his book (I had seen no other book published by him), and the main thing I had in mind was this: I appeared to have found in that book a certain brief synopsis of both Euclid and Regiomontanus on triangles. And so from these rules or figures, all of which I copied onto a single half-page, I had later computed, ignoring Euclid, the geometrical bodies which the basis of my construction required.[6] This is what I

---

3 See Ch. 1, pp. 12, 14–16.   4 *K.g.W.*, XIII, 341–8.
5 In his letter of 16 May 1599 (*K.g.W.*, XIII, 332).
6 Kepler may refer to the computation in Ch. 13 of the *Mysterium* of the ratios of the sizes of the spheres which circumscribe and inscribe the nested Platonic solids. It was presumably the sine-tables of Ursus' *Fundamentum astronomicum* that he had used. In the *Mysterium* the calculations are derived from propositions in Euclid, but Kepler indicates that they can more easily be carried out using sines.

derived from him in the construction I undertook. But I could never have anticipated that the letter would be published. Nor did I know that it was published, though I had written to Ursus three or four times, until I gathered from Maestlin that Tycho reproached me for it: that was last February, the same time as I received Tycho's letter, delayed somewhere for ten months.[7] So I am very grateful indeed to you for having sent the little book I have so long wished to see. For I could not remember all that I had said, nor what in particular had incited me to such highly improper praises (for there were contributory causes). You will understand how upset I was that Ursus should, in effect, score a triumph with that letter of mine (for I see no other reason for his having judged it worthy of publication; and as for the content of the letter, he knew and declared that my book had appeared); that, in addition, having laughably acquired it, he should abuse it to the utmost, namely, for the purpose of embellishing his calumnies; and that having decided on this he deliberately and knowingly did his utmost to render me (who had never injured him) liable to odium (though Tycho's humanity in forgiving me is no less than his candour in reproaching me); and that he is unashamed to expose my name by this untimely collation of authors to those of the learned who are grudging, whoever they may be, now at the very dawn of my reputation – you will, I repeat, understand how much this upset me. Consequently it is evident that I am not free from prejudice and bias. Nevertheless, since he virtually set me up as a judge in this dispute by seeking fame with the help of my letter, I think it permissible for me to pronounce judgement. What more can I say? This will be a private judgement and you play the advocate for both parties before me. So, though in law I play the fool, let me act as arbitrator.

His writing can be considered from various points of view. So I leave to your discretion the question of the charge and the sentence, whether delivered verbally or put on record. For if I say that he goes too far in calumny, the same will be said of my praise. And just as men fear savage reproaches, so, on the other hand, they despise adulations. So perhaps he and I have gone equally far in opposite directions, so that neither has the right to object to the other.

As for the question at issue, whether one stole the hypotheses of the other, I think this will become apparent most readily if the works of each are examined. For truth asserts itself and by its weight it crushes what militates against it. And something can, indeed, be discerned there. Ursus declares that these hypotheses are not held by himself alone, except insofar as he differs somewhat from Tycho. Tycho, on the other hand, maintains that the hypotheses which he sets out (in that part in which they differ from the Copernican hypotheses) he discovered by his own efforts, and that he nowhere saw anything similar. Ursus, on the contrary, says that both his and Tycho's hypotheses were derived from the hypotheses of Apollonius of Perga made public long ago around the time of Christ. He bases his assertions on some passages in Copernicus in which [Ursus claims] Tycho's hypothesis is in part explicitly posited by Copernicus himself and in part ascribed to Apollonius of Perga

7 This is slightly inaccurate. Though Kepler had not received Tycho's expostulatory letter until February 1599, he had first heard of Tycho's wrath from Maestlin in August 1598.

(f. Dii). But what Copernicus says in these passages could not be further from what Ursus maintains. For he neither sets out Tycho's hypothesis – Ursus is mistaken in his reference – nor does he ascribe it to Apollonius of Perga. Ursus declares: 1, that he derived his hypothesis from elsewhere; 2, that he found it set out in Copernicus and attributed to Pergaeus; and 3, that it is the same as that which Tycho holds. So he admits that he copied it from someone else, which is just what Tycho claims.

It should be noted that whoever revealed Ursus' hypotheses to him (lacking a little, which it is easy to make good) clearly held the same hypotheses as Tycho. But who is he? Ursus says he is Copernicus, and through him Pergaeus, and no one else besides. So no one apart from Pergaeus in Copernicus or Tycho is the author of the Ursine hypotheses. But I declare that neither Pergaeus nor Copernicus was the author of the Ursine hypotheses. So Tycho alone, and no one else, was the one who showed Ursus the way to his hypotheses. The confirmation of the case lies in a grasp of the true meaning of Copernicus – Ursus himself provides the rest by his confession. In Book 3, Chapter 25 of the *Revolutions*, where Ursus thinks the Tychonic hypotheses are expressed by Copernicus, he does indeed concede that some of the motions which he ascribes to the earth can be transposed to the sun. But he does not concede this for the annual motion, nor does he there deal with the principal motions of the five planets. Nor does the argument of that book, which is concerned with the single solar (or terrestrial) motion without regard to the motions of the other heavenly bodies, deal with this.

To make the matter clear I shall provide two figures to explicate Copernicus' opinion. In the first figure A is the centre of the world and of the solar body. B is the centre of the epicycle which takes the place of the eccentric (Copernicus turns the epicycle and great orb into an eccentric, but he himself says it comes to the same thing). And B goes round A uniformly in the period of a year. C is the centre of the epicycle which produces the variation in eccentricity. C likewise goes round B in a year, in the opposite direction. D is the earth which goes round C every 3,434 years.

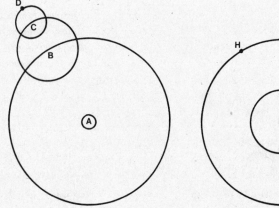

In the other figure F is the centre of the world and of the motion of the earth, H, which still goes round uniformly in an annual motion. E is the centre of the solar body which goes round the centre G in 3,434 years. G is the centre of the epicycle which bears the sun, producing the progression of the sun's apogee. Now according to Copernicus and the Prutenic Tables it goes round F under the fixed stars in 17,108 years. So the sun is indeed moved in its place by two motions, but very slowly and not much. For these little circles are very small in comparison with the circle in which in both figures the earth runs. But Ursus, who hoped that no one would understand these matters, supposes either through ineptitude or stupidity that Copernicus is talking about a kind of motion of the sun that would be similar to the annual Copernican motion of the earth, and that would produce some effect in the parallaxes of the five planets. But these two motions, assigned to the sun by Copernicus in the second case, are ignored in the treatment of the other planets, whether they are attributed to the earth or to the sun, although they produce some small effect in need of demonstration, which Tycho seems to have detected in his observations. So much for the first passage. In the remaining passages cited from Copernicus, Book 5, Chapter 3 and Chapter 35, Ursus has a yet worse case. Copernicus derives his reference to Apollonius of Perga and his lemmas from Ptolemy;[8] and I am amazed that Ursus did not read it in Ptolemy. Pergaeus' opinion is of an utterly different kind. For he is not concerned with treatment of the entire system of the five planets whose centre is in the sun, as are Tycho, Ursus, Copernicus and the ancients. He is concerned only with the following question. What should be the ratio of the epicycle of each of the planets to its principal orb, and the ratio of these to its motion of anomaly of commutation and its regular periodic motion, if the station and retrogression of the planet are to follow from that ratio? Thus his hypothesis is defective, in that Pergaeus thought that each planet varied only in the single anomaly that Ptolemy saved with an epicycle, not knowing that the centre of the epicycle proceeds unequally or, according to Ptolemy, in an eccentric. (Pergaeus is not commended by Ptolemy for his construction of hypotheses but for the subtlety of his demonstration, which is, indeed, very elegant. See the chapter on the stations in Reinhold's commentary on Peurbach's *Theoricae*.)[9] All expositors declare that this is the sense of the passage cited. But Ursus, either through blindness or trickery, would convince us by parenthetical interpolation of the word *mobilis*[10] that Pergaeus, or rather Copernicus on his behalf, spoke of that motion common to all the five planets in relation to the sun, their centre, as it is carried around in an annual motion, the motion which Tycho's hypotheses expound and which is attributed to the earth by Copernicus. And this is the confirmation of the case. Suspicion will naturally arise

---

[8] *Almagest*, XII, I.

[9] *Theoricae novae planetarum Georgii Purbachii…ab Erasmo Reinholdo Salveldensi pluribus figuris auctae, èt illustratae scholiis* (first publn, Wittenberg, 1542; revised edn, Wittenberg, 1553).

[10] Ursus interpolates *mobilis* into his quotation from *De revolutionibus*, V, 35, to misleading effect (*Tractatus*, sig. Dii, v).

from all the evidence in one who carefully considers these things. He declares that he is guilty of a philosophical theft, and that at Magdeburg he made public Tycho's opinion of the disposition of the world, but did so together with his own opinion derived from a source which does not exist, as our investigations have just shown.[11]

Thirdly, you ask if perchance this treatise contains anything useful or worthwhile. I cannot tell you what you especially wish to know. I reply that although his mastery of astronomical matters seems to be nil as far as his understanding of the masters Ptolemy and Copernicus is concerned (though one might say that it is strikingly bad), that does not preclude him from being a good geometer and a good arithmetician. The mathematical problems which he presents are ones which I, like him, very much wish to understand; but it may be that others besides have decribed his method of dealing [with them]. If, however, he does solve these problems I would declare him an astronomer,[12] nor would I spurn such a treatise. At present I cannot discuss the mysteries of angles and triangles, for I would need his *Fundamentum astronomicum*, and the owner of the copy (which I saw in town) is abroad. Now in geometrical matters there is, as you know, certain knowledge and each axiom we either know or know ourselves not to know. But in astronomy opinions hold sway in the matter of hypotheses. So in geometry nothing once discovered is rejected, but in astronomy it is necessary to reject some things. Thus it is that one who is confident in the cleverness by which he constructs proofs in geometry readily rejects and shows indifference to all astronomers – just what Ursus seems to do. For this is how I imagine his reasoning. 'I understand geometry, and I am fertile in solving problems. Why shouldn't I see the truth before the others in astronomy as well? Since astronomers contradict one another, I shall ignore them all and rest content with my own cleverness. For they are difficult to understand, and it takes time to read them carefully. I shall save myself the time and occupy myself more profitably.' Considering this it could well be that Ursus is inexperienced in the astronomy of the masters, but that he is nevertheless skilled and experienced in geometry. But Ursus wants to pass judgement on an astronomer on the basis of geometrical and arithmetical knowledge. Now it is true that he who lacks wings will never fly. But it is not true that one who has these Platonic wings thereby knows immediately how to fly.

Moreover he recalls certain authors known to him, Vieta, Birckerus, Rubeus and others, and they have made notable contributions to mathematics. But none of them is known to me in this corner of Germany, this solitude. I could perhaps ask amongst

[11] In the *Tractatus* Ursus quotes Tycho's letter to Rothmann of February 1589, in which the charge of plagiarism and theft is levelled, adding his own variously abusive and sardonic notes. Note 17 reads: 'I myself, unless I am mistaken, communicated to Rollenhagen, Rector of the School at Magdeburg, both my own hypotheses and the Apollonian or pseudo-Tychonic hypotheses, already clearly known to me from Copernicus.' Note 20 reads: 'Let it be a theft, but a philosophical one. It will teach you to look after your things more carefully in future' (*Tractatus*, sig. Fi, r).

[12] Evidently a slip; the context suggests 'mathematician' or 'geometer'.

the Jesuits what Clavius has published, and particularly about his work which shows how to represent the astrolabe with rule and compass.[13] But the time does not seem appropriate for even so slight a degree of familiarity with them. Nor do I recall any of our Jesuits who excels in knowledge.[14] But if your Lordship will tell me what works of the authors Ursus names are available, I shall at once undertake to obtain them for my use.

Finally I say this about Tycho's hypotheses. I believe him to be their author and inventor. And it could well be that he discovered them of his own accord, not guided

[13] The citation of Viète and Clavius here is unproblematic. Viète, referred to by Ursus as the greatest mathematician of the age (*Tractatus*, sig. H2, v), had in his *Canon mathematicus* (Toulouse, 1597) discussed prosthaphaeretic methods which anticipate those of Wittich, Bürgi and Ursus. Clavius is referred to in a letter of 1593 from Rubeus, quoted by Ursus (*Tractatus*, sig. Fii, v–Fiii, r). Rubeus reports that he has shown Ursus' *Fundamentum astronomicum* to Clavius, who intends to present Ursus' method of prosthaphaeresis in his forthcoming work on the astrolabe. Clavius duly did so in his *Astrolabium* (Rome, 1593), 178–80. The citation of Rubeus and Birckerus is more puzzling. The Rubeus whose letter Ursus quotes, Teodosio Rossi, was a jurist and published no mathematical works. Ursus cites him as Theodosius Rubeus Romanus. It is possible that Kepler's belief that he was a mathematician was encouraged by association with another Romanus, Adriaan van Roomen, whom Ursus cites as 'the greatest mathematician of our age after Viète', and to whose criticism of a method of quadrature of the circle, retailed in the *Fundamentum astronomicum*, he refers (*Tractatus*, sig. H2, v); the reference is to *In Archimedis circuli dimensionem expositio analysis...Exercitationes cyclicae contra J. Scaligerum, O. Finaeum, et R. Ursum...* (Würzburg, 1597). His misapprehension persisted, for in *De stella nova* (1606) Rubeus appears, alongside Galileo, as a mathematician (*K.g.W.*, I, 229). The citation of Birckerus may have a somewhat similar origin. For Ursus quotes a letter from Johann Georg Brengger (Birckerus), a physician who published no mathematical works (*Tractatus*, sig. Fiii, r–v). He also repeatedly cites Jost Bürgi (Byrgius or Birgius) as his teacher and mentor, and ascribes much of the credit for the new methods of prosthaphaeresis to him. It is unlikely that Kepler actually confused the two, for he already knew of Bürgi as a mathematician, and Brengger's letter actually mentions Bürgi by name. But Birckerus may well be a *lapsus* for Birgius resulting from this association of the names. (A couple of similar slips occur in the *Apologia*.)

Herwart replied to this puzzling request for bibliographical information as follows: 'Christophorus Clavius' work on the astrolabe I send forthwith...Prenkerus [*sic*] has published nothing. Rubeus I do not know. Franciscus Vieta or Vietaeus Gallus, preeminent in geometrical and algebraic matters, has published many things, most of which I have asked him for, but I have not yet been able to obtain them' (*K.g.W.*, XIV, 20).

[14] *Nec memini quenquam inter nostrates Jesuitas in arte excellere*. Perhaps Kepler means 'in the mathematical art', for the scholarly interests of the Jesuits at Graz were primarily in the humanities: see J. Andritsch, 'Gelehrtenkreise um Johannes Kepler in Graz', in P. Urban and B. Sutter, eds., *Johannes Kepler 1571–1971: Gedenkschrift der Universität Graz* (Graz, 1975), 159–95. Otherwise, though perhaps attributable to Kepler's resentment over the persecution of Protestants, the remark is puzzling. For Kepler had been introduced to the Catholic Herwart through the Jesuits and had been on good terms with them.

by Copernicus' opinion, but that nevertheless he constructed a world that can easily be derived by alteration of the Copernican opinion. Though as yet I know little of the special hypotheses that depend on Tycho's general hypothesis, yet from what I have so far been allowed to know I judge that the prospects are good for the Copernican hypotheses. Tycho has this single advantage, that he is freed from the immense size of the sphere of the fixed stars: all the other things which Tycho asserts on the basis of his most accurate observations can be, and have been, easily transferred to Copernicus...

This too occurs to me. Ursus says that Copernicus employs the hypotheses of Aristarchus of Samos. That is true, as is evident from Maestlin's note on Rheticus, f. 116 in my book.[15] And if Ursus had not read this, perhaps he would not have discovered it by reading Archimedes,[16] for he was preoccupied with reading historians. But he dissimulates. Certainly Copernicus had not read it.[17] But Ursus is also in error in thinking that Aristarchus was the first to teach it. In fact Aristotle too more than fifty years earlier knew something of this Copernican hypothesis,[18] and he ascribed it to the Pythagoreans and refuted their argument. But he refutes it in such a way that it is readily apparent that neither the opinion of the Pythagoreans nor their astronomy were sufficiently understood by Aristotle. We can, however, gather the following from Aristotle. 1) They supposed that the sun is the centre of the world (Aristotle says 'the fire', but they used that word to designate the sun). 2) The earth goes round the sun with an annual motion. 3) The moon (which I maintain is called 'counterearth' in Aristotle) is turned round the earth and causes eclipses. 4) The distance of the fixed stars is so great that the earth's entire orb is vanishingly small compared with the fixed stars. 5) The retrogradations of the planets appear because of the motion of the earth. That those whom Aristotle refutes thought this is proved by the clear words of Aristotle himself; and it is the same view as that of both Aristarchus and Copernicus. I adhere with the firmest allegiance – this by way of a conclusion – to that most ancient teaching about the world, whatever Ursus may maintain about the falsity of all the hypotheses that ever were. So enough of Ursus' treatise. I cannot imagine that you want more. But if there is anything else that you wish to know about this treatise and that I am capable of explaining, I shall be at your service.

---

[15] In the preface to the edition of the *Narratio prima* appended to Kepler's *Mysterium* Maestlin refers to the heliocentric system of Aristarchus (*K.g.W.*, 1, 84–5).

[16] Archimedes, *Arenarius* (*De arenae numero*), 1, 4–7, quoted in Ch. 5, fn. 159.

[17] Kepler is right in supposing that Copernicus was not aware of this passage in Archimedes, but wrong in maintaining, as he later did in his *Hyperaspistes*, that Copernicus was altogether unaware of Aristarchus' views; see J. Dobrzycki, ed., *Nicholas Copernicus on the Revolutions*, translation and commentary by E. Rosen (Warsaw and London, 1978), 25, for the relevant deleted passage from the autograph of *De revolutionibus*, and 361–2 for its sources.

[18] *De caelo*, II, 13. Kepler discusses this passage in more detail in Ch. 2 of the *Apologia*.

On Ursus and Tycho. Lord Herwart, nobleman and Chancellor to the Bavarian Court, a man most competent in every branch of learning, sent a copy of the book by that scourge of Tycho. In it I have at last seen my letter, which he slipped in by the following stratagem. His intention was to harass Tycho and Roeslin together with their discoveries and speculations. He reproaches Roeslin for [postulating] a size of the Martian orb in conflict with observations, and he sets out the method for ascertaining distances from the anomaly. He then adds: 'In this way one can find out whether or not the opinions, etc., of Kepler are entirely true (I know they are truer than Roeslin's hypotheses). This opinion he communicated to me in this letter, worthy of publication, etc.'[20] Since he mentions the fact that my little work was advertised in the Frankfurt catalogue, I do not know why he judged my letter to be worthy of publication, unless it was because it praised him. I do not know whether he removed anything from it; but what he retailed is all mine. But though I wrote several times, I am altogether convinced and hold it to be almost certain that right at the outset I mentioned Tycho. He betrays himself like a rat [by its squealing],[21] and in my judgement he is unmasked by his own confession in his own words. He says that he himself took his hypotheses from Ptolemy,[22] but that he changed some things in the opinion of Apollonius of Perga, who proposed a theory about the situation of the planets, and that Tycho derived his [hypotheses] from the same source, but denies it. Read what Ptolemy says about this and you will declare that Ursus is raving. It is indeed true that when the two opinions of Apollonius are set out by Ptolemy, the second of them points to some extent in the direction of Tycho's hypotheses. For he supposes that each of the three superior planets (for Ptolemy openly says that the second opinion concerns 'only those which have the property of complete elongation') has its own eccentric, which goes round with an annual motion in the order of the signs; then he says that in each eccentric there is turned an epicycle going round the earth, whose motion is in precedence equal to the motion of commutation of the planet. But that stupidest of men does not adduce this second opinion, but rather the first one, which clearly has nothing in common with Tycho. And to prove his assertion he cites Copernicus, Book 5, Chapter 35, and openly maintains that in that chapter Copernicus anticipated Tycho in his hypotheses. For he says explicitly that the Tychonic [hypotheses] are to be found in Copernicus. But Copernicus recalls the lemma that pertains to the first of Apollonius' opinions. And nothing could be more ridiculous than his wishing to extract these hypotheses of Tycho's from Book 3, Chapter 25 of Copernicus. Copernicus there discusses the

---

[19] *K.g.W.*, XIV, 44–5. [20] *Tractatus*, sig. Civ, v–Di, r.

[21] Cf. Terence, *Eunuch*, 5, 6, 2: 'I've betrayed myself like a rat by my squealing...'.

[22] A puzzling remark. Ursus does not cite Ptolemy as the source of Apollonius' constructions, and Kepler very plausibly suggests below, and in the *Apologia*, that Ursus had not read the relevant passage in *Almagest*, XII, 1.

question whether the earth is carried round the sun with an eccentric motion or whether the sun [is carried] around the centre of the great orb with a very slow motion, the earth [being carried round] that same centre in a concentric with an annual motion. Nevertheless, that man, an ass rather than a bear, thinks that Copernicus is talking about the transposition of the sun in the circuit of its great orb with the earth in the centre of that orb. And on reading this I came to the conclusion that Ursus neither read Ptolemy nor understood Copernicus.

CONCERNING THE DISPUTE ABOUT HYPOTHESES BETWEEN LORD TYCHO AND URSUS[23]

When a certain learned man asked me to write fully to him about my judgement of this issue, I replied almost as follows, as far as I can remember.[24] First, I did not conceal the fact that I was angry with Ursus, who had disgracefully abused my letter by publishing it without consulting me. So I asked him to consider carefully whether I would be free of bias. Further, I was unwilling to deliver my judgement about the insults before a holder of public office. But on the question which one stole the other's hypotheses, although I follow another hypothesis, I said that Ursus seemed under suspicion of guilt. The main points were as follows. Ursus declares that he exhibited a certain diagram brought from Denmark at Rollenhagen's house; he calls this a philosophical theft; and he tells Tycho that it will teach him to look after his things more carefully in future.[25] Likewise he says that he was not the first author of that hypothesis, except insofar as he made the earth mobile with a daily revolution. Finally, he says that he gathered the hypotheses from the same source as that from which Tycho also had gathered them, and moreover that they are to be found explicitly in Copernicus and Ptolemy.[26] But this last claim is false, for those hypotheses are not to be found in the places cited. So from this falsehood and the three preceding truths – that he was not the first inventor of the hypothesis; that he derived the hypothesis from the same source as Tycho; and that he stole a diagram of Tycho's in which this hypothesis was to be found – the following conjecture is, so to speak, compounded out of elements: Tycho's hypothesis was stolen by Ursus and claimed as his own. Moreover, it is clear from Ursus' distortion of the passages he cites, which we shall consider in order, that Tycho's hypothesis is not to be found in Copernicus and Ptolemy.

Book 1, Chapter 10. Throughout this chapter Copernicus, taking his cue from Martianus Capella, gradually and discreetly leads up to the exposition of his hypotheses and their highest degree of necessity. Now since Capella says that Venus and Mercury have paths turning about the body of the sun, it is apparent that he assigned the centres of their orbs to the sun.[27] And afterwards it follows, if one

[23] *K.o.o.*, I, 281–4.
[24] The reference is to Herwart and the assessment sent to him in May 1599, translated above.   [25] See fn. 11.   [26] See fn. 22.
[27] In Ch. 4 of the *Apologia* Kepler considers in detail the passages from Martianus Capella to which Copernicus alludes.

takes this cue, that should one (understand – should I, Copernicus) assign Saturn, Jupiter and Mars to that same centre, he will not be in error, as is shown by the reckoning of their motions and as is shown in that work of mine in which I derive that reckoning from such a hypothesis. Here Ursus pretends that Copernicus is talking about some mutation of Copernicus' own hypothesis, even though Copernicus has not yet expounded his own hypothesis, but rather with these words provides himself with a cue for expounding it. Now if Copernicus had realised as did Tycho that the sun can move with an annual motion, but nevertheless remain the centre of the five planets (but he could not see this because he believed in the reality of the orbs),[28] why did he say a little later that either one must concede that the ratio between the earthly orb and that of the fixed stars is negligible or one must concede, following the ancients, that infinite multitude of orbs?[29] For one who holds Tycho's opinion neither need be the case.

Book 3, Chapter 25. No one who understands Copernicus can doubt that here Copernicus either was not understood by Ursus or was misrepresented through malice. Copernicus is dealing with the model of the sun, to which that book is devoted. Ursus, however, thinks that he is talking about the totality of hypotheses and about the mutual relations of the parts of the universe. To make the matter clear I shall provide two diagrams. One will represent Copernicus' first opinion; the other will contain the variant of which Copernicus here speaks.

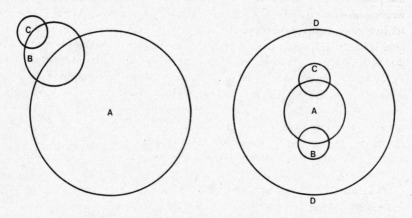

[28] Kepler implies that such a transposition was impossible for Copernicus because it would have involved interpenetration of the orbs of Mars and the mean sun. N. R. Swerdlow, '*Pseudodoxia Copernicana*: or, enquiries into very many received tenents and commonly presumed truths, mostly concerning spheres', *Archives internationales d'histoire des sciences*, 26 (1976), 136–7, appeals to this passage in support of his conjectures about the role of celestial orbs in the genesis of the Copernican system.

[29] But the size of the universe is so great that the distance earth–sun is imperceptible in relation to the sphere of the fixed stars. This should be admitted, I believe, in preference to perplexing the mind with an almost infinite number of orbs, as those who kept the earth in the middle of the universe were forced to do.

Based on Rosen's translation in Dobrzycki, *Copernicus on the Revolutions*, 20.

A in the first figure is the centre of the earth's eccentric, namely the sun itself. B is the epicycle which is returned to its starting point in an annual motion like the eccentric, but in the opposite direction and lagging behind a little in such a way that from the discrepancy the precession of the apogee follows. C is the centre of the second epicycle in which the earth goes round in 3,434 years, as a result of which the variation in eccentricity occurs. This, Copernicus claims in the words cited by Ursus, can be transformed as follows. Let A be the centre of the universe, but not the very body of the sun itself, around which the earth, in D in the second figure, goes in an annual motion (N.B. the earth still goes round). Let the sun be in C and be moved in the epicycles B and C, but so slowly that in a single year it makes no detectable progress. For epicycle B is turned under the fixed stars in 25,000 years, and epicycle C in 3,434 years. Now read Copernicus and you will see how greatly Copernicus differs from Ursus' opinion. The things they say are as different as onions and garlic.

Book 5, Chapter 35. In this passage Copernicus recalls Apollonius and says the same thing on the same topic as does Ptolemy in Book 12, Chapter 1. This is how that passage appears to me.[30] Ursus was advised by someone more intelligent than himself that something Tychonic is to be found in Ptolemy under Apollonius' name. Ursus, being not in the least expert in these matters, was content with what Copernicus takes from Ptolemy and seized on something that is clearly foreign to that passage. He overlooked what was most opportune to him for the purpose of weaving his calumny, because he did not take the time to go to Ptolemy. For there are two forms of the Apollonian lemma. The one of which Copernicus and, after him, Ursus speak contributes no more to Ursus' opinion or to Tycho's hypothesis than does Copernicus himself, above. Its meaning is explained in this figure.

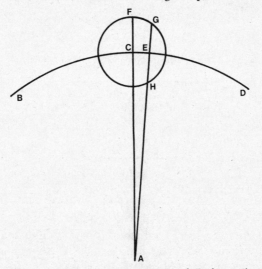

[30] Ch. 3 of the *Apologia* elaborates this defence of Tycho against the charge of anticipation of Apollonius of Perga. Tycho had called for such an elaboration in his letter of August 1600 (see p. 26).

A is the immobile earth; BD is the planet's concentric, which in the case of Saturn is revolved in 30 years, etc. Now Apollonius thought that there existed no inequality of the planets other than that which is related to the sun and brings about the retrogradations, and so he neglected to make BD an eccentric. FG, the epicycle, has a notion of commutation. Apollonius then demonstrated that if a line AG is produced from A, the centre of the earth, cutting the epicycle at H and G, so that EG is to AH as is the motion CE of the concentric to FG [the motion of] the epicycle, then the planet is made to stand still at the points H and G. This first form of the lemma to which Copernicus, Ptolemy and Ursus refer clearly has nothing in common with Tycho's hypothesis, concerning as it does very different matters. The other form is this, and had Ursus seen it he would have confounded heaven and earth in his insane joy.

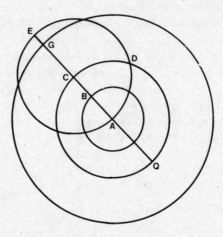

A is the centre of the earth, of the universe and of the planetary orbs; B is the sun in its orb; CD is the concentric of each planet going round together with B, the solar orb, and in the same direction. ED is a large epicycle of the planet, so that its centre C goes round the earth; the motion of its epicycle is counter to the order of the signs of the zodiac with a motion of commutation. On this assumption Apollonius again sought a line determining the stations, as before. This form is so constructed that a clever man might derive the Tychonic hypothesis from it, but not without great labour and with infinitely more difficulty than from the Copernican basis. For, firstly, there is no hint, no vestige, of the Tychonic [hypothesis] to be found openly and explicitly in it, since the author's concern was with demonstration of the stationary point and not at all with the disposition of the parts of the universe. Secondly, the centre of the planets is not placed in the sun, but in the earth. Thirdly, it does not deal with all the planets considered together, but with one at a time. Fourthly, each planet has as its orb a geocentric. Fifthly, what Tycho calls the orb of a planet is in Apollonius (even if a suitable adjustment is made), and is called, a large epicycle. Sixthly, Apollonius does not say that the sun and the centres of those large epicycles are carried in a single concentric round the earth, but rather says that they are carried together in a revolution of the same period. Seventhly, in Apollonius the orb of Mars does not intersect the orb of the sun. Eighthly, there does not appear in Apollonius, insofar as he bequeathed us writings, a basis on which to derive the ratios of the spheres from the annual parallax. Ninthly, the motion in consequence of that great sphere, according to Apollonius, or orb, according to Tycho, has in the former a much shorter period, being annual or nearly so. In Tycho it takes 30 years in the case of Saturn, 12 in the case of Jupiter, and about 2 in the case of Mars. From these considerations it is evident both that Copernicus, Ptolemy and Pergaeus have not been understood by Ursus, and that the Tychonic hypothesis cannot so readily be derived from Apollonius.

Master Johannes Kepler, Styrian Mathematician, wrote this account for the sake of the noble youth Franz Tengnagel, a member of Tycho's household, who wished to understand these matters more clearly. March 1600.

4

The scope and form of the *Apologia*

Tycho's letter to Kepler of September 1600, the letter he received just before his final departure from Graz in desperate circumstances, sets out his brief for the *Apologia*.[1] He is to refute those parts of the *Tractatus* 'which are mathematical and pertain to hypotheses' and he is especially required to defend Tycho's claim to originality in the formulation of new hypotheses.

The *Apologia* is incomplete and before reconstructing the strategy by which Kepler attempts to satisfy his brief we must consider the extent of this incompleteness. As it stands the work falls into two main parts. In Chapter 1 Kepler seeks to refute Ursus' account of the nature of astronomical hypotheses and to set in its place an adequate account. In the remaining three chapters Kepler rebuts, point by point, Ursus' 'history of hypotheses', offering detailed interpretations of the primary sources to which Ursus had, or in Kepler's opinion ought to have, appealed. From Kepler's three earlier assessments of the *Tractatus* it is clear that the incomplete fourth chapter would have ended with a detailed refutation of Ursus' interpretation of the passages from *De revolutionibus* on which he bases his extraordinary claim that the Tychonic hypotheses were explicitly anticipated by Copernicus. Kepler's intimation to Maestlin in February 1601 that he would welcome comment on the 'problems proposed for discussion' by Ursus – that is, presumably, those posed as a challenge to Tycho at the beginning of the *Tractatus* – could perhaps be regarded as an indication that he intended to rebut Ursus' claims to mathematical superiority over Tycho.[2] But there is nothing in the *Apologia* as it stands to substantiate this, and it seems likely that Kepler had come to realise, perhaps as a result of his meetings with Ursus, that Tycho's claims to priority in the invention of prosthaphaeretic methods and his charge of their plagiarism by Ursus were indefensible. What is quite certain, both from the preface to the *Apologia* and from the description of its contents offered to Maestlin, is that Kepler never intended to address any of the issues of Tycho's law-suit, the charges of theft, plagiarism and defamation of

[1] *K.g.W.*, XIV, 148; see Ch. 1, p. 26.
[2] *K.g.W.*, XIV, 165.

character. I conclude that with the addition to Chapter 4 of the relevant material on Copernicus to be found in his earlier assessments, together, perhaps, with a summary or peroration, the *Apologia* would be complete.

The way in which the three historical chapters of the work answer to the brief laid down by Tycho is clear enough. Tycho's claim to originality in the formulation of hypotheses is defended by showing that the evidence Ursus adduces for his anticipation by Apollonius of Perga, Martianus Capella and Copernicus is a tissue of misrepresentation and misinterpretation. Less clear is the role of the first chapter, a rebuttal of Ursus' sceptical views on the status of astronomical hypotheses in which Tycho figures only incidentally. This first chapter does, indeed, prepare the ground for the explicit defence of Tycho by discrediting Ursus' understanding of the nature of astronomy. But this is not, I think, its only role in Kepler's defensive strategy. Ursus had claimed that hypotheses are acceptable in astronomy provided only that they suffice to predict and retrodict apparent celestial coordinates. He had further claimed that contrivance of new and adequate astronomical hypotheses is an easy matter. In the first chapter of the *Apologia* Kepler rejects these claims. He argues that saving the phenomena is not the sole criterion of adequacy for astronomical hypotheses; they should in addition be confirmed by 'physical considerations'. And he points out that even if saving the phenomena were sufficient for the adequacy of astronomical hypotheses, contrivance of adequate hypotheses would not be the easy matter Ursus appears to suppose. As we have seen, Kepler had initially held Tycho's world-system to be a mere inversion of the Copernican hypotheses. But by the time he wrote the *Apologia* he had come to realise that Tycho's hypotheses included detailed planetary models designed to explain observations not in accord with Copernicus' models, and further that Tycho had used a variety of 'physical arguments' to support his system against its rivals. In the third and fourth chapters of the *Apologia* he emphasises Tycho's attempt to improve the predictive adequacy of astronomy and his concern with physical support of his system as tokens of his originality. Thus by arguing in the first chapter that both predictive power and physical support are required of astronomical hypotheses, and that neither is easily achieved, he prepares the ground for his detailed arguments in support of Tycho's claim to originality.

There is, I think, no doubt that throughout the *Apologia* Kepler writes sincerely. But the constraints imposed by his brief to defend Tycho do affect his presentation in a way which may easily lead to serious misunderstanding of the work. Throughout the work he insists that to fulfil his true aim, that of portraying the form of the universe, the astronomer must seek hypotheses that are both predictively adequate and physically plausible. But the few explicit appeals he makes to physical considerations – for example, his invocation of the capacity of the Copernican and Tychonic systems to explain why the superior planets are nearest the earth when in opposition to the sun – are prosaic when compared with the dazzling array of physical arguments he had offered in support of a Copernican world-system in his *Mysterium* and was later to offer in his *Astronomia nova*, *Epitome astronomiae Copernicanae* and *Harmonice*. It is no wonder that von Prantl was led to contrast the sober inductive methodology of the *Apologia* with the 'fantasies' of the *Mysterium* and *Harmonice*.[3] There are, as I shall show in Chapter 7, solid grounds for holding that no such contrast is valid; that the physical reasons to which Kepler alludes in the *Apologia* do in fact include the full range of architectonic, harmonic and dynamical considerations which Kepler deploys in these other works. The sparseness and prosaic quality of the explicit examples are to be explained by his decision to adduce only reasons that can be urged in support of the Tychonic system. For whilst he felt free to declare his adherence to the Copernican system, considerations of tact and relevance bound him to refrain from active promotion of Copernicanism.

Constraints of a very different kind arise from Kepler's choice of literary form. The modern reader may well be struck by two artificialities of presentation not to be found elsewhere in Kepler's writings. First there is the recurrence of rather strained legal terminology. For example, in the preface Kepler relates how he had intimated to Ursus that he would assume the role of judge in the dispute with Tycho; and later Osiander, whom Ursus had quoted in support of his case, is referred to as Ursus' 'witness' and is said not to be 'exempt from all contrary testimony'. Secondly there is the prevalence of *ad hominem* expostulation: 'I leave it to the thoughtful reader to judge...'; 'Why, then, you may well ask...'; 'Let the reader hearken to Pliny's words', etc. These and other puzzling features of the work are explained by the fact that the work is cast as a judicial oration, composed strictly in accordance with the strategies for conduct of a legal defence laid down by Cicero, elaborated by Quintilian, and codified in the many rhetoric and dialectic handbooks of the period.[4] This would be obvious were Kepler to have presented Tycho as his client. But given his resolve to keep clear

---

[3] K. von Prantl, 'Galilei und Kepler als Logiker', *Sitzungsberichte der philosophisch-philologischen und historischen Klasse der k.b. Akademie der Wissenschaften zu München* (1875), 394–408.

[4] The impact of Cicero on humanist educational theory is well documented; on the influence of Quintilian, whose more orderly and pedantic writing made his doctrines more readily assimilable into textbook form, see A. Messe, 'Quintilian als Didaktiker und sein Einfluss auf die didaktische pädagogische Theorie des Humanismus', *Neues Jahrbuch für Philologie und Pädagogik*, 156 (1897).

of Tycho's legal proceedings against Ursus, this course was not open to him. So with great artistry Kepler treats astronomical hypotheses as his client. As he informs us in the preface, he is out to refute Ursus' calumnies against 'the discoveries of mathematicians', that is, astronomical hypotheses. There is a famous parallel to Kepler's adaptation of the classical judicial oration. In Sir Philip Sidney's *Defence of Poesie* of 1595 it is poetry herself who appears as the client to be exonerated of the charges levelled against her by her detractors. In the following account of the formal structure of the *Apologia* I am indebted to Myrick's detailed analysis of the structure of Sidney's oration.[5]

Kepler's preface to the reader is the *exordium* of his oration.[6] Cicero and Quintilian, variously summarised and collated in the rhetoric and dialectic manuals of the period, tell us that an *exordium* should seek to render the judges well disposed to the advocate's case and eager for further information. It should be unassuming in tone, seeking to discredit the antagonist only by insinuation, not by denunciation. When the cause is an honourable one the advocate cannot do better than explain why he has accepted the brief. Kepler's polished preface adheres meticulously to these precepts. The next part of the oration is the *narratio*, in which the crucial facts relevant to the case are rehearsed. Detailed argument is not called for at this stage, but judicious selection and rhetorical devices may be used to present the facts in a light favourable to the case. In each chapter of the *Apologia* the *narratio* summarises those of Ursus' views that are to be refuted. In the first chapter it is but a single sentence telling how Ursus had denigrated hypotheses. In Chapter 2 it is again very brief, charging Ursus with having started with a reliable historical account in order to seduce the reader into acceptance of subsequent fabrications. In Chapter 3 the *narratio* is fuller, explaining by way of a detailed analogy between astronomy and architecture the division of labour between an author of astronomical hypotheses and the geometers and arithmeticians who provide him with materials for their construction. Only then is Ursus' misattribution of hypotheses to a mere geometer, Apollonius of Perga, retailed. In the fourth chapter it is again quite substantial, this time because not one but a whole series of historical misrepresentations is to be refuted.

[5] K. O. Myrick, *Sir Philip Sidney as a Literary Craftsman* (Cambridge, Mass., 1935), Ch. 2, '*The Defence of Poesie* as a classical oration'.

[6] The descriptions of the parts of an oration which follow are based on Cicero, *De oratore*, II, 76–81; Quintilian, *Institutio oratoria*, IV; and Melanchthon, *Elementa rhetorices* (Wittenberg, 1519), in C. G. Bretschneider, ed., *Philippi Melanthonis opera quae supersunt omnia. Corpus Reformatorum*, I–XXVIII (Halle, 1834–52; Braunschweig, 1853–60), XIII, cols. 431–6. Hereafter cited as *M.o.o.*

After the *narratio* come the *propositio* and *partitio*, both optional according to Cicero.[7] The *propositio* is supposed to indicate the main issue the advocate intends to address, and the *partitio* the way in which he intends to do so. Kepler has adapted the form of an oration to a work with chapters by using the chapter headings as his *propositiones*. The *partitio* of the first chapter announces Kepler's intention first to state the true nature of hypotheses, then to refute point by point Ursus' views on the subject. The second chapter lacks a *partitio*. In the final chapters the *partitiones* announce Kepler's intention to examine the primary sources on which Ursus' historical misrepresentations are based.

The heart of a judicial oration is the *contentio*. It has two parts: the *confirmatio*, in which the advocate's contentions are established, and the *refutatio*, in which the arguments put up by his opponents are refuted. In a defence the *confirmatio* comes first. Kepler follows the rule meticulously. In the first chapter the *confirmatio* sets out Kepler's own views on the nature of astronomical hypotheses and the *refutatio* demolishes Ursus' views point by point. In the remaining chapters the *confirmationes* offer Kepler's own interpretations of the texts on which Ursus has based his account, and the *refutationes* build on these to demolish Ursus' interpretations.

The rhetoric and dialectic handbooks of the period give precise, and to the modern reader extraordinarily pedantic, instructions on the way in which material is to be marshalled (*inventio*) and on the argumentative strategy (*argumentatio*) and rhetorical devices to be employed at each stage in an oration. Marshalling of material is guided by the so-called 'topics' or 'places of invention' (*loci inventionis*), general headings of discussion which vary in number and order according to the *status* of the case, that is, the type of question at issue. Thus where, as in the first chapter of the *Apologia*, the debate concerns the nature of something, Melanchthon instructs the orator to consider first its definition, then its genus and species, then its distinguishing characteristic or function (*proprium*), and so on.[8] Kepler follows this formula to the letter. When, as in the subsequent chapters, the debate concerns the meanings of texts, Melanchthon instructs the orator to attend to definition of words 'to make clear what is obscure', to *divisio* 'to resolve ambiguities' and, above all, to *circumstantia*, that is, the provenance of the text.[9] Again Kepler scrupulously adheres to these conventions.

Argumentation is the province of dialectic as the term was understood in the period. In his *Erotemata dialectices* Melanchthon recognises four main types

[7] *De oratore*, II, 19.
[8] *Erotemata dialectices* (*M.o.o.*, XIII, col. 663).
[9] *Elementa rhetorices* (*M.o.o.*, XIII, cols. 442–3).

of inferential strategy: syllogism, enthymeme, induction and *exemplum*.[10] Syllogism covers all deductive arguments in which the premises are made explicit; enthymeme those in which premises are suppressed. Induction is in modern parlance enumerative induction to the next instance. Thus in the *Apologia* we find the following argument identified as an induction: medicine, music and oratory were once more perfect; so why not astronomy as well? *Exemplum* is a wide category covering both support of a generalisation by citation of telling instances and a variety of forms of argument by analogy.[11] Such a variety of types of argumentative strategy is characteristic of the humanist 'deliberative' style, and is precisely what sets humanist logical theory and oratorical practice apart from its scholastic counterparts.[12] A careful reading of the *Apologia* shows that Kepler makes effective use of the whole range of such strategies. The strategy of a *refutatio* is guided in addition by an elaborate theory of fallacies, originating in Aristotle's *De sophisticis elenchis* and greatly enriched by the polemically inclined humanist dialecticians.[13] Melanchthon advises the orator in seeking to discredit his opponent to attempt to convict him of equivocation, *petitio principii*, false induction and false *exemplum*, use of unreliable testimony, etc.[14] Kepler's masterly refutations convict Ursus of the full range of such fallacies.[15]

Where the dialectic handbooks concentrate on the more formal means of persuasion the rhetoric handbooks concentrate on informal means, the various oratorical strategies and figures of speech that serve to insinuate rather than prove by rendering the advocate's presentation variously engaging, lively and apposite. It would be tedious in the extreme to detail Kepler's often effective use of *antonomasia, polysyndeton, paralepsis, anticipatio* and the rest. But one particular oratorical device deserves mention, for if unrecognised it is apt to engender confusion. In *concessio* the orator argues from his opponent's premises, or premises apparently favourable to him, in order to show that no matter how charitably one considers the opponent's case,

---

10 *Erotemata dialectices* (M.o.o., XIII, cols. 594–5).

11 In other handbooks argument by analogy is sometimes treated as a distinct form of inference and sometimes, following Cicero, as a kind of *inductio*.

12 On humanist dialectic the standard work is C. Vasoli, *La dialettica e la retorica dell'Umanesimo* (Milan, 1968). See also C. J. Armstrong, 'The dialectical road to truth: the dialogue', in P. Sharratt, ed., *French Renaissance Studies, 1540–70* (Edinburgh, 1976), 36–51; and L. A. Jardine, 'Lorenzo Valla and the intellectual origins of humanist dialectic', *Journal of the History of Philosophy*, 15 (1977), 143–64.

13 On the history of the theory of fallacies see C. L. Hamblin's fascinating *Fallacies* (London, 1970).

14 *Erotemata dialectices* (M.o.o., XIII, cols. 726–50).

15 Thus the first refutation of the first chapter convicts Ursus of equivocation, the fourth convicts him of *petitio principii*, the ninth of false *exemplum*, the tenth of false testimony and the eleventh of false induction.

it still fails.[16] Kepler makes repeated use of *concessio*. For example, in the first chapter he argues that even if saving the phenomena were (as Ursus holds) the sole function of hypotheses, Ursus would not have conclusive grounds for scepticism; and in the fourth chapter he argues that even if we consider the full range of classical texts which support Ursus' claim that geoheliocentric cosmology originated in antiquity, Ursus still does not have a good case against Tycho's originality.

Recognition of the dialectical structure of the work is important in another respect too. Whereas in a *refutatio* aggressive irony, *ad hominem* appeals, and even jocular facetiousness are quite proper, the tone of a *confirmatio* is supposed to be modest, confident and fully serious. It is in Kepler's *confirmationes* that we should expect to find, and do find, his considered expression of his own views.

None of Kepler's other works, not even his polemics against Roeslin, Fludd and Scipio Chiaramonti, in which such dialectical sophistication would have been appropriate, is composed in strict conformity to the classical rules of composition and delivery.[17] A number of factors may help to explain the formality of the *Apologia*. A high degree of stylistic polish is hardly surprising in what was in effect a commissioned work. The fiction of a legal defence, already adopted in Kepler's initial assessment of the work offered to Herwart, may have been suggested to him by Tycho's legal action against Ursus. And there may be an element of ironic mimicry in it; for though Ursus' *Tractatus* is stylistically a mixture of diatribe and satire, its preface presents it as a defence to be weighed up by Moritz of Hesse as arbitrator. Kepler's facility in the genre is likewise unsurprising. Instruction in the principles of classical oratory, using such textbooks as Melanchthon's *Elementa rhetorices* and *Erotemata dialectices*, was central to the first-year arts course in universities subject to humanist reforms. Kepler would have received such instruction at Tübingen;[18] and he had taught rhetoric at the Graz

---

[16] Cf. Quintilian, *Institutio oratoria*, IX, 2, 51.

[17] His vernacular *Antwort auff Röslini Discurs* (1609) answers Roeslin's work line by line and is jocular and mildly satirical in tone. His defence of the *Harmonice* against Fludd, *Pro suo opere Harmonices Mundi apologia* (1622), and his defence of Tycho's theory of comets against Chiaramonti, *Hyperaspistes* (1625), are more formal and severe, but again refute the opponent point by point.

[18] The distinguished humanist Joachim Camerarius the Elder had a major role in the reform of Tübingen instigated by Duke Ulrich of Württemberg in 1534; for an account of these reforms see H. Hermelink, 'Die Anfänge des Humanismus in Tübingen', *Württembergische Vierteljahrschrift für Landesgeschichte*, n.s., 15 (1906), 319–36. The place of rhetoric and dialectic in the first-year arts course is spelled out in the new statutes for 1536: see *Urkunden zur Geschichte der Universität Tübingen aus den Jahren 1476 bis 1550* (Tübingen, 1877), 386.

Stiftsschule.[19] What is remarkable is not his adoption of this genre but his success in exploiting it. Though he is no Erasmus, Kepler manages to make the rigid conventions of the classical oration the vehicle both for an effective defence of Tycho's claims to originality and for a series of profoundly original philosophical and historiographical insights.

[19] As there were few pupils for his mathematics teaching he taught 'Virgil and Rhetoric' to the upper classes (*K.g.W.*, xix, 8–9).

II

Apologia pro Tychone contra Ursum

5

Text and translation of the *Apologia*

The edition is from a microfilm of the manuscript, Pulkowo, v, f. 264r–299a, kindly provided by the Württembergische Staatsbibliothek, Stuttgart, through the good offices of Dr Martha List and Professor Robert Westman. For certain folios in which text was obscured by the binding, use has been made of photographs of the unbound manuscript kindly provided by the U.S.S.R. Academy of Sciences, Leningrad, through the good offices of Professor Owen Gingerich.

I have preserved Kepler's punctuation and capitalisation but I have not reproduced accents, except on Greek words. Ligatures are expanded, but not abbreviations. Because of the many insertions and deletions in the manuscript the paragraphing is on occasion conjectural. Where Kepler uses underlining or bold script for emphasis or quotation I have used italicisation. Two sections of the last chapter, lost since Frisch's use of the manuscript, are reproduced in the version of Frisch's edition.[1] There are (*pace* Frisch) few *lapsus calami*; in these rare instances the original is given in square brackets. Passages marked with corners, thus ⌐...⌐, are Kepler's interpolations. Most of the extensive deleted material consists of drafts of passages which appear in the final version. A few passages of independent interest are given in footnotes to the text.

The translation is fairly free, and I have not hesitated to break up Kepler's longer and more complex sentences. Kepler's use of the terms *sphaera*, *orbis* and *circulus* is far from constant. Often *orbis*, *sphaera* and *circulus* are used interchangeably to mean 'orb' or 'sphere'; *circulus* often means simply 'circle'; and both *circulus* and, more rarely, *orbis* may have the connotation of 'circular path'. My decision to render these terms constantly by 'sphere', 'orb', and 'circle', respectively, is motivated in part by recent controversy

---

[1] The missing parts are an entire folio between f. 295v and f. 296r (f. 270, l. 32–f. 271, l. 38 in Frisch's edition) and a glossed quotation from Vitruvius (f. 272, l. 31–40) to which Kepler refers at the foot of f. 296v: *Huc refer ex altero folio paragraphum.* Dr Martha List has informed me that the numbering of the folios is due to Frisch.

about the connotations of these terms in the period,[2] and in part by the wish to avoid difficulties that would otherwise arise in cases where the terms occur in passages from other authors quoted or glossed by Kepler. *Circuitus* is another awkward word: generally I have rendered it 'circuit', but on occasion 'period' seemed more apposite; I have avoided the tempting but anachronistic rendering 'orbit'. Difficulties in rendering other problematic terms, *artifex*, *calculus*, *commutatio* and *problema*, for example, are mentioned in the footnotes to the translation.

The parts of Kepler's oration are indicated in the margin of the translation.

[2] The dispute has mainly concerned the connotations of these terms in *De revolutionibus*. A useful note by A. M. Duncan on Kepler's usage is to be found in *Johannes Kepler: the Secret of the Universe* (New York, 1981), 14.

KEPLERI APOLOGIA PRO TYCHONE CONTRA URSUM[3]

Ad Lectorem praefatio.

Edidit ante triennium *Nicolaus Ursus, Caesareae Majestatis, dum viveret,* 264r
Mathematicus libellum *de Hypothesibus astronomicis*: in quo cum historiam
texuisset hypothesium, tandemque ad Helisaeum Röslinum, Medicinae
Doctorem, amicum meum singularem devenisset, et ad refutandam ejus
opinionem, adducta quadam Euclidis propositione: rationem tradidisset, ex
dedomenis necessarijs Geometrica methodo Planetariorum orbium ampli-
tudines dimetiendj (rem tritam apud Astronomos): subjunxit, ex ijsdem
fundamentis etiam examinarj posse mea placita, quae anno priore sub titulo
Mysterij Cosmographicj publicj juris feceram: eaque occasione capta, Epistolam
quandam, qua biennio ante super inventione mea ipsum consulueram, libello
suo inseruit.

Pessimam quidem hoc nomine gratiam apud me Ursus inijt. Scripseram,
utpote vigintj trium annorum juvenis, minime sane graviter. Nam et in Urso
laudando, et in me illj subjiciendo modum utrinque multum excessj.
Securissimus enim de hoc eventu, id saltem spectavj, ut responsum im-
petrarem. Et ijs ferebatur esse moribus Ursus, ut his armis opus esset ad
expugnandum ejus silentium. At si quibus de instituto meo non constat, ij
Ursj opera mihi merito succensebunt, quod immodicis Ursj laudibus,
injuriam ⌊multis in Europa⌋ doctissimis Mathematicis fecerim: mirabuntur
etiam, qui me propius norunt, me pro Maestlino, Ursum ⌊praeceptorem⌋
celebrare, illique transcribere, *hoc quantulum est cognitionis in Mathematicis*:
quem tamen, nisi in *fundamento* suo *Astronomico* loquentem nunquam
audiveram, nunquam videram. Sed haec ⌊adhuc⌋ levia viderj possunt:
graviora sequuntur. Nam etsi Ursus honorem meum egisse viderj vult,
publicata mea epistola, dum eam luce dignam praedicat: tamen quid
quaesiverit, non obscurum reliquit. Primum, in titulo libellj Nobili et Illustri
D, Tychonj Brahe dano, domino in Knudstrupp et Uraniburg etc // hypo- 264v
thesium novarum inventionem praeripit, eamque sibi arrogat. Putavit itaque
hac Epistola mea, in qua haec sunt verba (*Hypotheses tuas amo*) demonstratum

---

[3] In the margin of f. 264r is written in another hand: *Kepleri Apologia pro Tychone
contra Ursum una cum Documentis authenticis contra Ursum atque provocatis in epistolis
Tychonis.* The accounts of Ursus' suspicious behaviour at Uraniborg by Christian Hansen
and Michael Walther, which Tycho sent to Kepler in December 1599, are with the
Apologia in Pulkowo, v (*K.g.W.*, XIV, 469).

irj, esse et homines, qui Ursj bonitatem causae agnoscant ⌐eaque ratione me alienae litj immiscuit, et testimonium ab ignaro extorsit⌐. Deinde cum in calce titulj eruditionem suam in hoc libello, ceu ad agonem quendam olympicum emittere profiteatur, dum omnes et evidenter D, Tychonem Brahe de *palma*, *Magisterioque Mathematico* in certamen vocat: percommode meam epistolam in libello ⌐ceu cujusdam judicis loco⌐ introducit, quae talem fert sententiam, ut Urso *palma* illa citra controversiam debeatur. Denique feceram ego in eadem illa epistola mentionem Tychonis perhonorificam, Ursumque oraveram, ut hanc meam inventionem ad ipsum in Daniam perscriberet. Placuit igitur Urso, non aliter me de hac in Tychonem, adversarium suum, affectione optima ulciscj, quam si Epistolam meam, extrita mentione Tychonis, eum in libellum insereret, qui non ad aliud quam ad lancinandam famam Tychonis comparatus et editus est: ut ita, qui Tychonem privatim Urso laudaveram, poenae loco conviatorem Tychonis publice omnibus laudare cogerer. Tot nominibus, cum Ursum lubuerit abutj mea epistola, jam clarum est, cur luce dignam judicaverit. Nam quod inventionem meam attinet, si lucem ea meretur, jam eam, fatente Urso, viderat, extante in Nundinis integro libello, nec Epistolae hic opus fuit editione.

Quam graviter autem haec editio me haberet, cum literis ad D, Tychonem et Röslinum perscriptis testatum fecj: tum Urso ipsj superiorj Januario, primum atque Pragam venj, in faciem dixj, suppresso tamen meo nomine; ne si cum praesente Keplero sibi rem esse sciret, rem ad jurgia protraheret. Addidj et haec: quando lubuerit illj, me, qui discipulj sub persona scripseram, in judicis subsellia invitum protrahere: pateretur ergo me in illa lite literaria (de convitijs enim et injuriis ⌐politicis⌐ actionem Tychonj relinquere), exuta discipulj modestia judicis authoritatem assumere, et quod e re mathematica fore videatur, publice vicissim decernere. // Sicque, nomen tandem professus, pacifice ab ipso discessj.

Quo serius autem ⌐denunciatam Urso⌐ refutationem illius scriptj sub manus sumerem: multae ac praecipue domesticae occupationes impedierunt. Intervenit et mors Ursj, quae me nonnihil dubium habuit: ne esset qui hanc cum mortuo pugnam calumniaretur. Sed quia promissum meum D, Tycho Brahe urgebat, a cujus humeris, hoc onus scribendj ⌐pridem⌐ ultro in me susceperam: et quia per se res erat honesta, extirpare ex animis discentium erroneas opiniones, obliterare in Maecenatum persuasiones impressam inventionum Mathematicarum calumniam: nec inusitatum, de mortuorum, quanticunque fuerint, placitis et publicis monumentis, decernere superstites ⌐(quod ipse Ursus fol C 2 in demortuo Ramo taxando comprobat)⌐: tandem ergo diu agitatam hanc scriptiunculam animo sincero, et rej Mathematicae

265r

studio suscepj: putoque sic eam pertexuisse, ut et ad me defensae et ad lectorem agnitae veritatis utilitas et laus reditura sit.

Quid sit hypothesis astronomica.

In descriptione Hypothesium sic loquitur Ursus, quasi non ad aliud comparatae essent hypotheses, quam ad ludos hominibus faciendos. Quam ad rem non satis verborum invenit, quibus quam fieri potest, contemptissime de hoc genere inventionum loquatur: nec prius cessat orationem suam distinguere, subdistinguere, quam seipsum quodammodo superaverit.

Nos veram hypothesium naturam primum explicabimus; postea varie impingentem, seque ipsam conficientem Ursi sententiam particulatim expendemus: ne quid restet, quod imperitus lector Urso ⌊propter ingenij famam⌋ frustra credat.

Non statim cum ipsa caelj observandj consuetudine natus est mos iste, ut quam quisque philosophorum ex intuitu caelj, de mundj dispositione concepisset opinionem, ea nomen aliquod haberet, *Hypothesisque* diceretur. // Longinquitas temporis, philosophorumque successio ut caetera, 265v quae vocamus, secundae intentionis vocabula, sic hoc quoque in astronomia stabilivit. Prima vocis usurpatio fuit apud Geometras: qui ante natam logicam philosophiae partem, naturalj mentis lumine demonstrationes suas geometricas cum vellent expedire, soliti sunt a certo quodam initio doctrinam incipere. Quemadmodum enim in Architectura satis habet opifex si pro futura mole domus fundamenta infra terram extruat, neque sollicitus est, ut terra inferior vel ruat, vel in sese recedat usque ad centrum, ita in re Geometrica primj authores non ita stolidj fuerunt, ut, qui post secutj sunt Pyrrhonij, ut omnia vellent addubitare, nihil arripere, super quod ceu fundamentum certum et omnibus confessum caetera vellent superaedificare. Quae itaque certa et apud omnes homines confessa essent, ea specialj nomine ἀξιώματα dictitarunt, ceu sententias dicerent, quae haberent authoritatem apud omnes. Quaedam hujusmodj principiorum sic erant comparatae, ut quamvis non ab omnibus crederentur, ipsis tamen authoribus satis erant notae, ut non dubitarent eas alia demonstratione certas et veras ostendere. Id autem ut in ⌊sibi⌋ proposita demonstratione perstarent, vel nimis longum erat futurum; vel intempestivum, ut cum ejusmodj sententiae cognatione quadam cum alijs rebus, de quibus in Geometria non disputatur, innexae et confusae essent. Has dixerunt: αἰτήματα, utpote, quae a discente sibi concedj a principio postularent. ⌊Hoc tum maxime usu venit, cum in figuratione angulum vel lineam talem designare vellent, qualis manu praecise fierj nequit. Tunc enim dicebant, ἔστω *talis vel talis*. Est et aliud genus, cum quae vel esse nequeunt, vel non sunt, tanquam essent assumuntur, ut demonstratione patescat, quid secuturum foret, si illa essent. Hoc quoque genus tum locum habet, cum e geometria egressj in cognatas scientias demonstrationis Methodum transferunt.⌐ Utrumque genus, et quae alia brevitatis causa intermitto, quando ad actum demonstrationis veniebant, ὑποθέσεις seu *suppositiones* appellabant:

266r respectu scilicet ejus, quod demonstrare intenderent. // Supponebantur enim nota, et super illis extruebantur ignotiora, ostendebanturque discentj.

Haec origo vocis, hic usus apud Geometras. Statim in Geometricis figuris et numeris, nempe in re totius naturae clarissima, mentique humanae maxime apta, deprehenderunt rerum contemplatores lumen illud mentis nostrae, quod rectissime quidem in figuris, sed tamen etiam in alijs rebus omnibus generatim vigeret, et sine quo nihil esset, cujus cognitionem capere mens nostra posset. Itaque hanc demonstrandj rationem ex Geometria ceu exemplo desumptam excoluerunt, et in formam artis adeoque et scientiae redegerunt, quae logica dicitur. Vocabula et ut loquimur, terminos, quos in Geometria repererunt, retinuerunt, adeo crebro, ut negare nemo possit. Ita *hypotheseos* vox quoque usitatissima est in doctrina demonstrationum apud Aristotelem. Cum itaque Logicae jurisdictio per omnes pateat scientias, ex logica quoque in Astronomiam vox *Hypotheseos* introducta est, quae vel sine logica propter solas demonstrationes in astronomia geometricas, ⌐et sic genuino nomine⌐ usurpabatur.

Hypothesin autem in genere dicimus, quicquid ad quamcunque demonstrationem, pro certo et demonstrato affertur. Ita in omnj syllogismo hypotheses illae sunt, quas propositiones sive praemissas aliter dicimus. In longiore vero demonstratione, quae multos habet syllogismos subordinatos, primorum syllogismorum praemissae dicuntur hypotheses. ⌐Ita in Astronomicis, quoties ex ijs, quae in caelo diligenter et apte attentj vidimus, aliquid numerorum et figurarum ope de stella, quam observaveramus, demonstramus; tunc ea quam dixi observatio, in instituta demonstratione fit hypothesis, super quam praecipue demonstratio extruitur.⌐ Genuina haec est vocis *Hypothesis*, notio. In specie tamen, cum numero multitudinis astronomicas dicimus *Hypotheses*, facimus id more scholarum hujus saeculj: designantes summam quandam conceptionum celebris alicujus artificis, ex quibus totam ille rationem motuum caelestium demonstrat: qua in summa insunt cum physicae tum Geometricae propositiones, omnes, quotquot ad totum opus Astronomo illj propositum asciscuntur: sive illas aliunde ad suum commodum transtulerit artifex, sive antea ex observationibus demonstraverit, et jam via

266v reciproca, secum // demonstrata a discente sibi ceu Hypotheses concedj postulet; ex quibus ille [illi] et eos (quibus initio secum hypothesium loco

88

usus erat) observatos stellarum situs, et quos porro similiter apparituros sperat, necessitate syllogistica demonstraturum polliceatur. Dixi quid sit hypothesis, Dicam et qualem esse justam hypothesin sit necesse.

Cum in omni disciplina, tum in Astronomia quoque quae concludendo lectorem docemus, ea serio facimus omnino, ludos non agitantes. Itaque quicquid est in nostris conclusionibus, verum esse persuasum habemus. Porro ut verum concludatur legitime propositiones in syllogismo, hoc est hypotheses veras esse necesse est.[4] // Tum demum enim fine nostro potimur ut verum 267r lectorj ostendamus, cum hypothesis utraque undiquaque vera lege syllogistica in conclusionem directa fuerit. At si error vel in utramque vel in alterutram praemissarum hypothesium irrepserit, etsj verum interdum in conclusione sequitur: id tamen, ut jam pridem Cap: 1 mej Mysterij Cosmographicj allegavi, fortuito fit, nec semper; sed tum demum, cum error propositionis unius, occurrit aptae vel verae vel falsae alterj, ad verum eliciendum. ⌊Exemplj loco latitudinem ☽ Copernicus supposuit non sat magnam. Visa est illa olim cor Scorpij tegere. Copernicus falsa hac latitudinis hypothesi, confecit tamen eo tempore oportuisse tegi stellam a ☽. Sed alter error accessit; assumpsit enim latitudinem stellae tanto minorem justo, quanto minor justo latitudo Lunae constituta fuerat.⌋ Atque, ut in proverbio monentur mendaces, ut sint memores: ita hic falsae hypotheses, verum semel fortuito concludentes, in progressu demonstrationis, ubi alijs atque alijs fuerint accommodatae, morem hunc verum concludendj non retinent, sed seipsas produnt. Ita tandem fit, propter syllogismorum in demonstrationibus implexum, ut uno inconvenientj dato, sequantur infinitae. Cum itaque nemo eorum, quos Hypothesium authores celebramus, in periculo versari velit erroris ⌊circa conclusiones⌋, ut ante dixi: sequitur, ut neque quisquam eorum inter Hypotheses suas aliquid velit recipere sciens, quod errorj sit obnoxium. Adeoque non tantum de demonstrationum eventu et conclusionibus, sed saepe magis de assumptis Hypothesibus sollicitj sunt: adeo illas, quotquot adhuc fere fuerunt authores celebres, cum Geometricis tum physicis rationibus expendunt, et undiquaque conciliatas cupiunt.

[4] There follows a deleted passage of some interest:

Nam si alterutra sola praemissarum hypothesium falsa sit, certissimum, est, quod concluditur verum non necessario futurum ⌊, sed crebrius naturam praemissarum⌋. Haec ab Aristotele rationibus evidentibus artificiose demonstrata, vide apud ipsum authorem. Unus casus est, isque rarus admodum et temerarius, cum utriusque praemissae hypotheseos errores sic invicem concurrunt ⌊seque mutuo tollunt⌋, ut vera sequatur conclusio, etiam in forma syllogistica. Ut cum quis in Astronomia statuat hypotheseos loco *Omnes inaequalitates motuum caelestium provenire ex sola refractione aeris*. Quicquid verum lege syllogistica huic praemissa conjungat conclusio sequetur falsa. At si aliam falsam, huc de industria concinnatam subjungat, utpote, *Quod stellae justo maturius surgant, justo tardius occidant, inaequalitatem esse motuum ipsorum caelestium*, sive clarius, ⌊id esse πάθος motuum caelestium, aequale caeteris⌋ conclusio· vera fortuito sequetur, si necessitate syllogistica concludas, *Quod stellae videantur surgere justo maturius occidere tardius, id evenire ex sola refractione aeris...*

Quid igitur est dicas, quod cum omnes eosdem motus caelestes demonstrent, tanta tamen sit hypothesium diversitas? Non sane id inde evenire existimandum est, quod ex falso verum per accidens sequi soleat. Nam, ut ante dixj, in longo demonstrationum per varios syllogismos anfractu, quales solent in Astronomia existere, fierj vix unquam potest, ac mihi sane exemplum non occurrit, ut ex assumpta non sana hypothesj, exitus undiqua-

267v que sanus // et caelj motibus analogus, aut talis qualem vult demonstrator, sequatur. Sed nec semper idem efficitur per diversas hypotheses quoties imperitior aliquis, idem effici putat. Quae ex copernicj sequuntur hypothesibus, ea Maginus ex alijs similiter, quoad numeros, demonstravit hypothesibus, quae Ptolemaicis essent ut plurimum conformes. Num idem uterque praestat? Minime gentium. Copernicus quidem causam simul demonstrare voluit et necessitatem, cur planetae superiores in oppositione cum sole semper essent terris proximj praeterquam quod numeris eorum motus in posterum voluerit repraesentare. Maginus omisso illo quod Copernicus in Ptolemaicis desideravit et in suis suppositionibus non parvj fecit, solos numeros copernicj fuit imitatus. In copernico sequitur, Martem admittere parallaxin majorem sole. In Magino hoc minime sequitur. Multa hic praetereo. Ac etsi diversae aliquae hypotheses idem omnino praestent in Astronomicis, quod de sua Copernicanarum mutatione Rothmannus in epistolis ad D, Tychonem jactavit: differentia tamen ⌊saepe⌋ est conclusionum causa alicujus considerationis physicae. Sic etsj omnino eosdem Tycho numeros hypothesibus suis, exprimeret, quos habet Copernicus, qui tamen vitiosi sunt: in hoc tamen differet intentum demonstrationum Tychonis a Copernicanis, quod praeter praedictionem futurorum motuum, cupit etiam evitare immensitatem illam fixarum, aliaque, quae Copernicus admittit sua hypothesj. Ita conclusione mutata, Hypotheses varias existere necesse est. Inconsideratus vero aliquis, ad solos numeros respiciens, idem ex varijs hypothesibus, adeoque verum ex falsis sequi existimabit.

Qui ad hanc normam omnia expenderit, sane nescio, an illi ulla sit

268r occursura hypothesis, seu simplex seu composita, quae non // peculiarem et ab omnibus alijs separatam atque differentem sit exhibitura conclusionem. Nam si in Geometricis duarum hypothesium conclusiones coincidant, in physicis tamen quaelibet habebit suam peculiarem appendicem. Sed quia ab artificibus illa in physicis varietas non semper considerarj solet; illique ipsj saepius cogitationes suas intra Geometriae vel astronomiae terminos cohibent, inque una aliqua scientia de aequipollentia hypothesium quaestionem instituunt, posthabitis diversis sequelis, quae respectu affinium scientiarum ventilatam aequipollentiam dirimunt et destruunt: par est, ut nos quoque sermonem et responsionem nostram ad ipsorum loquendj morem accommodemus.

Quid igitur, an non alterutram hypothesium ⌞circa primum motum (ut exemplo loquamur)⌟ falsam esse necesse est, ⌞aut⌟ eam quae terram ait moverj intra caelum, aut eam quae caelum vult gyrarj circa terram? Sane si contradictoria simul vera nequeunt esse, neque haec duo simul vera erunt; sed alterum omnino falsum. At utroque modo idem de primo motu demonstratur? Eaedem sequuntur signorum emersiones, ijdem dies, ⌞ortus et⌟ occasus stellarum, eaedem qualitates noctium? Ex falso igitur ⌞aeque ac ex vero⌟ verum sequitur? Minime gentium. Non enim haec quae recensita sunt, et mille alia, vel propter terrae vel propter caelj motum eveniunt, quatenus is terrae vel caelj motus est: Sed quatenus intercedit inter caelum et terram aliqua separatio, super tractu quodam, qui legitime ad viam solis sit inflexus: utrocunque jam corpore illa separatio perficiatur. Demonstrantur igitur antedicta per duas hypotheses, quatenus illae sunt sub uno genere, non quatenus inter se differunt. Cum ergo causa demonstrationis unum sint, causa demonstrationis etiam non erunt contradictoria. Ac etsi illis adhaereat contradictio physica, illa tamen demonstrationem jam nihil attinet. Non igitur et ex vero et ex falso // verum sequi, hoc quidem exemplo probatum ⌞268v⌟ est. Sed acrius fortasse nos urgebit notissimum exemplum de concentrico cum epicyclo, qui aequipollet Eccentrico. An non si planeta, seu quicquid id est quod planetam incitat, centrum eccentrici ⌞in⌟ motu suo respicit, centrum epicyclj simul non respiciet? Quod si quis astrologus hoc affirmabit, alter illud, falsum ab alterutro dici nonne necesse est, cum tamen ⌞idem⌟ uterque demonstret, aequali scilicet manente planetae motu moram tamen ejus in uno semicirculo longiorem fieri⌟? Ex falso itaque ⌞aeque ac ex vero⌟ verum nonne sic sequitur? Minime gentium: Eadem enim respondeo, quae supra. Id quo conficitur, moras in alterutro semicirculo longiores fierj, ab utroque dicitur, et ab eo qui eccentricum ponit, et ab eo qui concentricum cum Epicyclo. Nam uterque generale hoc affert, majorem circulj planetarij portionem esse in illo semicirculo. Hoc genuinum demonstrationis et coaequatum medium cum insit in utraque hypothesi generaliter, nihil jam demonstrationem attinebunt, quae uterque dicit in specie, sive pugnet illa, sive non. Neque sane planetis oculos et discursum intellectus humanum tribuimus, ut sibi circino vel hoc vel illud punctum signent, haecque ⌞quae dixi⌟ specialia sui potius captus quam explicandae naturae causa introducunt artifices. Non est igitur digna nomine *hypotheseos* ⌞astronomicae⌟ vel haec vel illa suppositio, sed id potius, quod utrique communiter inest, dum scilicet assumitur et supponitur certa et dimensa portio circulj, quem planeta decurrit, quae sit in uno circulj

Zodiacj semisse. Haec inquam tandem est hypothesis justa, qua longitudo
morae planetae in illo semicirculo demonstretur. Itaque ut concludam: Omnis
in astronomia conclusio non nisi ab uno et eodem medio perficitur, et
269r uniformem praemittit hypo // thesin: etsj illa a seipsa differat, quatenus extra
hanc demonstrationem consideratur. Et vicissim, quaelibet hypothesis, si
accurate consideremus, propriam nec ulli alij hypothesi communem penitus
producit conclusionem: Nec fierj potest in astronomia, ut ex omni parte
verum sit, quod ex falsa primitus hypothesj fuit extructum: et proinde
hypothesibus hoc est proprium (si ideam fingamus justarum hypothesium)
ut sint undiquaque verae: nec est astronomj, scientem, falsas supponere, vel
ingeniose confictas hypotheses, ut ex ijs motus caelestes demonstret.

Jam Ursj absurdas sententias paulisper expendamus.

Primo Hypotheses Astronomicas ait esse effictam delineationem, imagin-
ariae, non verae et genuinae formae systematis Mundanj. Quibus verbis aperte
negat Hypothesin esse, quae non sit falsa. Confirmat hanc suam sententiam
paulo post, ubi ait, nihil aliud esse Hypotheses, quam *commenta* quaedam. ⌐Et
inferius ait, non fore *Hypotheses, si verae sint.* Item, *proprium esse hypothesium
ex falso verum sciscitarj*: quare secundum hanc sententiam, nec movebitur terra
nec quiescet. Utramque enim hypothesin Ursus fatebitur. Et mille talia. Sed⌐
hoc monstrum definitionis jam antea satis est refutatum, nec verbis dignum
est. ⌐Tota series errorum est ex sinistro intellectu ἐτύμου, ὑποτίθεσθαι, quod
Urso videtur idem esse, quod *fingere*. At supra diximus, quis ejus sit
significatus ap: Geometras.⌐

Deinde Hypotheses ait confingj ad observandos motus caelestes. Haec
sententia e diametro pugnat cum veritate. Motus caelestes observare possumus,
nulla plane imbutj opinione ullius dispositionis caelorum. Etenim observatio
motuum caelestium astronomos manuducit ad hypotheses legitime constitu-
endas, non e contrario. Quod enim in omnj cognitione fit, ut ab ijs quae
in sensus incurrunt, exorsj, mentis agitatione provehamur ad altiora, quae
nullo sensus acumine comprehendj queunt: idem et in astronomico negocio
locum habet: ubi primum varios planetarum situs diversis temporibus oculis
notamus, quibus observationibus ratiocinatio superveniens mentem in
cognitionem formae mundanae deducit, cujus quidem formae mundanae,
sic ex observationibus conclusae, delineatio Hypotheseon Astronomicarum
postmodum nomen adipiscitur. //

269v   Tertio, astronomos in id saltem elaborare vult, ut ex assumptis suis
hypothesibus rationem reddant motuum caelestium, doceantque, quales futuris
temporibus appariturj sint. Neque hic sine exceptione verum dicjt. Nam etsj
hoc, quod dicit, astronomj primarium est officium, non est tamen astronomus
e coetu philosophorum, qui de rerum natura quaerunt, excludendus. Bene

fungitur officio astronomj, qui quam proxime motus et situs stellarum praedicit: melius tamen facit, et majori laude dignus habetur, qui praeter hoc etiam veras de mundi forma sententias adhibet. Ille namque verum, quoad visum, concludit: hic non tantum visui concludens satisfacit, sed etiam Naturae penitissimam formam concludendo amplectitur, ut supra fuit explicatum. Neque hic solummodo Ideam aliquam bonj astronomj describo: exempla suppetunt. Non fuit necessarium ad motus planetarum praedicendos ut Ptolemaeus de ordine sphaerarum disputaret, et id tamen is diligenter fecit. Non fuit necessarium ad motus planetarum calculo exprimendos et praedicendos ut Copernicus, et post illum D. Tycho causas inquirerent, cur planetae ἀκρόνυκτιοι [ακρονυχιοι] fiant terris proximj. Poterant enim vel Ptolemaica forma caelorum usj, correctis dimensionibus, idem praestare. Sed fecit amor cognoscendae Naturae, ut astronomj, astronomicis argumentis hanc physices partem explorandam sumerent, nec injuria gloriantur de suarum hypothesium cum rerum natura conformitate. Quid Copernicus? cum animadverteret in ptolemaeo quandam inaequalitatem motus Epicyclj; non quia is motus visui et experientae seu observationj stellarum, sed quia rerum Naturae repugnaret, ideo hinc occasionem se adeptum profitetur, a Ptolemaeo discedendj. Et o miserum Aristotelem, miseram ejus philosophiam, qua parte divinissima censetur, si astronomj tam malae fidej testes in philosophicis controversijs sunt. Is enim, cum // eo devenisset, ut certum 270r intelligentiarum caelestium numerum et ordinem exprimere necesse esset: ad astronomos Eudoxum et Calippum nos ablegavit, et quamvis in quibusdam dissentientes videret, utrosque voluit, ob adhibitam diligentiam, et communicatam motuum doctrinam, venerandos et amandos, audiendum vero, qui quam rectissime rem expediret. Aliter ergo sensit Aristoteles de veterum illorum astronomorum (quibus qui secutj sunt, non ignobiliores fuere) laboribus et praeclarissimo proposito, quam Ursus.

Quarto videtur argumento pugnare pro sua sententia, non esse necesse, ut verae sint astronomicae Hypotheses. Nam crassissima in omnibus adhuc hypothesibus absurda residere. Verum id quod est ignotum, probat per aliud longe incertius. Verum quidem est, si absurdum et falsum id censerj debet, quod unj alicui hominum coetuj tale videtur: nihil erit in tota physiologia, quod non pro crassissimo absurdo haberj debet. Variae sunt hominum sententiae, varij captus ingeniorum. Non credidit absurdum ⌊esse⌋ Copernicus, ut terra moveatur, Fixarum tanta sit immensitas. Non existimat absurdum Tycho, ut planetae solem mobilem ceu centrum suum, quocunque id secedit, sequantur, nec tamen suos motus deserant. ⌊Qua de re vide disputationem Rothmannj cum Tychone in *Epistolis Astronomicis*.⌋ Non est absurda Ptolemaicis, inaestimabilis illa ultimj caelj pernicitas. Sunt haec, ut a sensibus, ita et a captu ingeniorum nostrorum remotiora, et varijs disputantur argumentis. Atque haec illa sunt crassissima absurda: quae quidem simul vera esse nequeunt, quo fit, ut ⌞neque⌟ plus una hypothesium forma vera esse possit. Quid autem ex his verum sit, quid falsum, penes vere philosophum est decernere. Certum tamen est, non omnia esse, ut Urso videntur, crassissima absurda. Mihi sane nihil falsum Copernicus dixisse videtur: quodque meis pro Copernico defuit argumentis, id admirabili solertia, experiundique industria supplesse videtur Gilbertus Gulielmus Anglus, in re magnetica.

Quinto desperat veras Hypotheses in Astronomia: Pyrrhoniorum more 270v omnia incerta // faciens. Esse quosdam adhuc naevos, etiam in optime constituta astronomia, proptereaque et in Hypothesibus, nemo negat: et id est, cur hodiedum in restitutione tantopere laboretur. Caeterum incerta propterea omnia, ne quis dicat. Invenit astronomiae opera Ptolemaeus, fixarum sphaeram esse altissimam, sequi ordine Saturnum, Jovem, Martem, Solem his inferiorem, Lunam infimam. Haec certe vera sunt et formae mundanae consona. ⌞Idem⌟ Ascensum et descensum Lunae nimis magnum praesupposuit. Ex hac hypothesj falsum et ab apparenti Lunae magnitudine dissidens quid sequi Regiomontanus et Copernicus deprehenderunt. Dixerunt itaque, Lunam non tantos facere saltus in altum et profundum. Et quis dubitat, quin hoc revera sic in caelo se habeat? De proportione Terrae solis et lunae ἐν πλάτει quis quaeso est, qui dubitet, nisi Melancholicus aliquis et desperabundus? Ita hodie nemo porro est qui dubitet, quin, quod est in Tychonis et Copernici hypothesibus commune, Sol nempe centrum motus quinque planetarum, id revera sic se habeat in ipso caelo, etsi interim de motu vel quiete solis hinc inde dubitetur. Talia multa sunt beneficio astronomiae hactenus in physica scientia constituta, quae fidem in posterum merentur, et revera sic se habent: ut Ursj desperatio sine causa sit.

Sexto plurimas ait hypothesium ⌞formas⌟ confingj posse. Adeo leve illi negocium est, hypotheseon constitutio. At ea plena laboris est et curae summae; quod testarj poterunt, qui eam degustarunt. Praesidium quidem habet haec sententia nonnullum in eo, quod Ursus non nisi fictitias esse vult hypotheses, sine veritatis necessitate. At enimvero nec hoc quidem Urso facile, talium, quales ipse describit ⌞formas⌟ plurimas confingere. Nam etsj non curat, veraene sint, vult tamen ut repraesentent calculum motuum: quod non cuivis obviae et fictitiae hypothesj contingit. //

271r Septimo, sufficere ait, ut hypotheses, caelestium motuum calculo respondeant, licet non ipsis motibus. Mirabile dogma, si quidem mens Ursj ex ⌞his⌟ verbis censerj debet. Quid est igitur, Urse, calculus? Si bonus est calculus, hoc est, Methodus computandj loca siderum, tunc sane motibus est consonus, et ex hypothesj extruitur, quae itidem est motibus consona. Sin quis calculus motus non repraesentat, aeque fallax habetur, ac hypotheses, ex quibus derivabatur. Ita si nobis nil curae est, utrum aliquis calculus motibus respondeat: nihil sane impedit, quo minus plurimas fingamus hypotheses, quod sibi facile dictitabat Ursus, totidemque calculorum formas constituamus, vitiosas nempe et a caelo dissonas. Macte Urse hac inventionum ubertate. Forsan et hoc agis, ut hypotheses in solas scholas Astronomicas concludas, ut quarum usus in hoc solo consistat, ut praeceptor monstret discipulo in delineatione hypothetica, quas ob causas singula praecepta in calculandis caelestibus motibus observare debeat. Qui usus etsi non alienus est ab hypothesibus, posterior tamen est, et inferiore dignitatis gradu. Primum enim in hypothesibus rerum Naturam depingimus, post ex ijs calculum extruimus, hoc est, motus demonstramus, Denique indidem vera calculj praecepta via reciproca discentibus explicamus.

Octavo, nil putat officere sacras literas hypothesibus contradicentibus. At, quas ipse jactat hypotheses, ex scriptura se demonstrasse putat. Quod si proprium est hypothesium, ut sint falsae: hypotheses ipse nullas habebit, quia, quas jactat, infallibili Scripturae authoritate munitas contendit. Ita suiipsius immemor est. Id magis etiam in eo, quod cum proprium fecisset hypotheseon, ut sint falsae, tandem aequior, astronomis licere inquit effingere hypotheses sive veras sive falsas.

Nono. Quae de falsitate hypothesium astronomicarum dixit, ait in plerisque alijs doctrinis fierj. Id quidem fit, sed male, et ab imperitis, nec verum aut bonum nisi rarissime et fortuito, sed falsum et pernicies justo ordine consequitur. // Ut in Medicina, si quartana quis laboret: et falsam hanc 271v hypothesin imperitus medicus conceperit, causam febris esse a sanguinis redundantia, is vena secta, non sane faelicius juverit aegrum, quam Ursus ex falsa hypothesi motus caelorum colligeret. Sed Ursum per exempla sua sequamur. Exemplj namque loco regulas Algebraicas producit, ubi ponitur is qui quaeritur, numerus, valere unitatem cossice denominatam, cum tamen is non sit unitas. Ex hac tamen falsa positione in cognitionem veritatis pervenirj. At, o Urse, nec quadrat exemplum, nec recte explicatur. Cum nesciatur numerus is, quem quaerimus, ignotum illi nomen damus, eoque nomine pergimus leges praescriptas, donec arte nomen id nobis innotescat. Quod igitur ponimus, unitatem esse denominatam, falsum non dicimus. Nam eo ipso quod nomen additur unitatj, innuitur certus quidam numerus, quem, is qui quaeritur, semel valeat. Hic ergo nulla falsa positio. At in astronomicis hypothesibus nihil simile. Quam enim quisque profitetur hypothesin, non aliter nisi ut notam et exploratam profitetur. In regula quam falsj dicunt, quo altero exemplo pugnat Ursus, non alio quam nomine gloriarj potest. ʟNon enim ideo falsus dicitur numerus positus, quia verum inde sequitur, quod si fieret, non hoc illi nomen esset: sed ideo, quia falsum quoque concludit: quod considerantj facile patet.⌐ Numerus quaeritur, qui legibus imperatis tractatus certum quid efficiat. Arripitur obvius quisque, tractatur legibus imperatis, si, quod jubetur, efficit, ipsus est, qui quaerebatur, sin minus, falsus est, Hoc idem in altero tentatur, utrique inter se comparantur, ⌐et⌐ utrorumque peccata, quorum intuitu verus tandem elicitur. ⌐Huic⌐ quid in astronomicis hypothesibus simile sit, ijsdem verborum vestigijs dicam. Hypotheses quaeruntur, quae caelj motibus respondeant. Alphonsinae peccare deprehenduntur, Copernicanae itidem. At sollers artifex collatione utrarumque instituta et subductis aberrationum rationibus, tertium aliquid statuit, quod errorem omnem circa praedicendos caelj motus effugit; eoque modo utrasque illas

272r hypotheses corrigit. Quid hic est, quod pro Urso ∥ faciat? Proponitur nobis in ejus exemplo modus unus, isque obliquus et ἄτεχνος, quo solent astronomj hypotheses investigare. At is ostendere volebat, quomodo ex ⌐jam constituta⌐ falsa hypothesj verum et caelj motibus analogon demonstretur. In regula falsj numerus falsus, falsum quid etiam concludit. Ursus contendit ex falsis Hypothesibus astronomicis verum concludj.

Decimo, postquam suam Ursus sententiam in aciem produxit, eam authoritate munit ridicula. Quid enim me moveat ignotus? Itaque Ursus authoritatj suae authoritatem ipse conciliat, authorem dicens esse pereruditum. Juvabo laborantem Ursum, fuit ejus praefationis author, si nescis, Andreas Osiander, ut Hieronymj Schreiberj Noribergensis (ad quem Schonerj quaedam praefationes extant) manus in meo exemplarj visenda testatur. Qui quamquam Ursj sententiae quam proxime suffragatur, tamen et aequior est eo, et dilucidius se explicat. Quid ipsum moverit, non est obscurum. Noribergae dominabatur, ubi tum edebatur opus Copernicj. Cum igitur amaret Mathemata, non potuit non amare tam praeclarum opus; at cum videret παραδοξότατον δόγμα de motu terrae, sic statuit infringendam cupiditatem in lectoribus rerum novarum. Atqui o Osiander, quid eo te desperationis adegit, ut ex Astronomia de vera Mundj facie nihil certj colligj posse diceres? An, qui peritissimus eras harum rerum etiamnum dubitabas de rebus evidentissimis? Incerta tibi etiamnum erat proportio Solis ad terram, prodita ab astronomis, et sexcenta hujusmodj? Si haec ars causas motuum simpliciter ignorat, propterea quia nihil credis, nisi quod vides, quid medicinae fiet, in qua quis unquam Medicus causam morbi vidit, intus latentem, aliter, nisi ex signis et symptomatis extra corpus in sensus incurrentibus, perinde ut astronomus ex stellarum situ aspectabilj de forma

272v motus eorum ratiocinatur. Sed ad Ursum ∥ revertamur: Cui hoc de teste suo dico, non esse omnj exceptione majorem. Vult in eadem praefatione

96

persuadere lectorj, non fuisse in hac sententia Copernicum, vere terram moverj. Statim ipsum in limine refutat ipse copernicus. Suggessit Funccio Chronologo, si fides illi viro tribuenda, non retrogressum esse solem in gratiam Ezechiae, sed tantum umbram Gnomonis. Quasi haec ab invicem separarj queant, cum constet ex lectione Sacrae scripturae universaliter animadversum fuisse miraculum. Graviora absurda, quae hoc etiam ingenium occupare potuere, relinquo alijs exquirenda. Quid mirum ergo, si in Mathesi quoque, quam amavit, non professus est, exorbitavit. Contra eum sisto Aristotelem, cujus existimationem de astronomis longe praeclariorem supra tetigj.[5] // Non est tamen praetereundum et notandum, viderj Osiandrum   273r simulate scripsisse, non ex animi sententia, usum nempe Ciceronianis de Rep: consilijs, cum Copernico metueret a vulgo philosophorum, ne absurditate hypotheseos a tam praeclaro opere lectores absterrerentur, ista quasi lenimenta operj Copernicano praemittere voluisse. Copernicus quidem Stoica rigiditate obfirmatus, animi sensa candide sibi profitenda putavit, etiam cum dispendio hujus scientiae. Osiander utilia magis arti secutus, verissimam Copernicj et seriam sententiam praefatione sua maluit occultare. Valuerit sane consilium hoc Osiandrj per hos sexaginta annos: tempus jam tandem est, hanc simulationem, ut mihi quidem videtur, ex privatis Osiandrj epistolis detegere. // Nam cum Copernicus anno 1540 ⌞Cal: Jul:⌟ ad Osiandrum   273v scripsisset; sic illi inter caetera respondet Osiander Anno 1541. 20 Apr:[6] //

De hypothesibus ego sic sensj semper, non esse articulos fidej, sed fundamenta calculj,   278r ita ut, etiamsj falsae sint, modo motuum φαινόμενα exacte exhibeant, nihil referat; quis enim nos certiores reddet, an solis inaequalis motus ratione Epicyclj, an ratione Eccentriciatis con // tingat. Si Ptolemaej hypotheses sequamur, cum id possit   278v utrumque. Quare plausibile fore videretur, si hac de re in praefatione nonnihil attingeres. Sic enim placidiores redderes peripatheticos et Theologos, quos contradicturos metuis. //

  273v

Ad Rheticum vero eodem die sic scribit Osiander.[7] //

---

5 A marginal note, *da khompt ein gantzer pogen ein*, here refers to the quarto sheet 273r, 276v. On the collation of f. 272v–f. 278 see M. List, 'Marginalien zum Handexemplar Keplers von Copernicus *De revolutionibus orbium coelestium* (Nürnberg, 1543)', in E. Hilfstein *et al.*, eds., *Science and History: Studies in Honor of Edward Rosen*, *Studia Copernicana*, xvi (1978), 454–5.

6 The cross-reference to the letter to Copernicus, designated 'A' on f. 278r–v, is: *da khompt ein was mit litera A bezeichnet*. A deleted note on f. 278r gives the further information: *Copernicus scripserat Cal: Julijs 1540. Receperat Martio 1541*. The letter to Copernicus is introduced: *Ad Copernicum ipsum eodem*.

7 The reference to the letter to Rheticus, designated 'B' on f. 278r, is: *Allhie einzubringen was mit litera B bezeichnet*. The letter to Rheticus is there introduced: *Ad Rheticum 1541. 20 Apr*.

278r Peripathetici et Theologj facile placabuntur, si audierint, ejusdem apparentis motus varias esse posse hypotheses, nec eas afferj, quod certo ita sint, sed quod calculum apparentis et compositj motus quam commodissime gubernet, Et fieri posse ut alius quis alias hypotheses excogitet et imagines hic aptas, ille aptiores, eandem tamen motus apparentiam causantes, ac esse unicuique liberum, imo gratificaturum si commodiores excogitet, ita a vindicandi severitate ad exquirendj illecebras avocatj ac provocatj, primum erunt aequiores, tum frustra quaerentes pedibus in authoris sententiam ibunt. //

273v His authoris verbis, ⌊praeter eam, quam dixi in praefatione usurpatam simulationem,⌋ manifesta ⌊quoque⌋ apparet aequivocatio vocis hypothesis. Supra namque dictum, quasdam esse quasj minutj aeris hypotheses, vix dignas hoc nomine, quasdam justas et vere Astronomicas hypotheses. Ut cum in exemplo Osiandrj determinamus et enunciamus portionem circulj planetarij, quae sit in uno Zodiacj semicirculo, digna nomine hypothesis est, nec variarj aut omnino falsa esse potest. At cum calculum ascensus et descensus planetae

274r in illis inaequalibus portionibus, extruimus, saepe id varia for // ma potest perficj, et hypotheses quidem ⌊alias⌋ ad efficiendam priorem illam et astronomicam hypothesin comparamus (alius posito centro circulj planetarij supra centrum mundj, alius Epicyclo in concentrium inserto). Sed quae non ipsae per sese astronomicae, verum Geometricae potius hypotheses sunt. Ita si quis astronomus dicat, viam lunae ovalem exprimere formam, hypothesis est astronomica, cum ostendit, quibus circulis hujusmodj ovalis figura conficj possit, hypothesibus utitur Geometricis. Et Ptolemaeus, cum diceret, planetarum motus in apogaeo lentescere, in perigaeo accelerarj, hypothesin fecit astronomicam; cum vero punctum aequalitatis introducit, id ut Geometra facit, calculj causa, ut ratio inirj possit, quantum lentescat ille planetae motus, quovis momento temporis. Hanc hypothesium diversitatem nemo consideratus

274v est, qui non agno // scat, manibusque palpet. At vero Osiander Copernicum ⌊vel potius ejus lectores⌋ aequivocatione ludificatus est, ea quae in Geometricis Hypothesibus vera sunt, ad Astronomicas transferens, cum diversissima sit utrarumque ratio.

Summa, tria sunt in astronomia, Hypotheses Geometricae, Hypotheses Astronomicae, et ipsi apparentes stellarum motus; propterea et duo distincta astronomi officia, alterum vere astronomicum, Hypotheses astronomicas tales constituere, ex quibus apparentes motus sequantur; alterum Geometricum, hypotheses Geometricas cujuscunque formae (variae namque saepe in Geometria esse possunt) tales constituere, ex quibus hypotheses illae priores astronomicae, hoc est, verj planetarum motus, non adulteratj visus commutatione, et sequantur et calculentur.

275r Hic me tenor ipse sermonis mei invitat, ut Osiandro // Patritium

adjungam, eique conatus astronomorum omnes deridentj, hypotheses astronomicas omnes rejicientj nonnihil respondeam.

Succenset is astronomis, quod conentur ex varijs circulis ⌐et solidis
orbibus⌐ planetarum motus apparentes efficere, eosque circulos, eas hypotheses, cerebrj sui figmenta, rerum Naturae imputare. De planetis ipse hoc
affirmat, inter fixas illos ⌐in liquido aethere⌐ incedere, quemadmodum et
videantur, ⌐liberos ab orbium solidorum compedibus qui nullj sint⌐ vereque
hujusmodj spirales et varie rursum prorsumque contortas lineas, nunquam
sibi ipsis per omnia similes, eaque motus inaequalitate, plane ut in oculos
nostros incurrunt, describere. Nec mirarj nos debere varietatem hanc, cum
planetae revera sint animalia rationis capacia, comprobante id authoritate
philosophiae barbaricae, et omnipotentiae divinae non fuerit impossibile,
tanta sapientia creaturas condere, ut imperatos hos motus ad finem usque
mundj exsequantur. //

Primum hoc illi facile concessero, solidos orbes nullos esse. In quo et 275v
Tychonis rationibus facile subscribo, et privatas etiam habeo. Patritius vero
Tychoni ridicule facit injuriam, quem ait fascinatum opinione solidorum
orbium novam confinxisse mundj formam; quam opinionem Tycho omnibus
scriptis et ipsa hypotheseos suae conditione impugnat. Neque nego, planetarum circuitus ratione summa administrarj. Sed et hoc dabo, si Deo
Conditorj placuisset, ut Planetae spirales hujusmodj et irregulares circuitus,
quales aspicientibus nobis apparent, conficerent, id illi effectu facile futurum
fuisse. Verum in hoc jam differt Patritius ab astronomis vere philosophantibus,
cum quaeritur, quidnam potius Deo placuerit, an ut planetae compositos
hujusmodj, perpetuoque a seipsis diversos et irregulares anfractus pervolitent:
anne, ut uniformem et quam fierj potest regularissimum circulum describant,
plane diversum // ab ea motus confusj forma, quae in oculos nostros incurrit; 276r
causae vero possint invenirj, quibus visus noster delusus aliter illos regulares
motus intueatur, quam revera comparatj sint. Hic nemo est e sobrie
philosophantibus, qui non hoc affirmet, illud omnino rejiciat, sibique simul
etiam astronomis maximopere gratuletur, si causas deceptionis invenerint, a
veris planetarum motibus accidentarios et ex phantasia visionis ortos se-
junxerint, et circuitibus simplicitatem atque ordinatam regularitatem asser-
uerit. Patricius vero ita philosophatur, ut qui illum audit, non pedem movere
possit, quin miraculum contingere fateatur. Spaciatus enim per campos, ubi
sepes et propinqua viae ⌐sibi⌐ occurrere, remotos vero montes se comitarj,
visu judice, putaverit: non utetur judicio rationis, sed ne sensum quidem
communem consulet: quin potius visui credet, montes ⌐non minus ac se
ipsum progredi⌐ dicet: adeoque si curru aversus Argentorato aut Vienna
egrediatur: Jovem lapidem jurabit, turrium, quae his urbibus praealtae sunt,

276v   ruinas in se vergere. // Olim quidam meorum contubernalium per fenestrae vitrum pascentes in suburbano prato boves prospiciebat: cumque pro fenestra penderet araneus, eum ille cum bobus, aspectu directore, confundens, nec intervalla distinguens, ⌐nobis deridiculo fuit, exclamans⌐ miraculum, bovem multipedem. Non sane major est hoc loco Patricij sapientia, dum planetas inter fixas, ita ut nobis apparet, cursus suos peragere contendit.

Hoc itaque Patritio dictum esto, vias planetarum apparentes, et historiam motuum libris promere, astronomiae potissimum Mechanicae ⌐et practicae⌐ partem esse: vias vero veras et genuinas invenire, opus esse astronomiae Contemplativae (frustra Osiandro et Patricio reclamantibus): at dicere, quibus circulis et lineis depingantur in papyro imagines justae verorum illorum motuum, ad inferiora Geometrarum subsellia pertinere. Haec qui distinguere didicerit, is facile sese a somniantibus abstractarum formarum captatoribus, qui materiam (rem unam et solam post deum) nimis secure contemnunt, eorumque importunis sophismatis expediet. Atque haec Patricio

Frisch, 248,   respond // ere volui, propterea quod Urso et Osiandro errantibus in
l. 13–15   quibusdam suffragatur. Ad Ursum redeo. //

272v   Undecimo ⌐Ursus⌐ ut penitus prosternat astronomiam nostrj saeculj, laudat antiquam, more querulorum senum, aut si quod aliud est genus hominum desperabundum et Melancholicum. Sed et argumento utitur ⌐ex op-tima, ut puto, forma⌐. Pleraeque inquit artes olim perfectiores fuere; Med-icina, Musica, Oratoria, quid ni et Astronomia? De medicina ⌐Thessalica⌐ sunto sane, qui divinius aliquid concipiant, quam ex residuis monumentis apparet: sunt vicissim non minoris nominis authores, Fernelius et alij, qui plus nostro saeculo tribuunt, majori verisimilitudine. De Musica judicium fert Urso ex eventu, qui non est vel in artis vel in artificis potestate. Credo multos admirarj veterum musicam, non aliam ob causam, quam quia qualis fuerit ignorant. Quibus ego, si deus vitam concesserit, nec ocium defuerit, in Harmonicis meis demonstrabo, quanto intervallo vetus Musica nostram sequatur. Quod autem clamitat Ursus, hodie non esse tantum Musices //

279r   imperium in animos, id Scythes ille, aeque antiquus hinnitu equi sui (quem optimae cantilenae praeferre solebat) facile refutat. Rarius audiebantur bonj cantores, vox, ut in Comoedijs et Tragoedijs, sic in cantu, quorsumcunque inflectebatur, sensus verborum aptabatur temporj et personis et loco ⌐verba clara articulata et vernacula⌐; magna gentis Graecae mollities, personae quoque Cytharistriae, Psaltriae, Lesbia Sappho [Lesbia, Sappho], vel lapidem moverint. Quae omnia et si quid aliud fuit apud veteres decorum, hodie quoque imitarj possumus: nec nunquam imitamur. Nostrum vero artificium veteribus plane fuit incognitum. Sed plura suo loco. De oratoria non contendo. Causa tamen magna est in natura ⌐et moribus⌐ gentis, quae mollissima facile cuivis cedebat affectui ∟gloriae vero cupidissima, id ejus adipiscendae causa facere non dubitavit, quod tamen illis laudabile, hodie, ap: omnes gentes detestandum est, ut quis sibi ipsi violentas manus afferat⌐. De astronomia vero res ita clara est, ut nihil supra. Veniant, quotquot sunt in Europa peritj rerum mathematicarum: non habiturus sit Ursus vel unum, qui hoc illi largiatur, perfectiorem fuisse veterum astronomiam. Itaque quantj sit faciendum, si quis praestet, quod pollicitus est Ursus, sc: restitutionem astronomiae veteris, consideratus lector censeto. Haec praecipua fuere, quae hoc primo capite de natura Hypothesium dicerentur.

CAPUT II

De Historia Hypothesium.

In historia hypothesium texenda Ursus artificio utitur non inscito. A veris enim exortus rationibus fidem sibi facit cognitionis historiarum: qua obtenta, nihil mutato filo orationis ad commenta sua delabitur, eaque lectori pro relationibus historicis obtrudit.

De Thalete, qui primus est in hac astronomorum serie in Graecia, ∟quamvis Linum et Musaeum antiquissimos, hunc inventorem Sphaerae, illum motuum Solis et Lunae celebret Diogenes Laertius⌐ si vera sunt, quae ∟de praedictione Eclipseos prodidere Plinius et⌐ Herodotus: certe equidem ∟hanc⌐ astronomiae ∟partem exacte⌐ calluisse necesse est. ∟Et Laertius quidem, primum astrologiae secreta rimatum, Solis cursum a solstitio in solstitium reperisse, anni quantitatem 365 dierum constituisse, de solstitijs et aequinoctijs scripsisse, anni tempora mensum esse, Solis et Lunae magnitudines comparasse, Solisque defectus praedixisse asserit.⌐ Neque tamen ∟Methodum Eclipses praedicendj a se ipso petere⌐ potuit, cum hoc non sit unius aetatis, observare sufficientes Lunae anomalias ad eclipsin solis praedicendam. // Credibile igitur, cum paulo ante eam ipsam Eclipsin, bellaque quae tunc gesta commemorat Herodotus, pax utique fuerit Lydos inter et Babylonios, Thales vero Lydis vicinus in Mileto habitaret, artem Eclipses computandj didicisse a Babylonijs. ∟Et Laertius quidem author est, AEgyptum adijsse. Ac ipse Thales apud authorem illum sic ad Pherecydem, de se. *AEgyptum penetravimus, ut isthic sacerdotibus et astronomis congrederemur.* Qui fuerint autem illi astronomj, Laertius in *Pythagora* innuit, quem eundem AEgyptum profectum narrat apud Chaldaeos conversatum. Erat enim AEgyptus sub Babyloniorum jugo. In fine vero epistolae jam allegatae, Thales addit, *At nos qui nil scribimus, Graeciam Asiamque peragramus.* Ita sive in Chaldaeam per continentem, sive marj in aegyptum iverit, Chaldaeorum utrinque copiam habuit: et astronomiam inde in Graeciam effere potuit.⌐ Etenim consulj solitos a Graecis Babylonios de

motu Lunae, illud persuadet, quod omnes pene Graecae nationes anno lunarj uterentur.[8] ⌐Huc refer verba Aristotelis lib. 2 de caelo Cap: 12. *Simili ratione de caeteris etiam stellis loquuntur* (quoquam quo tempore texerit) *qui earum rationem quondam plurimis annis observarunt, AEgyptij et Babylonij, a quibus multa fide digna de singulis sideribus accepimus.*⌐ Quas meas conjecturas hic cum Ursj conjecturis conferre libuit, quia is non sum, qui bene dicta carpere velim. Sed jam statim peccat Ursus, hypotheses, quas [quos] Physicorum esse ait, paulo post tempora Aristotelis demum constitutas existimans, nec ulla ratione scirj posse, quales nam fuerint temporibus Thaletis et Pythagorae, astronomorum hypotheses. Quo loco non videtur Plinium legisse lib 2. cap: 22. Etenim illae hodiernae physicorum, ut vocat Ursus hypotheses, seu illa sphaerarum ordinatio, quam Ptolemaeus est amplexus, Pythagorae usitata fuit. Audi lector verba Plinij. *Sed Pythagoras* inquit *interdum ex Musica ratione appellat tonum, quantum absit a terra Luna. Ab ea ad Mercurium, spacij ejus dimidium, et ab eo ad Venerem fere tantundem, a qua ad Solem sesquiplum, a Sole ad martem tonum, ab eo ad Jovem dimidium, et ab eo ad Saturnum dimidium, et*
280r *inde sesquiplum ad signiferum.* // Audis ordinem sphaerarum lector ex Plinio. Aut igitur Plinius malae fidej author est, aut Pythagoras de ordine sphaerarum ita censuit, ut Ptolemaeus, idque ducentis annis ante id tempus, quod harum hypothesium ortuj designat Ursus. ⌐Et id est quod Ptolemaeus lib 9. Cap: 1 innuit, cum ait, orbes Veneris et Mercurij ab ANTIQUIORIBUS quidem ⌐⌐(Pythagora)⌐⌐ sub solarj, a quibusdam vero posterioribus (Heraclito Platone, Eudoxo Aristotele) supra Solarem collocatos: et denique Veterum sententiam, ut magis probabilem sequitur, negans certj quid ex astronomia hic haberj posse. Quid, quod Patricius Zoroastrj Persae qui multis saeculis Pythagoram antecessit, eandem sententiam tribuat, ex his ejus verbis. Ἓξ αὐτοὺς ὑπέστησεν (aliter Ἒξ ἀνακρεμάσας Ζώνας) ἕβδομον ἡελίου Μεσεμβολήσας πῦρ. Si Sol in medio planetarum (Terram enim idem Zoroaster in eo medio, quod centrum est, collocat) Ergo ♀ ☿ infra solem erunt. Cumque Pythagoras a Zabraco Assyrio, gentilj Zoroastrj (Patricio referente, puto ex Proclo) hanc sapientiam hauserit: ordinem hunc indidem hausisse verisimile est.⌐ Quamvis qui post dictj sunt Pythagoraej, hunc magistrj suj ordinem deseruerunt. Illi namque, ut paulo post probabitur, eandem tenuerunt de mundi forma opinionem, longe ante Aristotelem, quam post aristotelem annis quinquaginta sub Ptolemaeo Philadelpho, Aristarchus

---

8 The following interpolation replaces an interesting deleted passage: 'Temporis posterioribus bella futura ipsas inter et Persas, rerum Babyloniorum dominos. [Vero?] quod Calippo et Hipparcho a Babylonijs petere, amplius non latuit, correctionem [anni?] exorbitantis, id a seipsis inceperunt petere, sicquidem nobis astronomiam constituerunt.'

Samius, et nostra aetate Copernicus, qui ut haec ex Aristotele ⌐et Plutarcho⌐
de Pythagoraeis, et ex Archimede de Aristarcho intelligere possemus,
inventionibus suis effecit. Quo loco alter Ursi error se prodit, qui Copernicanarum hypothesium originem primam ad Aristarchum refert, quae
multo est antiquior.

Ac de Philolao quidem Pythagorico, qui medius fuit Aristotelem inter et
Pythagoram primusque Pythagorica vulgavit, cum antea nisi a Pythagoricis
ipsis ignorarentur, de eo igitur non tantum obscure Laërtius asserit, sensisse
terram juxta primum circulum (intellige sub Zodiaco) *moverj*, sed multo
clarissime Plutarchus, cujus verba cum Copernicus ipse in praefatione operis
sui posuerit, mirum admodum est, ea non legisse Ursum, qui lectione
Copernici toties et tantopere gloriatur, eamque omissam alijs immerito
exprobrat. *Terram* vult Philolaus apud Plutarchum *in circulum moverj circa
ignem* (intellige Solem) *sub zodiaco in morem Solis et Lunae.* Aristoteles vero
lib 2 de caelo cap: 13. De Pythagoraeis in genere sic. *Ij qui Italiam incolunt,
qui ijdem Pythagoraej vocantur in medio ignem esse dicunt* (Hic Solem intellige,
nam et Thales Ionicae et Pythagoras Italicae philosophiae principes, occultanda
vulgo suasere philosophiae // mysteria) *terram vero quae sit unum astrorum*   280v
(planetarum) *motu suo, quo circum medium feratur, noctes et dies efficere.* Hic
Aristoteles non bene percepit eorum sententiam, aut certe defectus est, quem
sic restituo, *terram motu suo quo circum medium feratur, annum, quo vero circa
suum axem, noctes et dies efficere.* Pergit *Praeterea aliam terram contrariam
fabricantur, quam* ἀντίχθονα *vocant.* Hac voce Lunam significant, propter ⌐ea
quod cum terra eodem orbe circumvehatur, et propter⌐ multas ejus cum terra
communes adfectiones de quibus testantur opticj. Aristoteles vero qui hoc
non capiebat, absurditatem sententiae ominatus, subjicit, quod *non quaerant
rationes et causas, ad ea quae apparent, reddenda et efficienda, sed ad praeconceptas
suas quasdam opiniones et ratiocinationes ipsa phaenomena violenter accommodent,
et conciliare conentur.* Esse autem Ignem Pythagoraeis idem, quod solem ex
argumentis eorum apparet, quae sic proponit Aristoteles, *Viderj vero possit
multis etiam alijs, non dandum terrae locum medium, si probabilitatem aliquam non
ex apparentibus, sed ex rationibus coaptare velint. Ei enim, quod est praestantissimum,
honestissimum locum dari oportere, Ignem vero terra praestantiorem.* Hic si
expendas, non esse praestantius ullum corpus ipso Sole in toto mundo, vim
argumenti mirifice augebis, et quid ille ignis sit, intelliges *terminos vero
digniores interjectis* (Sunt enim quasi quidam fines, quos aliae [alios] non habent
quantitates. Finis vero semper praestantior eo, cujus finis est) *extremum vero
mundj et medium,* centrum, *esse terminos* omnium locorum intermediorum in
quibus planetae et Sol vulgo collocantur. Quare centrum et extimam
superficiem esse loca mundi honestissima. *Ex quibus concludentes, non putant*

281r   *terram in sphaerae meditullio loca* // *tam, sed Ignem* Solem. *Censent praeterea Pythagoraej, eo quod par sit, maxime omnium custodirj id, quod sit totius universi praecipuum: medium vero talis sit locus (quem et Jovis custodiam dicunt) ideo ignem* (Solem) *tenere hanc regionem,* centrum: *sicut, quod absolute medium dicitur, id est et magnitudinis et rej et Naturae medium.* Quae argumenta diluere conatur Aristoteles his verbis. *Atqui, ut in animantibus non idem est animalis et corporis medium, ita multo magis de toto caelo etiam cogitandum. Quare non opus est, ut pro universitate trepident, aut praesidium ad centrum collocent: quin potius medium illud quaerant, quale sit, et ubj. Medium enim et preciosum illud, principium est. At medium locj ultimo, quam principio similius videtur. Medium utique terminatur, extremum vero terminat. Praestantius autem, quod comprehendit, et finit, quam quod finitur. Hoc enim materia, illud essentia compositj est.* ⌊AEquivocatio in voce *Medium*, quod Pythagoraeis est centrum, aristoteli quovis alio modo sumitur. Quae si tollatur, coincidit Aristotelis argumentum cum adverso, proque Pythagoraeis est. Nam id *medium*, de quo illi, hic ap: Aristotelem *extremj* rationem obtinet.⌋ Haec ille contra Pythagoraeos, ad quae respondere nec est difficile, nec tamen hujus locj. Sed hinc apparet ipsorum sententia de loco potissimum terrae extra mundj medium. Jam et de motu ipsos ulterius audiamus. *Sed qui eam in media mundj sede locatam esse negant, circum medium eam volvj asserunt; neque hanc solum, sed eam etiam, quae ei adversa est* (Lunam manifestissime) *quemadmodum paulo ante diximus.* Et interjecta aliorum sententia, *Nam quia terra non est centrum,* punctum, *cum totum ipsius hemisphaerium absit,* (intellige, etiam illorum concessione qui terram in medium mundj referunt) *nihil putant prohibere, quominus phaenomena contingant, nobis extra centrum* (non tantum uno semidiametro terrae, sed quantum inter Solem et terram interest) *habitantibus, similiter, ac si in medio terra esset. Nam ne nunc quidem* ⌊(id est, ne quidem si terra in medio relinquatur)⌋ *quicquam notabile*

281v   *efficj, dum dimidia diametro a centro absumus.* // Haec de Pythagoraeis Aristoteles, fere quidem plura et explicatiora, quam de Aristarcho Archimedes. Quae si legisset Ursus, Copernicanarum originem non ad Aristarchum post Aristotelem, sed ad Philolaum et Pythagoricos longe ante Aristotelem retulisset.

    Neque tamen omnes Pythagoricj qui sequuti sunt, hanc Philolaj tenuere de mundo conceptionem. Eudoxus enim Cnidius ⌊jam Platonis et Aristotelis coaetaneus⌋ et ipse Pythagoricus, apud Aristotelem principis Pythagorae sententiam propius sequitur, in eo nempe, quod una cum illo statuit, moverj Solem. De quo et ejus correctore Calippo cum Ursus pro libertate sua affirmet, quibus usi sint hypothesibus, ignorarj, Et tamen paucis interjectis, eadem asseverandj temeritate, usurpatos ab illis concentricos a ⌊Pythagoraeis, et⌋ Hipparcho cum eccentricis permutatos fabuletur: videbimus ex Aristotele, quid certj de Eudoxj sententia haberj possit: praecipue cum hunc locum a caeteris aut neglectum ⌊aut male tractatum⌋ interpretibus, operae precium videatur, diligentius expendere.

Cum autem in gemina constituti simus difficultate, eo quod et Aristoteles, rerum mathematicarum imperitior ⌊ut multis locis apparet⌋, Eudoxj mentem vel sinistre perceperit, vel obscurius expresserit, et numerj manifeste a verbis dissideant: faciamus quod possumus, verborum involucra conjecturis sublevemus, numerorumque errores correctione verbis consentanea tollamus. Igitur cum lib XI Metaphysicorum ⌊Aristoteles⌋ in contemplatione versetur substantiarum simplicium et materia carentium; eoque sublimitatis evehatur accurata motus, et primae motuum causae disquisitione: invento ex simplici, uniformi et aeterno motu, quem primum dicimus, motore primo, ad intelligentiarum secundarum et primae subordinatarum numerum inve ∥ sti-   282r gandum Cap: 8. accedit: illud praecipue urgens, singulis mobilibus singulas praeesse intelligentias, nec introducendum quicquam in mundum, quod officio careat, ociosumve sit. Mobilium vero caelestium corporum numerum ex Motuum qui videntur in caelo, varietate, instituit. At motus ipsos, quinam et quot deprehendantur quaerit ab astronomis. Hinc ejus verba. *Ac multos quidem motus esse eorum, quae feruntur, ijs etiam evidens est, qui mediocriter in hac re versati sunt. Nam pluribus una motionibus singuli errones feruntur. Quot autem sint, nunc quidem ea, quae mathematicj nonnulli tradunt, cognitionis causa dicimus: ut animo certum quendam numerum possimus comprehendere: caetera partim nobis inquirenda sunt, partim ij qui ⌊haec⌋ inquirunt, interrogandj, num quid praeter ea quae dicta sint, ijs videatur, qui haec tractant: ac tum utrique amandj illi quidem erunt, eisque gratia habenda; sed ijs demum habenda fides, qui accuratius et diligentius tradiderunt.* Hoc prooemium audiat Osiander et Ursus, et quicunque astronomorum de rerum Natura conceptus parvj facit. Sequitur ipsa res. *Eudoxus Solis et lunae, utriusque lationem in tribus ponebat ⌊sphaeris⌋, quarum primam esse non errantium stellarum.* Hic in compendijs astronomicis vulgo depingitur amplectj totum alicujus planetae systema, idque facit concentricum. Movetur super polis aequatoris ⌊ab ortu in occasum⌋ et revolvitur una cum eo horis vigintj quatuor. Interdum ei aliud insuper officium, deferendj nodos, attribuitur, unde plerumque nomen habet. ⌊Sed hoc non ex Eudoxj mente.⌋ *Secundam autem per eum circulum ferrj, qui per medium Zodiacj est.* Apparet Eudoxum hunc orbem sic disposuisse ⌊in Luna⌋ ut ejus polj priorj orbi revolutionis diurnae insisterent, declinantes utrinque a polis illius in coluro solstitiorum spacio obliquitatis eclipticae. Motus vero ejus est sub priorj ab ortu in occasum seu in antecedentia, restituitur semel ad idem illius primj ⌊sive ad idem Zodiacj⌋ punctum annis novendecim. Quo spacio circumfert nodos. Qualis vero motus hujus circuli sit in theoria Solis, dicam jam, quantum dici potest, in tertio. *Tertiam autem per obliquum circulum ad latitudinem signorum.* Sic intelligo, tertium ad secundum in latum inflexum esse. Est ergo hic qui Lunam ipsam defert. Ejus poli secundo infiguntur, deflectentes utrinque spa ∥ cio latitudinis maximae lunaris: movetur in consequentia, seu ab occasu in ortum et restituitur ad primj, seu ad Zodiacj idem punctum spacio fere menstruo. ∟Ita fit, ut qui orbis propter motum longitudinis ponitur, is latitudini serviat, qui vero latitudinem stellae praestare debet, longitudinis motum efficiat.⌐ Sit ADG sphaera fixarum, et in ea ipse ADG colurus solstitiorum. AE aequator, C.G. ejus ∟axis et⌐ polj. HD Zodiacus. BF ejus axis et polj. Erit IKL primus orbis in Theoria lunae secundum Eudoxum: Ejusque polj I.K. NOQ secundus, ejus polj, N,O, affixj primo IK in punctis P.M. RST tertius, ejus polj R.T. affixj secundo in punctis V.X. Habes Eudoxj concentricos Urse, in theoria Lunae. I nunc et veterum astronomiam extolle. Theoria ∟haec⌐ non constat nisi concentricis, ∟sane⌐ quia caeteras Lunae inaequalitates Eudoxus ignoravit.

Porro quare Eudoxus etiam in Theoriam solis latitudinis circulum introduxerit, diu mecum quaesivj. Nam etsj hoc procul dubio et D, Tycho Brahe faciat, qui mutationem aliquam latitudinis fixarum deprehendit: hoc tamen tam subtile negocium nec nisi post multa saecula deprehendj potest, nec si Eudoxj tempore potuisset animadvertj, Eudoxo fuisset cognitum qui multo evidentiora praeterijt. ∟Interpretatio vero Fracastorej locum non habet. ∟∟*Haec igitur* inquit, *a secunda per Zodiacj longitudinem ducebatur. Motu autem suo faciebat, ut stella non in eadem maxima declinatione semper maneret, sed Luna modo in uno signo modo in alio videretur, Sol magis et minus declinaret ab aequinoctiali.*⌐⌐ Quid enim opus est pro varianda Solis declinatione novum et inducere circulum, cum id officium tertio quï sub Zodiaco incedit sit proprium? Et verba ipsius, non sunt astronomj. *Luna,* inquit, *secundum maximae declinationis ab aequatore punctum, nunc in uno signo, nunc in alio cernebatur.* At Luna Fracastorj nuspiam plus ab aequatore declinat, quam in ♋. ♈ non minus quam Sol.⌐ Adeoque jam fere concludebam mecum, voluisse Eudoxum motum fortasse fixarum demonstrare per secundum NOQ circulum, aliter revolutum, et super alijs polis: cum occurrebant sequentia. *In majore autem latitudine inflexum esse eum in quo luna, quam in quo sol feratur.* Quibus verbis omnis plane diversitas inter solis et lunae theorias tollitur, hoc uno excepto, quod OV, vel NX minor sit in sole, quam in Luna. Tandem incidit, Eudoxj intentum non tam esse astronomicum, quam philosophicum. Etenim et Eudoxus ∟astronomus⌐ inter philosophos, et quidem Pythagoricos a Laërtio recensetur. Consilium igitur hoc fuisse Eudoxo, quando ut lumine sic motibus plurimum Sol et Luna differe videantur a caeteris, aequalitatem omnimodam inter luminaria instituere, Solique ad lunae normam tribuere latitudinem, sed insensibilem ⌐aut certe insensibiliter variabilem⌐, quod ijs innui verbis existimo, quibus majorem lunae quam solj tribuit latitudinem. ⌐Haec enim aequalitas rationibus ejus philosophicis forsan admodum erat consentanea. Nec dubium, quin eandem ob causam ⌐⌐inter caeteras⌐⌐ et solem proxime supra lunam infimam collocarit, idque ab eo ut caetera Aristoteles hauserit.⌐ Hoc pacto si schema superius fiat Theoria solis, manebit primj circulj ILK situs et motus. Secundus NOQ motu carebit fere. Tertius RST Solem vehet et annuo spacio ad idem primj vel Zodiacj punctum restituetur. // ⌐Quamvis non negaverim fierj potuisse, ut cum Eudoxus in ⟨283r⟩ Sole delinearet colurum solstitiorum, in quo demonstraret maximam ejus declinationem, Aristoteles eum colurum pro sphaera mobili censeret. Sed nec eam conjecturam rejecerim quam Fracastorej verba suppeditant, cum ait. *Sol vero in maxima sua declinatione videbatur nunc magis ab aequatore distare, nunc minus.* Eudoxus itaque seu observationum solstitialium vitio ⌐⌐et varietate⌐⌐ seu veterum de obliquitate Zodiacj traditionibus (erant enim diligentes in hoc inquirendo propter Olympiadum initia) cum sua observatione collatis, ⌐⌐seu suspicarj seu vere⌐⌐ deprehendere potuit diversitatem aliquam declinationis solstitialis, cui efficiendae Solem in latitudinis circulo constituebat. De Iphito quidem communis est fama, causam illi fuisse astronomicam instituendarum Olympiadum. Et ipsa forma annj illius clamitat, notam fuisse Iphito rationem solstitij, adeoque declinationis maximae Solis, ut qui annum ab eo plenilunio incipiebat, quod proxime solstitium sequebatur. Ab eo vero ad Eudoxum sunt annj 400. Vide et quae supra de Thalete, qui medius fuit inter Iphitum et Eudoxum, item de Lino antiquissimo ex Laertio retulimus.⌐

Sequitur: *Planetarum vero aliorum cujusque in quatuor sphaeris, quarum prima et secunda eadem est, quae et inerrans, ferens omnes, et quae sub ea posita est, quae per medium Zodiacj motum habet communem omnibus,* habere omnes planetas unum orbem, qui sphaeram inerrantium in motu et situ observet: alium item, qui sub Zodiaco movetur, in contrarium primo, et ex obliquo. Is fuit supra NOQ, super polis N, O, incedens. Sed hic alia ipsius ratio est. Nam rationibus Eudoxj consentaneum est, ut hic jam non vehat nodos, sed revolvatur in Saturno quidem annis 30, in Jove annis 12, in Marte mensibus 23. In venere, et Mercurio mensibus 12. Sit rursum ABC portio fixarum sphaerae et in ea colurj solstitiorum, cuj respondeat DEG primus Eudoxj orbis, quem habent quinque errones, qui volvatur super DH polis et axe, axj primi motus coincidente. Sit TIV secundus orbis, cujus polj T. V, figantur in punctis F. G. primj, et sit DF obliquitas Eclipticae. Super ijs polis planeta movetur in consequentia in Saturno annis 30 ad idem primj vel Zodiacj punctum rediens. KML sit tertius orbis, cujus polj K L sunt in superficie secundj per media signa euntis TIV. Super ijs movetur ⌐superiorj parte⌐ in consequentia secundum tractum circulj maximj YIX, et sic movetur, ut requirunt nodj. ⌊Quodsi YXZ, pars lineae repraesentantis Eclipticam explicetur ut fiat circulus, colurus TIV fiet in pictura linea, similiter et KYL, et K L poli cum centro coincident.⌋ PRO sit quartus, cujus polj P O fixi sint in tertij punctis M N, super quibus movetur itidem in consequentia secundum tractum sui circulj maximi RS [IS]. Secundum quam viam stella circumfertur tempore eo, quod postulat anomalia commutationis. Haec Eudoxj philosophi, philosophica opinio de 5 Errantibus, qui alteram inaequalitatem ignoravit, quae est respectu orbis planetae.[9] //

285r   *Tertiae vero omnium sphaerae polos esse in eo circulo, qui per medium signorum transit: at quartae conversionem esse per circulum, qui obliquus est ad medium circulum hujus.* ⌊*Esse porro tertiae sphaerae polos reliquarum quidem stellarum proprios Veneris autem et Mercurij eosdem.*⌋ Ut hunc vere Gordium nodum cum ratione et via certa adoriamur: sic est nobis agendum. In Luna vidit Eudoxus tres motus, primum diurnum, secundum menstruum in oppositum primo, et obliquum, sub Zodiaco, tertium qui digressionem ejus ab Ecliptica variaret. Ad numerum hunc ⌊motuum⌋ totidem etiam ponebat orbes. In planetis caeteris quatuor observabat motus admodum evidentes: ⌊primum⌋ diurnum sub aequatore in antecedentia, secundum sub Zodiaco in consequentia: Tertium latitudinis ab Ecliptica: quartum, quo planetae illi, prae luminaribus, hoc peculiare obtinerent, ut fierent retrogradj. De primo et secundo motu quod ij sint omnibus communes, ipsa verba Aristotelis satis testantur. Quod vero et tertius latitudinis Lunae cum his 5 erronibus communis sit, etsj ex verbis Aristotelis non sequitur: non latuisse tamen Eudoxum, ratio persuadet. Nam si non plus diligentiae in observando adhibuit, quam quantum ad deprehendendas retrogradationes suffecit: non potuit non saepius in anno respicere ad planetarum conjunctiones, quorum unus alio vel borealior vel australior, et sic diversae ab altero latitudinis esset, nec ad unguem per medium signorum ferretur. Et in descriptione quidem quartj, cum eadem sint verba, quae supra in descriptione tertij: *medium circulum quartj obliquum esse ad medium tertij,* tanto minus est, cur dubitemus, alterum ex his latitudinj planetae, alterum retrogradationj deputarj. Itaque non est hic audiendus Fracastoreus, qui latitudines planetarum ab Eudoxo neglectas, ambobus vero his circulis unam retrogradationem repraesentatam asserit, Eudoxum ad sua Homo-centrica, adque formam motuum a se sane perquam ingeniose inventam violenter accommodans. Et obtinet ⌊ille⌋ quidem propositum, fit planeta retrogradus et velox et stationarius, positione duorum orbium terram in centro habentium, quorum alterius polj in Ecliptica sint, motus versus polos

---

[9] Cross-reference to passage headed 'B' on f. 285r: *Vide B.*

eclipticae, alterius vero polj ab hoc circumferantur distantes a polis hujus, quantum quantitatj retrogradationis sufficit, super quibus orbis hic interior in contrarium exterioris aequalj tempore restituatur. At hoc pacto planeta semper in Ecliptica, aut semper in aequali lat // itudine extra eclipticam 285v manet ⌐motu librationis cis et ultra Solem excurrens⌐, aut si fuerit inaequale tempus restitutionum, fiet ut olim planeta latitudinem habeat aequalem retrogradationj. Et ut demus, neglectam ab Eudoxo latitudinem planetarum, non tamen est verisimile, usum esse Eudoxum tam recondita forma hypotheseos ad retrogradationem efficiendam: cum Epicyclj conceptus sese ipsum quasi offerat ultro, retrogradationem considerantj: et in disciplinarum constitutione simplicissima et facillima [facilima] quaeque primo sequamur, necessitatem perpendentes, non ingenium, ut Fracastoreus. ⌐Neque tantum vero simile non est, sed plane contra omnem rationem, Eudoxo visum esse hunc retrogradationis motum in concentrico fierj, motu librationis, ita ut eandem a terra distantiam planeta conservet: cum etiam vulgi oculos perstringat Martis acronychij ingens magnitudo, qui jam ad solem accedens a tertiae magnitudinis stellis superatur, ex quo manifesta est ejus a terra diversissima utrinque distantia. Quam librationem si etiam statuisset Eudoxus, fuisset a Callippo reprehensus, et proinde orbium Eudoxj numerus et ordo mutatus, cujus contrarium affirmat Aristoteles. Cum itaque Callippus diligentior non haberet quod hic in Eudoxo reprehenderetur, non decet nos Eudoxo tantam ignorantiam tribuere.⌐ Sed ad Eudoxum redeamus. Ex tertio et quinto orbe diximus alterum latitudinj servire debere, alterum retrogradationj. Quod primo loco tenemus. Deinde, cum distincte narret Aristoteles de primo et secundo orbe, quod is omnibus planetis communis sit, de tertio vero vel quarto, quinque planetarum non affirmet, alterutrum cum tertio luminarium esse eundem: hinc recte concludj existimo, etsj horum alteruter idem cum tertio luminarium habet officium, vehendj planetas in latitudinem, positione tamen plurimum differe. Tertio considerandum nobis est, quid ex eo sequatur, quod tertiae sphaerae polj in Venere et Mercurio sint ijdem. Cum igitur hisce duobus commune nihil sit (in usitata quidem de mundo conceptione) praeter lineam medij motus, quae per centrum epicyclj trajicitur: latitudo enim diversa est, retrogradationis et tempus et quantitas itidem diversa: sequitur igitur, tertij orbis polos pendere a linea medij motus. Et amplius, cum certum sit, latitudines Veneris et Mercurij in diversis Zodiacj locis perficj, nec semper aequaliter ab invicem removerj haec loca: retrogradationis vero in utroque ⌐planeta⌐ loca media semper esse conjuncta: sequitur igitur tertium orbem retrogradationj servire, quartum latitudinj, quod quidem et verba Aristotelis non obscure innuunt. Hoc ita tamen verum est, ut et supra in Luna fuit: nempe tertius nodos vehere, quartus et retrogradationem et latitudinem una, beneficio tertij efficere potest. Quarto denique consideremus quod affirmat Aristoteles, polos tertij esse in circulo per medium signorum. Id multifariam fierj potes. Aut enim oppositj polj in oppositis Zodiaci locis cernentur, cum sc: terra in hujus tertij centro est: Aut uterque polorum in eodem Zodiacj loco, cum sc: terra extra orbem hunc constituitur, Ita ut orbis epicyclus sit, et linea ex terra per Epicyclj centrum coincidat cum ejus axe. Aut denique axis hujus Epicyclj secundum // Eclipticae longitudinem situs, utrumque polum in vicinis Zodiacj locis constituet. Primum Fracastoreo placet, sed ob causas supra dictas esse non potest. Nam ejusmodi orbis non est ita comparatus, ut retrogradationem efficere possit. Secundum, si verba praecise urgeamus, esse nec ipsum potest. Nam sic quidem planeta retrogradus fieret, at simul latitudinem aequalem efficeret retrogradationj. Nec tertium esse potest, Nam, planeta in latum veheretur, cis et ultra Eclipticam. In his angustijs, aliter statuere non possum quam quod aut Aristoteles ad Eudoxj delineationem respiciens, qui lineam ex terra per centra Epicyclorum educebat, illam axem sphaerae ⌊seu Epicyclj⌋ existimaverit, ideoque polos illa puncta, quibus linea epicyclum secabat. ⌊Aut certe quia ratione optica tale Episphaerion in justo situ ad planum Eclipticae visui perpendiculariter objectum pingj nequit, quin poli et centrum, adeoque totus axis in unum punctum coincidant, ipse etiam Aristoteles centra et polos confuderit, cum videret, lineam ex terra per Epicyclj centrum actam in Eclipticam ducj.⌋ Itaque factum ut existimaret, motum sphaerae (ut ⌊is⌋ episphaerium [epicyclum][10] appellat) esse circa illam lineam. Necessitate itaque adactj hoc de Aristotele suspicemur, et sic explicemus Eudoxum, tertiae sphaerae (quae Epicyclus fuerit, seu, ut Aristotelj placeamus, Episphaerium) centrum et planum medium in planum Eclipticae competijsse, ⌊ut lineae ex terra per centrum hujus Epicyclj ejectae in singulis planetis singulae fuerint, in Venere vero et Mercurio coinciderint et una linea fuerint.⌋ In ⌊hac vero tertia⌋ aliud Episphaerium fuisse constitutum, cujus polj a polis tertiae declinarint tanto spacio, quantum sufficiebat latitudinibus. Et quarti [quartus] quidem motu in consequentia planetam et in latum digredj, et retrogradum in imo Epicyclj fierj, tertij vero motu id efficj ut planeta sit in maxima latitudine, jam apud solem, jam in ejus opposito, jam in stationibus etc.

Sequitur in Aristotele: *Calippus autem positionem quidem sphaerarum eandem cum Eudoxo posuit, tum et intervalla eadem: porro ej quae Jovis et Saturnj eandem*

---

[10] This emendation is confirmed by a deleted passage on f. 283r: 'Tertius et quartus sunt ij quos Epicyclos dicunt. Sed Eudoxus ex ijs sphaeras facit. Itaque loco vocis Epicyclus, vocem Episphaerion nobis suppeditat,...' Since neither term occurs in Aristotle, Kepler's thought must simply be that since *epicuclon* literally means 'epicircle', its use to denote a sphere is inappropriate.

sphaerarum multitudinem // dedit. In Saturno et Jove penitus nihil mutavit, 283v
Nam inaequalitates eorum causa suj orbis, propter tardum eorum reditum,
nec ipse quidem, quamvis professione Astronomus, animadvertere poterat.
Soli vero et Lunae duas insuper adjiciendas existimavit, si quis praestare debet, quae
apparent: reliquis vero planetis unicuique unam. Callippus non philosophorum
alicuj sectae addıctus (non recensetur enim a Laërtio inter philosophos), sed
ut dixj, professione astronomus, astronomica sola // tractavit, ʟeaque⌐ 287r
accuratius quam Eudoxus philosophicas quaerens concinnitates. Animad-
vertit igitur observatione diligentj Solj, Lunae, Martj, Mercurio, Venerj,
plures ʟin⌐esse inaequalitates, quam Eudoxj ferunt hypotheses: eam, scilicet,
quae respectum habet ad parts Zodiacj, quam eccentricitatem dicimus.
Itaque orbem unum singulis addidit, qui utrum epicyclus fuerit, an deferens
eccentricum (utrumque enim fierj potuit, ʟmagis tamen hoc⌐) ex his verbis
decernj nequit. // ʟHic iterum incongruam Fracastoreus affert interpreta- 284v
tionem, primo in Sole et luna ait binos utrinque additos a Callippo, qui sunt
singulj, eosque sic comparatos, ut tarditatem et velocitatem, quae in utroque
apparet, quamque positione Eccentricitatis nos efficimus, inveheret luminaribus, et ordinet sphaeras ad modum Homocentrorum suorum. Quasi modo
tam difficili fuerint usj veteres, isque silentio a recentioribus fuerit sepultus.
Verum est orbes hos esse propter Eccentricitatem, sed singulj sufficiunt, quod
videtur desperare Fracastoreus, singulj sc: Epicyclj, aut circellj circa centrum,
si simul progressum Apogaeorum efficere ʟʟEudoxus⌐⌐ voluit. Nec de binis
ʟʟtrinisque⌐⌐ Aristotelis verba sonant. Deinde illum in caeteris ʟʟFracastor-
eus⌐⌐ putat additum, propter earum ab aequatore declinationem, ad normam
solis et Lunae; Quasi non aeque de Marte ♀ ☿, ac de ☉ et ☽ vel Eudoxo
prius constitisset, de digressione horum evidentissima ab aequatore; Et quasj
non Zodiaco circulo singulos pro singulis planetis Eudoxus subordinavit, qui
declinationis illius varietatem efficerent: aut quasi neque Callippus animad-
verterit, Saturnum et Jovem aeque atque caeteros quinque diversis temporibus
inaequaliter ab aequatore distare. Itaque cum nihil pro se Fracastoreus habeat
praeter numeros secuturos, eosque varie in exemplaribus depravatos, hoc in
causa manere existimo, duos tantummodo Solj et Lunae orbes a Callippo
adjectos, non quatuor, hoc est, cuique unum; additum vero singulis ex
quinque, non declinationis aut latitudinis, sed Eccentricitatis efficiendae causa.
ʟʟVideant alij an sola numerorum concordia sufficiat, ut statuamus, cognitam
fuisse Callippo geminam in ☽ inaequalitatem, ad cujus motuum similitudinem
etiam Solis inaequalitatem geminam fecerit.⌐⌐ // ʟManent igitur hac ratione 287r
Saturno et Jovj orbes quaternj, Martj, Mercurio, Venerj, veniunt orbes quinj,
Solj et Lunae infimis (in Aristotelico sphaerarum ordine) itidem quaternj.

Hinc lux affertur obscuro Aristotelis loco lib 2 de caelo cap 12. Ait ibi Aristoteles, *qui planetae medium teneant, eos pluribus motibus cierj.* Nempe ⌞hic⌟ Mars, Mercurius, Venus, medij secundum Aristotelem quinis motibus cientur, Saturnus et Jupiter summj, Sol et Luna imj, quaternis tantum. Quo vel solo loco nostra interpretatio contra Fracastoreum munitur.⌟

Hactenus ergo unam partem propositj sui expedivit Aristoteles, dum supra sibi proposuit *interrogandos, qui haec inquirunt.* Jam sequitur altera pars; dicebat namque *partim sibi quaerendum esse.* Itaque addit *si autem omnes simul positae debent apparentibus satisfacere, necesse est secundum unumquemque planetam esse alias sphaeras una pauciores, quae revolvant, et ad idem restituant semper primam sphaeram stellae infra ordinatae: hoc enim pacto solum continget, planetarum motus praestare omnia quae apparent.*[11] // Hic me aliorsum inclinantem Fracastoreus in viam reduxit: quamvis in quibusdam ab ipso etiamnum dissentiam, quorum hoc primum est, quod quae hic ab Aristotele ipso subnectuntur philosophica magis, quam astronomica, ipse ⌞etiam⌟ Callippo tribuit, quod inconsiderate factum. Itaque Eudoxus et Callippus totidem ponebant in singulis planetarum sphaeris orbis, quot motuum varietates deprehendebantur in stella. Aristoteles philosophus utitur axiomate philosophico, superioris motum sphaerae communicarj cum inferioribus, idque raptu quodam, ut quotcunque motibus cieretur aliqua superior, totidem et omnes illas quae infra hanc sint, cierj, nisj renitantur, proprio et insito quodam motus principio. Itaque quia, exemplj gratia, imus Saturnj orbis annis triginta circumit, rapturus est primum jovis, ut is fixarum motum nequeat observare. Quapropter oportet orbem primum Jovis in contrarium tendere, ut annis triginta semel imum Saturnj in oppositum percurrat, et sic eodem loco maneat. At hoc primus Jovis praestare nequit, habet enim aliud officium, ut sequatur fixas, et terram circumeat horis 24. Itaque Restituente hic opus est, quae loco primae Joviae, id quod necessarium esse diximus, efficiat. Porro autem hac restituente ab unico motu imae Saturniae liberatur, caeteris Saturniae sphaeris etiamnum obnoxia. Quapropter totidem restituentibus opus erit proxime infra sequentj Jovj, ejusque primo orbj, quot // omnino orbibus superior Saturnus fertur: una minus. Nam ubi prima Jovis per restituentes liberata fuerit ab alijs Saturnj motibus obnoxia tamen adhuc primae Saturnj, non opus est illam etiam ab hujus motu liberarj: cum utriusque idem motus sit et communis sphaerae fixarum. Quemadmodum igitur prima Saturnj motum nullum efficit insito principio ⌞activo⌟, sed rapitur a fixis ⌞24 horis⌟, ita prima Jovis rapitur a prima Saturnj, eoque ordine semper inferioris primam prima superioris, usque ad Lunam.

284v

284r

[11] Cross-reference to passage headed 'C' on f. 284v: *Vide C.*

Porro quae hic Fracastoreus interpretandj causa affert, ea si cum hac interpretatione conferantur, per se casura puto: ⌊Verum⌋ et ipsa Aristotelis haec ratio constituendj caelestes orbes, ⌊non tantum⌋ in astronomia non[12] videtur, ⌊sed⌋ neque naturalibus undique rationibus est consentanea. Primo quod sine discrimine superiorum motus in inferiores derivarj asserit, et eorum qui amplectuntur inferiores, et qui non amplectuntur Epicyclj dictj. Deinde quod, cum ex promiscua summa orbium concludat Aristoteles aequalem summam intelligentiarum motricum, septem tamen e numero, qui primj stellarum dicuntur, motu carent. Quid igitur illis opus intelligentia motrice? nisi forte passiva, cujus vi obedientiam praestent superiorj trahentj ejusque nutum ultro sequantur. Tertio cum ut dictum est septem e numero motu careant, cur non vicem restituentis quaelibet supplet et ita suam ipsa sphaeram ab una superioris motione liberat? Nisi forte servanda aequalitas inter primam Saturnj et primas reliquorum planetarum, ut quia hoc ocium est nacta, hae quoque eodem privilegio gaudeant.

Sed pertexamus Aristotelis orationem: *Cum igitur ferentium sphaerarum aliae octo, aliae quatuor et vigintj sint.* In alijs exemplaribus legitur 25. Octo quidem sunt Saturnj et Jovis singulorum quatuor. Reliquorum si quilibet etiam Sol et Luna secundum mentem Fracastorej habent 5. Summa quidem 25 conflatur. Vereor igitur ut haec correctio 25 pro 24 sit Fracastorej. Si Sol et Luna quaternas tantum habent, summa posteriorum erit 23, non 25. ⌊Vide an non sic intelligendum. *Aliae octo*, id est Solis et Lunae, *aliae 23*, id est reliquorum 5. Ut ita sit evidentior causa hujus divisionis in 8 et 23.

Et harum non necesse est habere restituentem solam illam, quae astrum ultimum fert. // Omnes sex superiores, sui ordinis imae, quae astra sua ferunt, habent   287v
sub se restituentes, quia habent et sub se aliud astrum. Sola Lunae, ultimi astri ferens et ima non habet sub se restituentem, quia nullum astrum sequitur, cujus prima restituatur. *Restituentium quidem sex erunt, duorum primorum.* Nam Callippus Saturno 4 dederat, totidem Jovj, primis planetis. Aristoteles addit utrinque totidem, uno minus, ut supra denunciaverat. Sunt itaque bis tres, quos addit, sc: sex. *Sedecim vero posteriorum quatuor.* Nam Eudoxus dabat Marti, Mercurio, Venerj Solj, 5. 5. 5. 4. Aristoteles addit totidem, semper uno minus, sc: 4. 4. 4. 3. quod summam facit 15. Sin autem et Sol secundum

---

[12] My interpolation of 'non' is prompted as follows. Kepler originally started the sentence: 'Sed et ipsa Aristotelis haec ratio constituendj caelestes orbes, etsj in astronomia videtur, tamen si...', and arrived at the final version by deletion and insertion. Both the original sentence and the final version require a 'non' before 'videtur', for Kepler has just insisted that Aristotle's counteracting spheres have no astronomical function, and hence that Fracastoro must be mistaken in attributing them to the astronomer Calippus. A similar cumbrous construction is to be found on f. 285v.

Fracastorej mentem habeat 5 ferentes, habebit et 4 restituentes, eritque summa
sane 16. *Quare omnium numerus, et ferentium et restituentium erunt* 55. Siquidem
illic Fracastoreo largiaris 8. 25. Hic 6. 16. At si 8. 23, et 6. 15 colligamus,
288r summam conficiemus 52. // *Si vero soli et Lunae non addiderit quis, quos diximus
motus, omnes sphaerae erunt novem et quadraginta.* Sive enim Callippus binos
addiderit, cadent primo 4 Ferentes, dein 2 restituentes Solis quia 2 ejus ferentes
cecidere. Sex igitur si auferas a 55, restant 49. Aliqua tamen exemplaria habent
vitiose 47. Sive singulj a Calippo sint additi, cadent primo 2 ferentes una ☽,
altera ☉, indeque ejus una restituens, quare 3 de 52 ablatae, relinquunt iterum
49. Nescio quid monstrj dant affectatj numerj, si tamen genuinj sunt. Est
binarij cubus 8, Numerus ferentium ♄ et ♃. Est binarij quadratum 4,
Numerus stellarum post duas illas, quae restituentes sub se habent. Est primus
perfectus 6. numerus restituentium ♄ et ♃. Est quaternarij quadratum, 16,
Numerus restituentium in 4 ultimis. Est quinarij quadratum, numerus
ferentium in 5 ultimis. Est denique septenarij quadratum 49 summa omnium,
demptis etiam Callippi additionibus in ☉ et ☽, nec patente causa ulla, sed ne
suspicione quidem, cur Aristoteles hanc additionem in ☉ et ☽ supervacuam
putaverit, non item in caeteris, ut videatur Aristoteles solum numerum
quaesivisse. Nisi toties Aristoteles pythagoricos explosisset numeros, cred-
erem, illum de industria conquisitis ratiunculis affectare quadratum septenarij
quem deorum numerum forsan ex alicujus religionis mysterio persuasum
habuerit. Nunc liber a superstitionum suspicione Aristoteles, an haec ita
scripserit, jure dubito. Summa haec. Praeclarum ejus institutum fuit, per
astronomorum oculos et observationes ad obscuram rerum naturam con-
tendere: vituperandum vero, quod astronomicis observationibus sua miscuit
ratiocinia philosophica, toto genere diversa. Quid enim commune 31 orbibus
Callippi totidem motuum, cum 21 Aristotelis, qui novum nullum motum
inferunt? Et sic se habet celebris iste locus Aristotelis, ex cujus explicatione
apparet, quod dixj, confidentj Ursum oratione usum in sua de hypothesibus
historia, dum ait Eudoxum et Calippum concentricis usos, quod nulla ratione
ex textu deducere potuit. ⌐Fracastoreus vero hujus rej author est non idoneus.
Concentrica cum scripserit, Veteres ad se traducere nititur. De caetero fatetur
ipse nihil de illis sciri nisi ex obscuro loco Aristotelis.⌐

Quod vero ait, Pythagoraeos [Pythagoraeus] post Platonem explosis Eudoxj
et Calippj non sufficientibus concentricis, eccentricos assumpsisse, tribus
nominibus temeritatem narrationis suae prodit. Primum id Callippus ipse fecit
in Eudoxj hypothesibus corrigendis, ut defectus ipsas argueret, orbemque
unum, qui maxima verisimilitudine Eccentricitatem invexisse credj potest,
adderet. Non fuit igitur necesse, ut hoc Calippj correctioni accideret post
platonem, nisi forte in Saturno et Jove, quorum motus non bene ⌐ipse⌐

exploraverat. Deinde, dum Eudoxum a Pythagoraeis ait explosum, non ita
bene suam orationem historiarum notitiae attemperat, uti illius tenore
ʟhancʝ jactat. Eudoxus enim et ipse fuit unus de secta Pythagoraeorum:
referente Laertio. At quis haec ex Ursj oratione colligeret? Denique velim ab
illo discere, quos appellet Pythagoraeos? Si quoscunque, qui aliquod
Pythagorae dogmatum amplectuntur, Pythagoraeos habet, jam non loquitur
ad consuetudinem historiae, cujus cognitionem sibj arrogat. Sin eos, cum
historijs Pythagoraeos dicit, qui sectae huic addictj, placita ejus, disputation-
ibus, instituta vitae, moribus suis exprimunt: audiat ex Laertio // sectam il-   288v
lam in Epicurum desijsse. Natus autem est Epicurus, jam demum inclarescente
Aristotele, qui eodem fere tempore Platonem audiverat, quo Exodus. Ecce
autem Eudoxj placita longo post tempore, jam adulto Epicuro, Aristoteles
celebris philosophus usurpabat, inque libros suos de caelo, et Metaphysica
referebat. Epicurus vero, Pythagoraeorum ultimus, quid in astronomia
scripserit, qui mundum ex atomis constituebat? Itaque et haec narratio
vanescit instar fumj.

Sed Hipparchus forte pro Calippi concentricis eccentricos resumpsit? De
calippj concentricis fabulam supra expedivj. Hipparchj vero quosnam
eccentricos (praeterquam in demonstratione apogaej unius Solis, demonstrationis facilitate schema eccentricum suadente) Ptolemaeus innuat, nesciunt
ejus lectores. ʟEt Cardanus fidem hujus assertionis ad authorem Fracastoreum
remittit.ʝ Hoc potius apud ptolemaeum lib: 9. Cap: 1 inveniunt, Hipparchum
quidem erroribus manum nullatenus admovisse, solas vero observationes
reliquisse conscriptas. Minus igitur tentavit Hipparchus in astronomia,
angustioribusque pomoerijs cogitationes et curas suas inclusit, quam vel
Eudoxus vel Callippus. Quae est ergo illa tam splendide asseverata immutatio
hypothesium quam Calippicis Hipparchus intulit?

Venimus usque ad tempora Hipparchj, qui posterior Aristotele fuit, quo
tempore Ursus ait introductas physicorum Hypotheses, hoc est eum
sphaerarum ordinem qui hodiedum vulgo creditur. Id autem totum, merum
esse somnium, supra probavj, cum dicerem, Pythagorae ipsj sic esse visum
de sphaerarum ordine, uti postea Ptolemaeo. ʟEt quid multis? Quomodo-
cunque locentur haec tria corpora Solis Veneris et Mercurij, inter Martem
et Lunam, semper (in usitatis quidem hypothesibus) res eodem redit in
astronomia. Nam hac via quidem ingressos observationes destituunt. Solae
physicarum rationum verisimilitudines, authores, ut quisque hanc vel illam
potissimum sequitur, in contraria rapiunt, in caeteris consentientes. Quo
pacto, sive ante sive post Eudoxum disputatum sit de moderno sphaerarum
ordine vulgarj, non censendae sunt horum et Eudoxj diversae hypotheses
astronomicae. Cum enim inde a Pythagora hucusque vulgo astronomorum

in capite consenserint, dubitatio haec de trium horum corporum ordine semper mansit, nec dum a physicis est sublata.[13] Heraclitus Solem fecit penintimum, eodem cum Aristotele usus argumento, quod caetera astra minus calefaciant. Plato Pythagoricam Socraticam et Heraclitiam philosophiam miscens referente id laertio, locupletavit hanc opinionem optica conjectura: si ♀ ☿ infra Solem sint, diminuta olim facie, ut ☽, visum irj.[14] Ptolemaeus addit ex illorum sententia et hoc argumentum, quod nunquam Solem obtenebrent. Quae argumenta declinantes modernj, pellucida illa corpora faciunt, contra quam Lunam, et vel Solj propinquantj viam cedere, ut Ptolemaeus vult, vel etiam subtercurrere, ut modernj ex Proclo et Averrhoe exemplis dubijs contendunt. At contra penetrationem orbium Solis Veneris Mercurij et Lunae nondum amolitj sunt; ut melius stet Alpetragius, qui unum tantum Solj et Lunae interponit. Et tamen conjecturis physicis contendunt, ☿ infra ♀ ⌐quem supra Crates⌐, quia motibus pluribus moveatur, motuumque similitudine ad ☽ proxime accedat: Utrumque vero sub Sole, ut is possit esse medius, et spacium inter ☉ ☽ impleatur. Ita disputatio haec physica est, non astronomica. Sed ad Ursum redeo.⌐ Et loquitur quidem Ursus ubique ita, ac si temporibus illis, ut philosophicae, sic astronomicae quoque scholae fuissent, Artificumque, ut philosophorum, continua successio, dogmatum astronomicorum vel defensio, vel publica ⌐et solennis⌐ immutatio: quod longe aliter se habet. Semper enim haec fortuna fuit 289r astronomiae, ut paucj essent, qui suas in eam curas conferrent. // Ac etsi sane non desunt exempla, ubi unus alterum audivit in astronomicis: magis tamen tum temporis agebatur negocium astronomicum intra privatos parietes, nec ulla celebris et frequens auditoribus schola astronomica ex historijs nota est. Adeoque non omnes invicem astronomi se internoscebant, quamvis coaetanej essent, aut libros scriptos reliquissent, typographiae, puto, defectu. Philolaus enim quid senserit supra dictum: cujus Astronomica perfectiora si fuissent Eudoxo ⌐quem vivendo Philolaus attigit⌐ cognita, quid opus fuisset, ut propria cura, minus perfectam commentaretur astronomiam. Etenim ex Archimede constat, definisse Eudoxum diametrj solis ad Lunae proportionem, quae est 9. ad 1. Aristarchum vero, qui eadem cum Philolao principia

---

[13] There is a smudged marginal note here, of which I can decipher only the words *Crates et Metrodorus*.

[14] A draft of this passage on f. 283v is clearer on this point:

Quamvis viderj possit uterque, et Eudoxus et Aristoteles, hanc Solis infra quinque planetas collocationem ex communi praeceptore Platone hausisse: is vero ex Heraclito. Tradit enim Laertius, miscuisse Platonem tres philosophiae formas Pythagoraeam Socraticam et Heraclitiam. Heraclitus vero Solem terris prope admovet eadem cum Aristotele usus verisimilitudine, quod caetera astra minus calefaciant, itaque remotiora. Platonicj vero, quod et Copernicus lib. 1 cap: 10 refert, opticam causam afferunt, Si ♀ et ☿ infra solem sint dimidiata facie ut ☽ conspectum irj.

sequebatur aliam, sc: fere, quae est inter 21. et 1. Quorum illud longe est a veritate alienum, hoc ⌊ej⌋ proximum. Itaque eam quisque privatim hypothesium rationem est secutus, quam vel praeceptorj suo acceptam tulit, vel physicis rationibus consentaneam censuit, vel demonstrationibus suis aptissimum expertus est: nulla vero ita ut Ursus vult, publice *invaluit*, aut solenniter *introducta* ⌊fuit⌋ hypothesium forma.

Jamque tandem ad celebrem illum Aristarchum Samium temporum ordo nos deducit, cujus quas recenset Archimedes hypotheses, Pythagoraeorum et Philolaj quoque fuisse ante Platonem jam sat nobis constat. Accommodavit ille quidem sermonem suum demonstrationibus, et circulum, quo terra volvitur, dixit eam habere proportionem ad fixas, quam habet centrum ad circumferentiam. Quod explicans Archimedes ijsdem propemodum verbis utitur, quibus supra Pythagoraej suam apud Aristotelem mentem explicaverant: quod deprehendet, qui conferre utraque volet. Idque est, quod tanto magis confirmat, quod supra saepius iteratum est: easdem utrisque et Pythagoraeis et Aristarcho fuisse Hypotheses. Quamvis Archimedes de Aristarchj sensu conjectura nonnihil impingit. Vult Archimedes, sic intelligendum Aristarchum, quae sit proportio terrae ad sphaeram solis, eandem esse sphaerae solis (seu terrae) ad sphaeram fixarum. // At haec adhuc 289v proportio non sufficiet Aristarchj demonstrationibus. Nam ita est sensibus conprehensibilis terrae ad sphaeram solis proportio, ut parallaxin ⌊solis⌋ minutorum aliquot efficiat. Totidem igitur minutorum parallaxis inter fixas et sphaeram, qua Tellus vehitur intercederet. At Aristarchus omnem plane sensum excludit, dum infinitam, quoad sensum, proportionem introducit. Sed haec obiter. Non est autem impunis habenda Ursj dubitatio de aetate, qua claruit Aristarchus, ne, qui sciunt, Ursum etiam historijs operam dedisse, cum ipso citra causam dubitandum sibi existiment. Dubie, inquam Ursus Aristarchum refert sive sub Ptolemaeum Philadelphum, sive sub Philometorem. Intolerabilis incertitudo, cum hi duo reges centum annis invicem distent. At facilis labor hanc dubitationem tollendj. Syracusas expugnavit Marcellus consul Romanus anno nono Ptolemaej Philopatoris: ea in expugnatione Archimedes interfectus est. Scripsit igitur libellum de arenae numero, ⌊in⌋ quo de Aristarchj Hypothesibus ⌊narrat⌋, ante nonum annum Philopatoris; et forte sub finem imperij Euergetis, quem proxime Philadelphus antecessit. At Philopatorj, Epiphanes, et huic demum Philometor successit. Ita conficitur, Aristarchum, cujus libros Archimedes legit, sub Philadelpho claruisse, nec ulla causa de Philometore suspicandj. ⌊Habet historicam confirmationem historicus: audiat astronomicam astronomus. Observavit Aristarchus aestivale solstitium authore Hipparcho, ut Ptolemaeus lib 3 Cap 1 refert, anno 50 primae periodi Callippicae. Is fuit annus 6 Philadelphj.⌋

Tot nominibus in historia sua impingens Ursus, malum insuper agit interpretem Aristarchj et copernicj, et imputatam alijs in Copernico intelligendo tarditatem in sese manifeste transfert: dum, quam Copernicus Axis terrenj (qui axj aequinoctialis respondet) *inclinationem* ad planum Eclipticae statuit, hanc ille *nutationem inter planetas, ad instar navis in marj, jam in proram jam in puppim incumbentis* interpretatur. At toto caelo Copernicano aberrat. Nam etsj axis Terrae in solstitijs inclinis est ad lineam e Sole in centrum terrae ductam, in aequinoctijs vero rectos cum ea efficit angulos: Non tamen id fit annutu vel abnutu, sed fit translatione centrj terrae seu φορᾷ ejusdem per sui circulj quadrantem, manente axe terrae semper sibi ipsj in diversis situbus parallelo, mimime vero nutante. // Nec simplex haec asseveratio vera est, angulum hunc variarj. Nam eadem semper manet axis terrae ad planum Eclipticae inclinatio, sive in aequinoctijs, sive in solstitijs (de mutatione, quae post multa saecula perquam exigua contingit, jam non loquor) sed non semper solem respicit; transfertur namque Tellus cum sua inclinatione, quae semper in easdem mundj partes vergit. Itaque inclinatio haec, eadem manens cum plano ipso, diversos tamen angulos cum diversis planj lineis constituit, ut ex Geometrica solidorum doctrina constat. ⌊Quaere in Euclide⌋.

290r

Adeoque qui apte sibj voluerit imaginarj primum et tertium Terrae motum secundum Copernicum (quamvis tertius ⌊hic⌋ per se motus non est) Is Cylindrum binis planis parallelis oblique secet, ad angulum 66½, et quia hoc pacto, communes sectiones ⌊seu bases⌋ fiant Ellipses, per corollar 16 libri 1. Serenj de Cylindrj sectione: fingat is interim, esse circulos perfectos. Quae igitur est in circumferentia ⌊basium⌋ superficies alicubj recta, inclinata alicubj, talem superficiem axis terrae inclinate circumlata describit. AC, BD, bases cylindrj illa inferior haec superior. CD, superficies aequealta axi ⊕, hic est inclinata, ut et in AB recta in E, F. Ita CD, AB, solstitiorum sunt loca, E, F, aequinoctiorum.

Nec faeliciter Ursus Graecas voces tetigit κύλισις [κύλησις] δίνησις, potestque vel hoc solo Graecae linguae perito, Copernicum obscurum reddere. Tibi, Urse, terra uti quidem statuis, in jactatis a te hypothesibus, δινεῖται, in Copernico, orbis qui terram vehit ⌊si quis palpabilis esset⌋ δινεῖται. In usitatis hypothesibus, omnes orbes, et nominatim mundus totus et sphaera extima δινεῖται, in torno globus δινεῖται quiescentibus polis, et manente ⌊mobili⌋ in suo loco. Terra vero in Copernico, si solum annuum motum spectes, mente separans diurnum, simpliciter περιφέρεται, sin tu dicas terram copernico δινεῖσθαι, non ego intellexero de motu annuo, sed de diurno, cum ab eo mente separamus annuum. Sin autem dicas κυλίνδεσθαι [κυλίσκεσθαι] jam utrumque complexus es, et motum diurnum et annuum. Etenim κύλισις [κύλησις] si nescis, compositus est motus ex φορᾷ [φορᾷ] et δινήσει. Et κυλίνδεται [κυλίσκεται] globus ille lusorum, quo ad stantes cuneos evertendos collimant. Nec mehercule concinnius est exemplum motus Telluris compositj, quam si quis fingat, Globum terrae, instar ejusmodj globj lusorij, obliquo tamen axi provolvj in directam lineam, sed quae non sit in plano, verum in globo, Ut si terra in superiorj schemate superficie EAFC provolveretur via EGF, et sic tres Copernici motus ⌊in⌋ una hac provolutione insunt.

Perduxit Hypothesium historiam Ursus, usque ad // tempora Christj, 290v
qualicunque successu. Quod sicubi a scopo erravit, error crimine tamen carere viderj potest. Per ignorantiam enim omnes multifariam labimur. At jam ubi ad Apollonium Pergaeum devenit, adeo dura fronte est, adeoque nugacissimam sciens volensque narrat fabulam, ut nescias, rideasne hominem potius, an irascaris. Circumstantijs omnibus narrationem suam instruit, ut, cui de re ipsa non ante constat, jurare ausit, non esse nihilj rem, qua de loquitur. Qua in rem cum praecipuum causae suae firmamentum ponat, praecipuam etiam fraudem committat; visum est peculiare caput in ea consumere; in cujus initio, quae de Apollonio nobis constant ex fide dignis authoribus, indicabimus; postea nugas Ursj cum illis conferemus.

CAPUT III

Nullas extare Apollonij Pergaej hypotheses astronomicas.

Quod in Architectura usu venit, ut non ex una officina petantur omnia ⌊calx, caementum, clavj, serae, fenestrae⌋, nec idem architectus et lapicida sit et lignarius, et ferrarius, et arcularius: idem et in Astronomico negocio usu venit. Etenim qui omnj animj cogitatione in id intentus est, ut ex ijs quae in caelo diversis temporibus eveniunt, ⌊argumentatione derivata⌋ Hypotheses aliquas, mundj formam exprimentes constituat, hunc architectj loco habeo: cujus quidem tanta est occupatio, ut impossibile sit, ipsum omnia a seipso petere. Accipit igitur observationes ipsas, vel omnes vel aliquas, ab alijs; sunt qui tabulas illi construant, usui quotidiano necessarias. Alius arithmetices praeceptis clarus, nonnihil habet in sua supellectilj, quod in suum usum seligat artifex; Alius operandj per numeros compendia comminiscitur, alius doctrinam sinuum artificiose tradit, alius triangula docet aestimare facilius; Denique sunt, qui Geometricis problematis demonstrandis, laborem principalis Artificis, non mediocriter sublevant. Et tamen, ut qui clavos cudit, fenestras construit, lapides imperatam in formam excidit, non ideo est habendus pro Architecto, etsj ejus opera utitur architectus, domusque
structura non huic sed // Architecto adscribitur: ita et in astronomia, non 291r

quia Astronomus ⌐sine⌐ Geometria et Arithmetica geminis alis astronomus esse non potest: propterea statim astronomus est, qui bonus Geometra bonusque est Arithmeticus: nec si quis doctrina triangulorum tradita Artificem in condendis tabulis egregie juvit, proptérea tabularam vel Hypothesium author est.

Haec cum ita se habeant, vehementer initio miror, Apollonio ⌐ab Urso⌐ tribuj Hypotheses astronomicas peculiares, qui quod astronomica attinet, nihil omnino scriptum reliquit, praeter Geometricum problema (professione namque geometra fuit), quo Ptolemaeus in stationibus planetarum demonstrandis in parte nempe domus exornanda usus est.

Ut autem clarissime pateat, quid Apollonius demonstraverit, quid vero Ursus securissime fabuletur, age totam ejus intentionem e fonte suo, scilicet e Ptolemaeo, (quo misso ⌐quod vehementer miror⌐ Ursus ad rivulum inde deductum, scilicet ad Copernicum provocat) in has pagellas derivabimus et transponemus. Equidem, quod Ursus nulla Ptolemaej mentione facta, qui clarissime mentem Apollonij exposuit, in Copernico, qui obscurius rem exprimit, magnum causae praesidium ponit, id ejusmodj est, ut magnam suspicionem concitet, nunquam visum Urso Ptolemaeum, sed extitisse monitorem aliquem, qui Urso subjicerit, tale quid Apollonij problema sapere, qualia Tychonis in se continent hypotheses, et illum quidem ⌐problema⌐ bipartitum esse. Tantum enim nec plura nec distinctiora Ursus de Apollonio, cum haec scriberet, sciverat. Sed ad Ptolemaeum.

Cum ea aetate in magno precio essent Geometricae demonstrationes, et flocci penderentur vel Mechanicae descriptiones vel propinquae vero numerationes, artificio demonstrationum destitutae; quod ex illa longa Ptolemaej excusatione lib. 9. initio facile colligj potest; disputatum etiam varie fuit, quonam pacto punctum certum Geometrice designarj queat, in quo planeta stare videatur: cum haec res non aliter, nisi prope verum, constitui, et numerando colligj posse videretur. Extitit igitur inter caeteros et Pergaeus Geometra, problema Geometricis accommodavit demonstrationibus, eaque sibi sumpsit, quibus concessis expedire demonstrationem speraret: quae quidem demonstratio admirabilj profecta ingenio, et inventione solertissima, plane Conica ejus ⌐authoris⌐ sapit. //

291v   De eo sic Ptolemaeus. Supposita una saltem inaequalitate motus, ea nempe, quae ad Solem restituitur, geminam instituisse formam problematis. Primo si hanc inaequalitatem quis in epicyclo statuat accidere, ita ut Epicycli centrum in concentrico moveatur in signorum consequentia, stella in Epicyclo, similiter in consequentia ex parte superiorj. Quibus positis, sc: F Terra, et circum eam concentrico OE, inque eo Epicyclo ABI circa centrum ⌐Et Sole semper in linea FEA versante⌐ ducit Apollonius lineam FIOB talem, ut OI dimidium portionis intra circulum ad IF, residuum ab oculo F ad I convexam epicyclj partem sit in ea proportione in qua est Epicyclj velocitas, ex E in O, ad velocitatem stellae ex A in B: demonstratque, si planeta in id punctum I incidat, stare visum irj. Sin autem quis malit hanc inaequalitatem ⌐quae respectum habet ad Solem⌐ per Eccentricum administrare valituram hanc rationem non nisj in tribus superioribus, qui omnj aspectuum forma cum sole configurentur; et oportere centrum hujus Eccentrici circa centrum Zodiaci, aequaliter Soli moverj in consequentia. Stellam vero in Eccentrico, circa centrum ipsius in praecedentia, inaequalitatis ⌐illius⌐ motuj aequaliter. Quibus iterum positis, sc F Terra, N, E centro Eccentricj circa F annuo motu volubilj via NEQ, A vero planeta in eccentrico versus BCT, motum commutationis (Copernico dictum) perficiente; ducit rursum per F terram, lineam talem, ut medietas OT lineae BT ductae, proportionem eam habeat ad FT partem minorem, quam habet velocitas eccentrici ex E per Q in N, ad velocitatem stellae ex B: et demonstrat, quando stella in T punctum propius Terrae F incidit, stare viderj. Atque haec summa est et intentionis et laboris Apollonij illius; praeter quae nihil plane, quod in astronomia scripserit usquam comperj. Quae ejus demonstrata, quomodo Copernicus ad suas transtulerit Hypotheses, in ipso vide Copernico, locis ab Urso citatis. ⌐Nihil enim ⌐⌐aliud⌐⌐ dixit de Apollonio Copernicus, quam Ptolemaeus, ex quo ille haec in suum opus transtulit.⌐ Quamvis ne haec quidem, per sese ingeniosissima inventio admodum utilis fuerit artificibus, propterea, quod Apollonius, ut exitum in demonstrando reperiret, ⌐in⌐inaequalitatem alteram in stellarum motibus, quae ad partes Zodiacj restituitur, exclusit. Itaque ea demonstratione nisj ad singulos casus ⌐inaequalitatis alterius⌐ variata utj Ptolemaeus et Copernicus non potuere. // Adhuc ergo nullas habemus Apollonij Hypotheses astrono- 292r micas, quae quidem sic dicj et inter caeteras formas recenserj mereantur. Quae enim ille supponit in demonstratione sua, non ita supponit, tanquam revera ita sese habere statuat (quod solent artifices) sed ita, ut patescat, quid circa stationes futurum esset, si ⌐illa⌐ ita haberent. Propterea gemina utitur forma principiorum demonstrationis suae, quia sc: nihil decernit ut astronomus artifex, sed merum agit geometram in motuum genere.

Jam vide quam suaviter Ursus de hoc Apollonij invento nugetur. Primo ait, ab ipso mutatam esse universam formam Aristarchianarum, perinde ac si professione fuisset astronomus, ut Aristarchus et alij. Mathematicum Ptolemaeus appellat, at certe nuspiam ej astronomj titulus ascribitur. Deinde, cum obversaretur ej memoria duplicis formae demonstrationis Apollonianae, sed memoria tenuis, tanquam rej per somnum visae (nam supra conjectatus sum, non ipsum Ursum legisse haec in Ptolemaeo, sed ab alio qui haec fortasse legerit, admonitum): ipse quoque geminam Aristarchianarum immutationem

Apollonio affingit. Primam ait institutam ideo, ut ingens illa vastitas fixarum, quam Aristarchus apud Archimedem introduxit, tolleretur. Ideoque sic esse institutam illam immutationem, ut centrum fixarum, una cum terra circumvolveretur annuo motu. At hoc totum merum esse figmentum, et sphingem ex Ursj cerebro natam, facile videt, qui cum hac Ursj interpretatione Ptolemaej verba supra allegata ⌊de priorj demonstrationis Apollonianae forma⌋ conferre voluerit. Major equidem est candidj et atrj coloris cognatio, quam Ursj et Apollonij. Ac si quis ad τὸ ἦθος suppositionum Apollonij respiciat, illi non potest ullo modo verisimile esse, tantam Apollonio curam fuisse de illo hiatu, illaque vastitate fixarum. Alteram immutationem Ursus talem ab Apollonio fingit institutam, ut cum antea Terra (centrum fixarum) volveretur annuo motu circa Solem centrum planetarum, quiescentem: jam circa terram, et centrum fixarum, quiescentia Sol et centrum planetarum

292v eodem annuo tempore circumierit // quae quidem hypothesium forma ⌊a⌋ Tychone Brahe, (cujus exagitandj gratia Ursus haec confinxit) propter causas ab ipso authore passim inculcatas, et ex occasionibus non contemnendis suscepta, inventa et constituta est. Habet haec quidem assertio verisimilitudinem aliquam; si quis eam cum Apollonij suppositione conferat. Sit enim ⌊juxta mentem Apollonij et in superiorj schemate⌋ ⌊F⌋ Terra circa eam ⌊NE⌋ circulus, quem centrum Eccentrici annuo motu describit, ita ut linea ex ⌊F⌋ terra per ⌊N.E.⌋ centrum Eccentrici ejecta semper eadem sit cum eâ quae ex ⌊F⌋ terra per Solem ⌊sive is jam in ampliorj circulo DG, sive in angustiorj KL moveatur.⌋ Et sit ⌊H⌋ stella in Apogaeo erit itaque juncta solj. Transferatur Apogaeum ejusque linea ⌊FH⌋, centrum sc: eccentrici ⌊N⌋, ex eo loco in E, sitque locus Apogaej in ⌊A⌋ iterum cum ⌊G vel L⌋ Sole. Stella vero moveatur a linea apogaej in partem contrariam motu commutationis, ex ⌊A⌋ scilicet in ⌊B⌋. Patet, quod quotannis fere semel in perigaeum incidat, etsj non ad oppositas partes Zodiacj evadat. Qua ratione fit, ut ad accessum Solis ⌊ad se⌋ tollatur in altum, cum decedente sole, sese simul demittat. Eadem vero Tychonis quoque hypotheses habent. Sic cum Apollonius dicit ⌊D vel K; G vel L⌋ solem, et ⌊N E⌋ centrum Eccentricj esse in eadem linea, Tycho dicit, ⌊N E⌋ centrum Eccentrici esse ipsum Solem. Sic cum haec Apollonij sententia locum aliter habere non possit, nisi justa sit proportio ⌊FE⌋ eccentricitatis ad ⌊EA⌋ radium Eccentrj (aliter justa retrogradationum quantitas non conficietur) fiat utique in stella Martis, ut ⌊EP⌋ eccentricus, secet ⌊FQ⌋ circulum a centro Eccentricj descriptum. Id quidem ex anteposita

293r suppositione Apollonij firmiter deducitur. // Cum enim BT recta ducta sit extra centrum E, et alia AP per centrum E. Quare per 3 tii[15] AP longior

---

[15] The reference is to Euclid; here, as in the following citations, Kepler evidently left the space to be filled in later with a reference to the relevant proposition.

erit quam BT. Est autem AP divisa bifariam in E centro, et BT in O ex
hypothesj. Quare per communem conceptionem dimidium EP dimidio OT
quoque majus est. Amplius, cum a puncto F extra centrum circulj CP
ducantur binae, una ex centro veniens FP, altera secus FT, incidentes in
circumferentiam PT. Erit per ⌊3 tii⌋ FT major quam FP. Jam supra dixerat
Apollonius, OT sic esse ad TF, ut velocitas ⌊centrj⌋ Eccentricj (Martis) ⌊NE⌋
ad velocitatem stellae, seu motum commutationis, AB. Hoc est ut 59 ad 28
fere. Cum igitur ejusdem ⌊totius⌋ ad majorem ex duabus partibus minor sit
proportio, quam ad minorem ⌊Per 5 ti⌋, Erit major proportio OT, ad FP,
quam 59 ad 28. Et rursum cum ex duabus majoribus quae longior est,
majorem proportionem habeat ad minorem, quam quae brevior per 5 li;
major igitur erit proportio EP ad PF, quam OT ad PF. Prius vero et OT
ad PF major erat proportio quam 59 ad 28. Multo igitur major est proportio
EP ad PF quam 59 ad 28. Cumque 59 ad 28 proportio sit major dupla, tribus
igitur nominibus EP proportio ad PF major est proportione dupla.[16] Si EP
plus est quam duplum PF, ergo PF minor est quam dimidium EP
semidiametrj Eccentrj. Ablatur igitur PF minor dimidia relinquit FE vel FQ
aequalem majorem dimidia parte. Itaque FP, distantia ♂ perigaea a Terra
minor est FQ semidiametro circulj qui centrum eccentricij vehit. Idem autem
asserit Tycho, de circulo ♂ secante circulum ☉. Ex hisce igitur apparet ea,
quam dixj, cognatio hypothesium Tychonis, cum Geometrico problemate
Apollonij. // At ex his tamen nondum sequitur, id quod in causa versatur, 293v
Apollonium hypotheses edidisse astronomicas legitime et proprie sic dictas,
multo minus vero id, quod sine omnj haesitatione affirmaverat Ursus,
Apollonium secunda immutatione Aristarchianarum facta, hanc formam
systematis mundanj, quae hodie Tychonj placet, expressisse. Primum enim
⌊ut initio⌋ non fuit Apollonius Astronomus professione sed Geometra,
quodque demonstravit ex astronomia transsumptum problema, id non ipse,
ut astronomj requirit officium, ad usum adhibuit, ut hac assumpta hypothesj
planetae ullius motus ex observationibus deduceret et demonstraret, sed
demonstrationem mere Geometricam ad astronomos tanquam propola clavos
aut securim ad Architectos attulit, si quis esset, cujus usibus ista servirent.
Deinde, cum hypotheses astronomicas supra cum Urso definiremus certam
aliquam de mundi forma, corporumque caelestium dispositione, conceptionem: Apollonij haec demonstratio plane nihil tale redolet. Nam et liberum
astronomis relinquit, sive Concentrico cum Epicyclo velint utj in omnibus
quinque, qui stationes faciunt, sive Eccentrico in tribus superioribus: et nihil

---

[16] By a deleted draft of the next few sentences is a scrawled marginal note: 'H[?]
qu[?] quidem ego, de Tych: scri[?] Potuisse[?]: idem forsan inferius diligens ac bona
fortuna usus. Sed non facile est.'

illj curae est orbium diversorum causa situs vel magnitudinis mutua comparatio. Respicit enim ejus demonstratio unum solum planetam, eumque non aliter, nisi propter unicum stationum et regressuum φαινόμενον. Non est igitur Apollonij problema pro Hypothesibus astronomicis habendum. Adde, quod ne quidem ad unius planetae motus demonstrandos sufficientes assumpsit hypotheses, cum eam motus inaequalitatem quae ad eundem Zodiacj locum restituitur, de industria praeteriret. Quo manifestum fecit, nihil aliud se quaerere, quam Geometricam conclusionis ex assumptis deductae certitudinem, quae vehementer ingeniosos contemplatores delectare solet. //

294r   Jam, ut concedamus, pro hypothesibus justis habenda demonstrata Apollonij, quod jam refutatum est, desumptas tamen ∟aut transpositas⌐ ex Aristarchianis, quis Urso credet? Etenim magna probabilitate contendit, vel ignoratas vel non intellectas esse Aristarchj conceptiones Apollonio. Etenim cum sciret Apollonius, quibusdam ex Astronomis placere Eccentricos, alijs concentricos cum Epicyclo: problema suum utrisque accommodavit. Quod si de Aristarchiana terrae volutione scivisset, credo equidem, aeque facile ∟ad⌐ illam suum problema conformasset, ac post eum Copernicus. Deinde si Aristarchum exprimere voluisset: quid quaeso opus illi fuisset asserere, quod centrum eccentricj aequaliter cum sole moveatur? quanto majus fecisset operae compendium, si, quod hodie Tycho, asseruisset, omnium planetarum centra esse in Sole. Nunc, cum separet centra planetarum a sole et ab invicem, certas, et a Tychonicis longe diversas metas extruendae formae mundanae praefixit ijs, qui hoc voluerint praestare. Nam post orbes Lunae Systema sequetur Mercurij cum Epicyclo, supra hunc Venus cum Epicyclo longe amplissimo: Tum Sol. Et jam supra Solem L (si ad proximum schema respexeris) E centrum Eccentricj Martis tam alte sublatum circumibit, ut EC (quae certam habet proportionem ad EF) longior sit, quam EP, ne orbis solis aliquando Marti cogatur hospitium praebere; quod omnibus illis impossibile viderj necesse est, qui palpabiles et adamantinos orbes cum Aristotele credunt. Hoc enim est, quod permagnam facit differentiam inter veras Apollonij hypotheses, et eas quas Ursus illj affingit, (cum Tychonis sint) quod utrinque quidem, ut supra demonstratum, stella ♂ in Eccentrico duobus locis incurrit metas centrj Eccentricj: Tychoni vero eaedem metae idem circulus est et Solis ipsius, itaque Mars illi longe intra Solis propinquitatem ad Terras descendit: Apollonio minime, vel certe non necessario: Tychoni ∟hoc⌐ ex observationibus compertum, diversitate aspectus rem subtilem patefaciente: Apollonio nulla observata, quibus contrarium evincat. Sic ergo Mars in Apollonio

294v   nunquam humilior sole fingendus. // Ut autem ex Apollonij demonstratis centrum eccentricj Martis supra solem circumducendum esset, ita Saturnj

centrum infra solem ⌊omnium vero trium vel longissimis intervallis supra solem circumire⌋ nihil prohibeat. Quo modo nulla certa ratio erit proportiones
orbium, ut apud Aristarchum et Tychonem, investigandj, nisi haec unica,
quae a contactu orbium, et occursatione mutua duorum planetarum cavenda
desumitur: qua ratione nihil aliud docetur quam certae metae, infra quas
planetae non descendunt: at quam alte supra has metas attollantur, pervestigarj
hinc nequit. Ecce vero schema hujusmodi integrae hypotheseos, si quis illam
ad normam Apollonianae demonstrationis cupiat constituere. Usque adeo
magna est inter utramque hypothesin differentia, dum altera centra orbium
⌊tantum⌋ cum sole movet, altera in ipsum insuper Solem ea collocat. Nec
debet movere quemquam, quod supra dixj, planetarum circuitus ⌊in
Apollonio et vero et ficto⌋ ad solis praescriptum et cum eo situm mutare:
ut propterea Urso fidem adhibeat asserentj, Apollonium ex Aristarcho loqui.
Nam si utramque formam problematis Apollonianj inter se conferas, videbis,
occasionem alterius formae excogitandae, non ex visis Aristarchianis, sed ex
prioris formae ⌊quae Ptolemaica est⌋ transpositione fluere.

Cum enim jam tum inter astronomos seu Geometras convenisset, omnia
quae per Eccentricos demonstrantur, posse itidem et per concentricum cum
epicyclo: Apollonius jam in priorj problematis sui forma inaequalitatem 5
planetarum per epicyclum in concentrico conficj ab astronomis acceperat et
assumpserat. // Statuunt enim Ptolemaicj, in orbibus trium superiorum tres 295r
esse Epicyclos, qui superiori parte in eandem plagam moveantur cum
Eccentrico, ea lege, ut cum centrum epicyclj est cum sole, planeta sit semper in
apogaeo epicyclj. Hanc ergo formam transposuit Apollonius in alteram, quae
⌊in⌋ eccentrico consistit: utque id recte fieret, oportuit ⌊omnino⌋ motum centrj
eccentricj, qui est pro motu Epicyclj, ad solem alligare, quia prius et Epicycli
circa suum centrum conversio ad solem alligabatur. Haec occasio hujus
hypotheseos, non certe alia. Demonstrationis causa: describatur centro M, et
eccentricitate AM, Eccentricus EFH. Cum ergo centrum M loco non
movetur per circumferentiam MN, circulus perfectus describetur per EFH.
Videbitur motus stellae ex A tardus circa E apogaeum, velox circa P
perigaeum. Idem accidet si describatur centro A concentricus BCD, et in eo
Epicyclus, cujus radius BE, CF, vel DH sit aequalis AM Eccentricitatj priorj:
motus vero sic sunt comparati, ut et orbis ME prior super M, et AB posterior
super A aequaliter et in partes easdem incedant, linea vero ex E F H planeta
per B, C, D centrum epicyclj ducta lineae MA eccentricitatj perpetuo
aequidistet dum circumfertur in BCD concentrico. Qua ratione efficitur ut
Epicyclus in sese, et respectu totius universj non minus quiescat, quam M
punctum quiescit. Sed Geometrae ad E. G. I apogaeum Epicyclj respicientes,
dicunt, planetam in epicyclo non quiescere, sed moverj ⌊in oppositum

eccentrico sc⌐ ex G in F tanto arcu, quantus est BC via centrj Epicyclj. Esto jam ut M centrum Eccentrici moveatur, et aliter quidem quam Eccentricus: moveatur modo in plagam eandem in quam et Eccentricus, sed velocius illo, ut Eccentricus quidem angulum QLK, centrum vero Eccentricj angulum MAL conficiant. Fiet hoc pacto, ut stella ab E in K velociter videatur

295v descendere, nec circulum perfectum describat. Idem fit per con // centricum cum epicyclo, ubi concentricus hic illius Eccentrici motum suscipiat, Epicyclus vero, sive in eo stella K, motum illic centrj Eccentricj, numerandum a linea DH aequidistante AM, ut HDK et MAL sint aequales. Ex his apparet occasio genuina, quae Apollonio hanc assumptionis formam suppeditaverit. Nam quod accessum et recessum planetarum ad nos, proportionalem conversationj solis attinet, eundem etiam Ptolemaicae hypotheses exhibent. Itaque ⌐adhuc⌐ nihil est, quo sese tueatur Ursus, obtineatque, Apollonium esse inversarum Aristarchianarum, quaeque Tychonj debentur, authorem.

His adde et hoc, quod cum inaequalitas illa motus, quam Apollonius a sua demonstratione sejunctam esse voluit, negligj nequeat; itaque apogaeum et perigaeum in linea HN, CE, ⌐in schemate abhinc tertio⌐ semper sibj ipsi parallela constitui debeat; ex eo evenit, ut AE apogaej lineam dicere jam porro non possimus, totaque nomenclatura Tychonis differat ab Apollonio. Quare neque motum planetae in praecedentia fierj statuit, ex A in B, sed in consequentia ex C in B, contra quam Apollonius. In tanta rerum termin-orumque diversitate frustra laborat Ursus abstractam a Tychone formam hypothesium ad Apollonium deferre.

CAPUT IV

Non extare ullibi mentionem hypothesium, quas Tycho
sibj asserit, Ursus Apollonio affinxit.

Nescio, quo affectu stimulante, id unum agit Ursus, ut Tychonis de mundo conceptiones antiquitati asserat, manifestam veritatj vim afferens. Primum ex veteribus in aciem supra produxerat Apollonium. Si hoc defendendum suscepisset, cognationem quandam intercedere Apolloniano problemati cum Tychonis hypothesibus, poterat tolerarj partim, in caeteris excusarj. Nunc quia plane utrumque exaequat, apparuit proximo capite, eos non plene ⌐Urso fuisse⌐ cognitos, minime vero vel lectum vel intellectum Apollonij

Frisch, 270,   problema. // Cognatio autem illa hypothesium Tychonis cum Apollonii
l. 32   problemate, quam et supra demonstravi et jam concedi posse dixi, jam quidem cernitur, postquam Tycho suam mundi formam diversissimis inductus argumentis et constituit et prodidit. Absque hoc esset, haud equidem scio, an quisquam adeo felici futurus fuisset ingenio, ut ex solo Apollonii

problemate inspecto in Tychonis cogitationes incidere potuisset; adeo obscura est haec cognatio coecusque transitus.

Alter ex veteribus Tychoni ab Urso tanquam aemulus ob oculos addicitur Martianus Capella, in quo multo adhuc imbecillius stat Ursus quam in Apollonio. Etenim is auctor, cum honestam e literis animi recreationem peteret (vir enim politicus erat et praeses Africae), poëticen encyclopaediae conjungere statuit, fabulaque pereleganti conficta de nuptiis Mercurii et Philologiae, cum philosophiam Platonicam tum in specie septem artes (ut appellamus) liberales ita tetigit et percurrit, ut non tantum summam cujusque brevissime exponeret, sed etiam quidquid in singulis ipse deprehendebat insigne, abstrusum, admirandum, aut tale quod ipsum potissimum afficiebat, id cum diligentia cincinnatoque stylo inculcaret. In astronomia Macrobium, Plinium et Vitruvium potissimum secutus est, Romanus Latinos scriptores idque adeo, ut ne quidem ab erroribus eorum discesserit, // quin in suum   *Frisch, 271* opus eos transferret. Itaque de eorum sententia Solem centrum fecit orbium Veneris et Mercurii. Sed Ursus, qui Mart. Capellae dogmatis notitiam uni Copernico acceptam fert, ceterorum mentionem nullam facit, Martianum primum auctorem existimans: magno incommodo causae suae. Quantum enim fecisset ad speciem, quam in auctoribus et antiquitate studio quaesivit, si Martiano Macrobium, si Macrobio Plinium, si Plinio Vitruvium ante-posuisset. Nam et hic in astronomicis Plinio in praedam cessit, ut Plinius Martiano. Quin imo ne Vitruvius quidem auctor est astronomicorum dogmatum, quae recenset in opere suo. Quid enim architecto cum astro-nomia? Non agit astronomum ex professo Vitruvius, sed ad commendationem et jucunditatem operis occasione captata cum alia rara tum astronomica aliunde petita interspergit. Macrobius vero et vetus ad Bedam commentarius hanc opinionem ipsi Platoni tribuit, utcunque eam successores ejus Platonici quemadmodum alia multa corruperint. Imo idem Macrobius eandem traditionem ad ipsos Aegyptios refert, ut ita pene cum ipsa astronomia nata fuerit. Hanc itaque praedam Ursus e rictu amisit, dum vix conspecto Mart. Capellae nomine in Copernico, in hoc unum intentus est, ne quidem Copernici indicationem expendens. Quaesisset enim saltem, quinam essent illi alii Latinorum, quos idem cum Martiano *percaluisse* Copernicus affirmat; quaesisset cujusnam e Latinis auctoribus essent illa verba: *conversas absidas* quae Copernicus e Plinio transscripsit.

Placet vero de hac circa Venerem et Mercurium hypothesi verba transscribere, ut quae jam dixi luculenter appareant. Capellae verba clara sunt ex lib. VIII.: *Tria item ex his cum Sole Lunaque orbem Terrae circumeunt,* ♀ *vero et* ☿ *non ambiunt Terram.* Et paucis interjectis: *nam licet ortus occasusque quotidianos ostendant, tamen eorum circuli Terras omnino non ambiunt, sed circa*

Solem laxiore ambitu circulantur. Denique circulorum suorum centron in Sole constituunt, ita ut supra ipsum aliquando, infra plerumque propinquiores Terris ferantur. Infra vero, cum prius de Luna et Sole mutuisque in sese et in Terra affectibus egisset, ordinis ratione ad eos pervenit, *qui circa Solem peragratione mundana volvuntur,* ♀ et ☿, quorum *circulos epicyclos esse, superioribus,* ait, *memorasse,* i.e. *non intra ambitum proprium rotunditatem Telluris includere, sed de latere Terrae quodammodo circumduci.* In quo epicyclo quomodo moveantur, infra in Venere declarat his verbis: *in suo posita circulo eum varia diversitate circumdat, quia aliquando eum transcurrit, aliquando subsequitur nec comprehendit, aliquando superfertur, nonnumquam subjacet.* Haec verba et sententiae quomodo desumpta sint ex aliis jam patebit. //

296r   Plinij mens in ambiguo est, cum verborum obscuritate, tum interpretum in hoc negocio authoritate. Valido probatur argumento Ex Cap: 8. lib: secundj, Capellae mentem Plinio non fuisse perspectam. Eo namque loco, cum ex professo sphaeras ordinaret, *Infra solem* inquit, *ambit ingens sidus, appellatum Veneris.* Et paulo post *Proximum Veneri sidus Mercurij, inferiore circulo fertur.*

Retinuerunt haec verba Collimitium Zieglerum Milichium [Mycillum],[17] ⌐interpretes Plinij⌐ non immerito, quo minus inferius, cum in specie Plinius de Venere et Mercurio verba facit, in Plinio Capellam agnoscere vellent; Sunt autem haec ex Cap: 17. *Primum igitur dicatur, cur hi duo nunquam longius a Sole abscedant, saepe ad Solem reciprocent? Conversas habent utraeque absidas, ut infra solem sitae; tantumque circulis earum subter est; quantum superne praedictarum; et ideo non possunt abesse amplius, quoniam curvatura Apsidum non habet ibi longitudinem maiorem.* Obscurum esse Plinium in his verbis omnes fatentur,

296v   Collimitius // et ex hoc Zieglerus interpretationem unam afferunt, et simul rejiciunt, ut impossibilem: Zieglerus alteram superaddit, Milichioque [Mycilloque] suppeditat, quam negat ⌐Lipse⌐ esse Plinianorum verborum genuinam. Sed Copernicus ⌐communiter mihi cum Urso receptus in hoc negocio testis, is ergo⌐, posthabito cap. 8 emendato vero contextu, quod crebro in hoc authore necessarium occurrit, efficit, ut jam in Plinio Capellae vocem exaudias. Sic nempe scripsit ex mente Copernicj Plinius. *Conversas habent utraeque apsidas, ut circum Solem sitas; tantumque de circulis earum subter est, quantum superne.* Ut vox *Infra* sit adulterina, *praedictarum* vero surreptitia. Movit Copernicum, ut hoc Plinium dixisse crederet, illud potissimum, quod et Vitruvius ante plinium, et Capella Plinij Metaphrastes, cum eadem de planetis dogmata, easdem sententias, eosdem errores, eadem saepe verba cum praecedentia tum sequentia ⌐Luna cum Plinio⌐ usurpent, posterior a priore

---

[17] For explanation of this emendation see fn. 192 below.

mutuo. In hoc loco, ubi de Veneris et Mercurij cum Sole conspiratione agitur, illi quidem hunc sensum, quem Plinio Copernicus restituit, exprimunt. Cum enim Plinio fuerit alter magister et praemonstrator, alter discipulus; uterque pro interprete obscuritatis Plinianae ⌐jure⌐ adhiberj viderj potest. Ac mihi sane, dum quid Plinius octavo capite dixerit, quid jam cap: 17 dixisse credendum sit considero, videtur vir ille rationes astronomorum et Vitruvij ⌐verba⌐ non satis percepisse, sed dum raptim torrentis instar omnes authores percurrit, omnia convolvit; ubertate scientiae tumidus, ⌐multa semicruda, nec plene secum ipsi percepta⌐ profert, obscuritate verborum se simul et lectorem decipiens. Hoc illi sane crebro alias usu venit; ubi secretiores ab usu quotidiano disciplinas attingit. Itaque non est in hoc authore absurdum, credidisse hunc ordinem sphaerarum, qui est octavo capite, et nihilominus cum vitruvio sensisse cap: 17, quod Solem ambiant Venus et Mercurius.

Sed audiamus et Vitruvium. Is cum motum[18] // primum, qui fixarum et signorum zodiaci est, explicasset jamque in genere dixisset, quomodo *Luna, stella Mercurii, Veneris, ipse Sol, itemque Martis, Jovis et Saturni* (hoc enim usitato ordine planetas recenset, quo Plinium forte decepit) contrariam fixis viam sub zodiaci signis peragrent:[19] accedit ad explicationem singulorum eo ordine, quo proposuerat. Primo Lunae motus et spatia menstrua describit; dein de Veneris et Mercurii stella sic ait: *Mercurii autem et Veneris stellae circum Solis radios, Solem ipsum uti centrum itineribus coronantes, regressus retrorsum et retardationes faciunt.* Postea ad Martis, Jovis, Saturni stellas accedit earumque motus explicat. Clara est itaque verborum ejus sententia. //

Platonis quidem, ipsius ⌐verba⌐ obscura sunt, sed tamen indicia quaedam praebent ejus sententiae, quam illi tribuit Macrobius. *Cum* inquit *effecisset Deus corpora singulorum, ea posuit in circumlationes, quas cujusque circuitus requireret, septena in septenas, Lunam quidem in primum circa terram, Solem vero in secundum supra terram, Vesperum vero, et quam Mercurio sacram dicunt in tales, qui celeritatis quidem ratione, aequecurrentem Solj orbem eunt, contrariam vero ej vim sortitj sunt; quare et comprehendunt, et comprehenduntur a se mutuo, et secundum eadem, Sol et Mercurius, et Venus.*

Primum videmus, celeritatis ratione eundem tribui tribus hisce orbem, sed propter evagationes Veneris et Mercurij dividj in tres. Id autem fit si Sol teneat Eccentricum, Venus et Mercurius singulos in eo Epicyclos. Deinde

Frisch, 272, l. 31

297r

---

[18] Kepler refers to the following material on the lost folio: *Huc refer ex altero folio paragraphum.*

[19] Frisch has 'contrarium...peragrent' as quotation. But this is not in any edition of Vitruvius that I have seen, and the context suggests that it is Kepler's gloss. I am indebted to Professor Bruce Eastwood for drawing my attention to this discrepancy.

vide mihi Platonem, quasj ex Martianj schemate recensentem ordine Solem Mercurium et Venerem: non contra Solem Venerem Mercurium ⌐Mercurius enim hac ratione propior fit soli quam Venus⌐. Tertio phrasin loquendj *comprehendere*, pro antecedere, videtur Plato Capellae suppeditasse. Quarto et Plinius suam vocem *Conversas Apsidas*, ex platonis voce *Contrariam Vim* deduxisse videtur. Itaque duo hi, Plinius et Capella Platonem imitatj viderj possunt, suamque ex hoc hausisse sententiam. Omnium vero manifestissime Macrobius, qui inter utrumque vixit, quinto post Christum saeculo; is enim se manifestum Platonis interpretem profitetur. *Ciceronj* inquit *Archimedes et Chaldaeorum ratio consentit Plato AEgyptios omnium philosophiae disciplinarum parentes secutus est, qui ita Solem inter Lunam et Mercurium locatum volunt, ut rationem tamen deprehenderint et edixerint; cur a nonnullis Sol supra Mercurium,*

297v *supraque // Venerem esse credatur. Nam nec illj, qui ita existimant, a specie verj procul aberrant.* Uno verbo hoc dicit Macrobius, utrumque ⌐quodammodo⌐ verum esse, et quod Sol supra venerem et Mercurium sit, et quod infra. Sequitur. *Opinionem vero istius permutationis hujusmodj ratio persuasit.* Permutationis voce videtur illa Platonis verba perstringere, *Comprehendunt se mutuo, et comprehenduntur,* Quod, inquit, Plato opinatus est, situm horum duorum Erronum et Solis ⌐interdum⌐ cum contrario permutarj, causa haec fuit. *A saturni sphaera, quae est prima de septem, usque ad sphaeram Jovis, a summo secundam, interjectj spacij tanta distantia est, ut Zodiacj ambitum superior triginta annis, duodecim vero annis subjecta conficiat. Rursus tantum a Jove sphaera Martis recedit, ut eundem cursum biennio peragat.* His ⌐et sequentibus⌐ verbis habes interpretationem ejus, quod plato dicit, *posuisse deum corpora in ambitus eos, quos cujusque periodus* nota scilicet temporis restitutorij quantitas *desideret.* Pergit Macrobius *Venus autem est tantum regione Martis inferior, ut ej annus satis sit ad Zodiacum peragendum. Jam vero Venerj ita proxima est stella Mercurij, et Mercurio Sol propinquus; ut hi tres CAELUM SUUM* (caelum in singularj dicit) *parj temporis spacio, id est, anno, plus minusve circumeant.* ⌐Ideo et Cicero hos duos cursus, comites Solis vocavit, quia in parj spacio longe a Se nunquam recedunt.⌐ Iterum particulam de textu Platonis supra posito explicuit. Sequitur. *Luna autem tantum ab his deorsum recessit; ut quod hi anno, viginti octo diebus ipsa conficiat. Ideo neque de trium superiorum ordine, quem manifeste clareque distinguit immensa distantia, neque de lunae regione, quae ab omnibus multum recessit; inter veteres aliqua fuit dissensio. Horum vero trium sibi proximorum Solis, Mercurij et Veneris ordinem vicinia confundit, sed apud alios. Nam AEgyptiorum solertiam ratio non fugit, quae talis est.* Si quem obscuritas,

298r quae in hoc quoque authore // nonnulla est in dubitationem adducit de ⌐ejus⌐ sententia, is haec jam posita ⌐verba⌐ probe ponderet. Sic enim est intelligendus sequens textus, ut inde ratio pateat ordinis non tantum inter solem et duos ejus comites, sed etiam inter ipsos hos. Itaque interpretatione ùnius verbj in sequentibus opus est. *Circulus, per quem Sol discurrit* (intellige Epicyclum in Concentrico, quo eccentricitatem salvare solemus; secus feceris, Macrobij verba in concordiam non rediges) hic inquam *a Mercurij circulo, ut inferior* (malim interior) *ambitur, illum quoque superior circulus Veneris includit. Atque ita fit, ut hae duae stellae, cum per superiores circulorum suorum vertices currunt, intelligantur supra solem locatae; cum vero per inferiora commeant circulorum: Sol eis superior existimetur*; subintellige, cum ⌊tamen⌋ idem caelum circumeant, uti supra dicebat. ⌊Explicat autem illa verba Platonis, *Comprehendunt sese mutuo et comprehenduntur, idque in plagas easdem, scilicet, Ante, Pone, Supra, Infra.*⌋ *Illis, ergo, qui sphaeras earum sub sole dixerunt, hoc visum ex illo stellarum cursu, qui nonnumquam, ut diximus, videtur inferior; qui et vere notabilior est, quia tunc liberius apparent. Nam cum superiora tenent; radiis magis occuluntur. Et ideo persuasio ista convaluit; et ab omnibus pene hic ordo in usum receptus est. Perspicacior tamen observatio meliorem ordinem deprehendit etc.* Haec ex Macrobio [Vitruvio] lucis gratia transcribere volui, ut Platonis verborum sensus intelligeretur; quem, Commentarius ad Bedam vetus, qui Martiani Capellae aetatem proxime secutus est, ⌊capite 14⌋ non obscure confirmat, dum ipse quidem Epicyclum veneris inter Solem et terram locat, ejusdem tamen Epicyclj decursum Platonj causam ait praebuisse, cur *aliquanto quam Solis est elatiorem luciferj globum astruxerit.* ⌊Et quidem⌋ si haec illius interpretis verba considerem *proximum Solj viderj excelsiorem, proximum terrae, humiliorem*, conjicio id, quod dixj, Epicy//clum Veneris totum infra Solem ab illo collocarj. At si schema ⌊alterum⌋ (est autem editus liber coloniae anno 1537) ⌊aspicias⌋, sane ipsissimam capellae sententiam refert. Nam punctus Solis seu litera κ ponitur intra complexum Epicyclj Veneris, nec verba modo allegata repugnant, si modica interpretatione juventur; et illud *proximum solj* referatur ad visum, *proximum terrae*, ad centrorum distantiam. Porro schema prius aut vitiosum est, aut ad repraesentandam eorum sententiam aptatum, qui Epicyclum veneris supra solem constituerunt: posterius inscitia descriptorum depravatum sic restituo: itaque hoc confer cum illius commentj lectione, si lubet. Obiter enim te juvare voluj.

Sed ad rem redeamus, et illud de nostra liberalitate Ursi copijs per sese imbecillibus subsidio mittemus: sententiam hanc, quod Sol sit centrum orbium veneris et mercurij, non tantum Capellae, sed etiam Commentatoris ad Bedam, Macrobij, Plinij, Vitruvij, Platonis, adeoque ipsorum AEgyptiorum fuisse. Quid hinc sequitur? Num hoc quod Ursus extruit? Hypotheses Tychonicas esse ab his quos modo nominavj longe ante Tychonem usurpatas? Minime. Nam primo, ut supra quoque in Apollonianj problematis consideratione sum argumentatus, haec una tantum pars est

formae Mundj Tychonianae, quare pro justa et legitima hypothesj astronomica, quae distincta sit a caeteris, haberj non potest. Ac ne quis suspicetur, Capellam, Vitruvium, Plinium, Macrobium, vel quemquem alium, caeteras etiam hypotheseos Tychonianae partes adjunxisse huic conceptionj, provoco ad ipsorum opera quae extant, ubi in caeteris omnibus partibus contrarium a Tychone, et idem cum Physicorum hypothesj (ut cum Urso loquamur) verbis evidentibus, claris et indubitatis sentire se profitentur: quae quidem
299r verba nimis longum foret allegare, sufficit digitum intendere. // Stat itaque assertio ursj firma et immobilis ut currus unica rota scilicet subnixus. Nam si partem totum; si duo, vigintj esse persuaserit: hoc quoque, quod contenderat, obtinebit.

Denique sicut antea Apollonium negavj professione fuisse astronomum, quare neque pro authore novarum hypothesium habendum contendj: ita hic idem multo quidem efficaciorj argumento urgeo. Nam etsj Apollonij problema de solis stationibus planetarum fuit, horum vero authorum cogitationes in tota machinae mundanae forma vagantur: ille tamen Lsuae inventionis⅃ (quantum sibi Geometra proposuerat) author primus merito celebratur: hi alienas de mundo conceptiones de industria consarcinant, ijsque sua opera exornant: nec sibi arrogant: si pauca in Plinio excipias, quae ille a Vitruvio mutuata non magna ejus injuria dissimulat, cum uterque erraverit. Apollonius laudem ingenij et admirationem omnium subtilissima et verissima speculatione meruit: hj scatent erroribus, et gratiam quidem habent transmissarum ad nos et conservatarum illarum speculationum: peculiarem vero gloriam nec sibj nec magistris suis redemerunt. Male nobiscum agatur si etiamnum astronomia in tanta incertitudine versetur, si tantum etiamnum in illa desideraretur: si latitudinum si evagationum Veneris et Mercurij cognitio non certior: si Solis circuitum etiamnum inflexum et contortum putemus: si causas rerum astronomicarum non meliores habeamus, quam ab illis didicimus.

Ac nisi Tycho his quas sibi merito vindicat hypothesibus astronomiam nobis perfectiorem tradiderit, quam illi reliquere: non deprecatur, quamvis totum ille mundum complexus sit, quin meditationes suae inter Pliniana figmenta referantur, et indigna habeantur, quibus hypotheseon astronomicarum nomen tribuatur. //
299v Sed alios insuper producit Ursus authores Tychonianae hypotheseos; non solos illj Geometras, physicos, Architectos, historiae naturalis scriptores, Poetas, Grammaticos, astronomiae imperitos objicit: quos contra dissimilitudine professionis excipiat: sequitur et astronomus, astronomorum omnium post Ptolemaeum celeberrimus, nec scientia tantum sed et eventu clarus, edito Revolutionum caelestium opere, unde tabulae derivatae sunt, quibus hodie

in usus Ephemeridum receptis priorum tabulae antiquatae sunt. Age Urse, virum te praesta: rectam enim viam insistis. Dic age, fuitne Copernicus Tychonj praemonstrator suae hypotheseos? Sed cum dixeris, memineris, tecum mihi rem futuram ⌞tuisque cum verbis⌟, non cum alijs. Nam ut supra quoque innui: post vulgatas Copernicj hypotheses, non nego esse et fuisse multos, quos inter se Rothomannus profitetur: quibus copernicanae hypotheseos contemplatio digitum intendit ad eandem cum Tychone viam ingrediendam: quorum neminem sua laude privandum existimem, etsj quis in hoc, Tychone praeceptore non fuerit usus. Quamvis Tychoni cum suam quaereret hypothesin, non fuerit in animo Copernicanarum ad terrae immobilitatem traductio; utj alijs, sed Martis in oppositione solis ad Terram appropinquatio, Epicyclorum ad solem alligatio, et in summa, eadem pene quae Copernico prius fuerant proposita. Ac cum ipsi viam ad suam hypothesin intersepsisset orbium falso subjactata soliditas, commodum intervenere Cometae, qui transcursu suo illas solidorum orbium nebulas discuterent et impedimenta Tychonianae hypotheseos tollerent. Adde quod non sola in Tychonicis mutatio et transpositio ⌞Copernicanarum⌟ est, sed multa nova peculiariaque, ex apparentijs ipsis deducta, quae si Copernicus usurpare velit, suam ipse hypothesin corrigat et immutet, necesse est. ⌞quod ego quidem, qui Copernicum in capitalibus sequor, faciendum censeo.⌟ Sed tu // quid ais Urse: tu quid affers ex Copernico, quo probatum reddas id 299a quod aiebas: Hypotheses motuum caelestium a Tychone jactatas, expresse descriptas esse in relictis monumentis Nicolaj Copernicj? Si cognationem inter utrasque intercedere, si facilem transitum a Copernico ad Tychonem contendisses: meum suffragium adjungj jam vidistj. Expresse vero ubi in Copernico descriptae sunt?

KEPLER'S DEFENCE OF TYCHO AGAINST URSUS

Preface to the reader

[*Exordium*]   Three years ago Nicolaus Ursus, until his death Mathematician to His Majesty the Emperor, published a little book, *De hypothesibus astronomicis*. In it, when he had composed a history of hypotheses, he eventually came to Helisaeus Roeslin, Doctor of Medicine, a very good friend of mine;[20] and in order to refute [Roeslin's] opinion he took a certain proposition of Euclid and retailed a way of calculating the amplitudes of the planetary orbs from the requisite data by a geometrical method (a commonplace business amongst astronomers).[21] At this point he added that my opinions, which I had submitted to public judgement the year before under the title *Mysterium cosmographicum*, could also be assessed on the same basis. And seizing the opportunity, he inserted into his little book a certain letter in which two years previously I had sought his opinion of my discovery.

On this score Ursus certainly earned my gravest displeasure. I had written in the manner of a young man of twenty-three, not in the least bit seriously. For I went much too far both in praising Ursus and in subordinating myself to him. But in all confidence I expected that I would at least provoke an answer. Ursus' character was indeed reputed to be such that these weapons would be needed to overcome his silence. But if there are people to whom my intention is not evident, they will, thanks to Ursus' action, be justifiably angry with me on the grounds that with my immoderate praises of Ursus I have insulted many most learned mathematicians in Europe. And those who are better acquainted with me will also be astonished to hear me honour Ursus instead of Maestlin as my teacher, and to hear me ascribe to him my 'modicum of knowledge in mathematics', despite the fact that I had never seen him and never heard him except for what he says in his *Fundamentum astronomicum*. But everything so far can be regarded as trifling – more serious matters are to follow. For though Ursus wishes to appear to have had my reputation in view when he made my letter public while declaring it to be worthy of notice, he left us in no doubt what his purpose was. Firstly, in

---

[20] Kepler's acquaintance with his compatriot Roeslin went back at least to 1592, when Roeslin cast his horoscope (*K.g.W.*, XIII, 391). On Roeslin's life and works see C. D. Hellman, *The Comet of 1577: its Place in the History of Astronomy* (New York, 1944), 159–73.   [21] *Tractatus*, sig. Civ, v.

134

the title of his little book he snatches the discovery of new hypotheses from the noble and illustrious Dane Lord Tycho Brahe, Lord of Knudstorpp, Uraniborg, etc., and appropriates it to himself. And so he imagined that it would be shown by this letter of mine, which contains the words 'I admire your hypotheses', that there really are men who acknowledge the worthiness of his case. And in that way he involved me in a dispute that did not concern me and extorted testimony from one who was unaware. Next, at the end of the title of his little book Ursus declares his erudition, like one who sallies forth into some Olympic contest, at the same time challenging everyone, and explicitly Lord Tycho Brahe, for the 'palm and mastership in mathematics'. So he most expediently introduces into his little book, as if in the role of a judge, my letter which expresses the view that indisputably that palm is due to Ursus. Finally, in that same letter I had referred to Tycho most respectfully and had begged Ursus to communicate this discovery of mine to him in Denmark.[22] So Ursus decided to take revenge on me for my good opinion of his adversary Tycho, simply by inserting my letter, with the reference to Tycho deleted, into his little book, a book which he wrote and published for the sole purpose of damaging Tycho's reputation. In that way I who had praised Tycho in private to Ursus would be forced as a penance to praise Tycho's detractor to everyone in public. Given that Ursus saw fit to abuse my letter in so many ways, it now becomes clear why he had judged it worthy of publication. As regards the question whether my discovery deserved publication, Ursus had on his own confession already seen it as an entire book appearing in the catalogue;[23] so there was no need for publication of the letter there.

And in my letters to Lord Tycho and Roeslin I have indeed testified how seriously this publication concerned me;[24] and I also said it to Ursus himself, to his face, as soon as I came to Prague last January. However, I concealed my name for fear that if he knew that the matter was being raised with Kepler in his presence he would proceed to pick a quarrel about it. I added this as well. Given that he had seen fit to drag me, who had written to him as a disciple, against my will into the seat of judgement, it would be permissible for me to cast off the modesty of a disciple, to assume the role of judge in the scholarly dispute, leaving the legal action for defamation and civil damages to Tycho, and publicly to deliver judgement on that part of the

---

[22] Cf. Kepler's formal apology of February 1599 to Tycho (Ch. 1, pp. 18–19).

[23] I.e., in the Frankfurt Book-Fair catalogue: cf. *Tractatus*, sig. Di, r.

[24] The letter to Roeslin is not extant; it may well be the letter to which Roeslin refers in his *Historischer, Politischer und Astronomischer natürlicher Discurs von heutiger Zeit Beschaffenheit...* of 1609 (see Ch. 1, fn. 39).

issue which evidently pertained to mathematics.[25] And having thus finally revealed my name, I parted from him peaceably.

However, later many concerns, mainly domestic, prevented me from taking in hand the refutation of his work that I had intimated to Ursus. Besides, Ursus died, which caused me no little hesitation in case there should be some who would deprecate this quarrel with a dead man. But Tycho Brahe, from whose shoulders I had long ago voluntarily undertaken to remove the burden of writing this work, pressed for the fulfilment of my promise. And in itself the undertaking was an honourable one: to root out the erroneous opinions from the minds of students, and to obliterate the calumnies against the discoveries of mathematicians impressed on the minds of patrons. Nor is it an improper practice to pass judgement on whatever has come down to us of the opinions and public pronouncements of the dead (something which Ursus himself endorses on f. Cii of his book by censuring the late Ramus).[26] So for these reasons I at last undertook this long-deliberated little work with a clear conscience and out of devotion to mathematical studies. And I trust that I have so composed it that the renown and benefit of the truth will accrue to me, who have defended it, and to the reader who has acknowledged it.

CHAPTER I

[*Propositio*]   What an astronomical hypothesis is

[*Narratio*] In his description of hypotheses Ursus writes as if hypotheses had been established merely for the amusement of mankind. Accordingly, he cannot find words enough with which to speak as contemptuously as possible about this kind of invention. Nor does he stop elaborating and further elaborating his discourse until he has virtually refuted himself.

[*Partitio*]   We shall first make clear the true nature of hypotheses. Then we shall weigh up point by point Ursus' discordant and self-contradictory opinion, lest there should remain any matter on which the inexperienced reader might trust Ursus because of his reputation as a clever man.

[*Confirmatio*]   This custom of naming the opinion about the form of the universe that each of the philosophers had derived from his inspection of the heavens, and of calling it 'a hypothesis', did not arise at once with the habit of observing the heavens.[27] Just as the passage of time and the succession of philosophers

---

[25] Cf. the account offered to Maestlin in his letter of December 1598 of what he would say were he to meet Ursus (see Ch. 1, p. 13).   [26] *Tractatus*, sig. Ci, v.

[27] Kepler opens his *confirmatio* conventionally with a definition of the disputed concept. The account which follows is evidently intended to refute Ursus' equation of *hypothesis* with *aitema*, which has as one of its Aristotelian senses 'illegitimate or false assumption' (see Ch. 2, fn. 38).

established other words which we call 'words of second intention',[28] so also
they led to the adoption of this word in astronomy. The first use of the word
was by geometers. Before the birth of logic as a part of philosophy, when
they wanted to expound their demonstrations by the natural light of the
mind, they used to start their teaching from some established beginning. For
in architecture the builder is content to lay down foundations below the
ground for the future mass of the house, and he does not worry that the
ground below might shift or cave in.[29] Just so in the business of geometry
the first founders were not, like the Pyrrhonians who followed later, so obtuse
as to want to doubt everything and to lay hold on nothing upon which, as a
foundation, sure and acknowledged by all, they would wish to build the rest.
Those things that were certain and acknowledged by all they used, therefore,
to call by the special name 'axioms', that is to say, opinions which had
authority with all.[30] Certain principles of this kind were so formed that,
although they were not believed by all, they were sufficiently evident to their
proponents themselves for them not to doubt that they would show
themselves to be certain and true by some other demonstration.[31] However,
if they were to continue with the demonstration in hand, that would take
too long or be out of place. Hence opinions of this kind became by
association confused and entangled with other matters about which there
is no dispute in geometry. These were called 'postulates', being things they
would postulate the learner to concede to them at the outset.[32] (They made
use of this mainly when they wanted to represent to the imagination a line
or angle of the kind which cannot be precisely drawn.[33] They would then

[28] I.e., words which refer to other words.

[29] A source of the comparison between hypotheses and the foundations of a house
is Demosthenes' *Second Olynthiac Oration*, commonly used in the period as an
elementary Greek text.

[30] The classification of types of hypotheses which follows is partly based on Proclus,
Commentary on the First Book of Euclid's Elements, a work which Kepler had probably
first read shortly before he composed the *Apologia* (see his letter of September 1599
to Herwart, *K.g.W.*, xiv, 63). Thus the distinction between axioms and postulates
which Kepler draws here is close to the third of three possible ways of
distinguishing them given by Proclus (*Commentary*, 76 and 182–3). Though Proclus
reserves the term *hypothesis* in the strict sense for definitions, he mentions its use
in a generic sense to cover axioms, postulates and definitions (*Commentary*, 76), and
often uses it himself in this way. And like Kepler he treats the premisses of *reductio*
proofs as hypotheses (*Commentary*, 255). The account of the nature of postulates,
aitemata, echoes passages from Aristotle's *Posterior Analytics* and *Rhetorica* (see
below).

[31] Cf. Aristotle, *Posterior Analytics*, I, 10, 72b, 23–4.

[32] Cf. Aristotle, *Rhetorica*, xix, 1433b, 17–18.

[33] This evidently echoes the gloss on *hypotheses* given in Aristotle, *Posterior Analytics*,
I, 10, 76b, 39–77a, 3: 'assumptions which serve to bring the truth of the conclusion
home to the student but where truth is not required by the proof; e.g., the

say: 'Let such and such be the case'. There is another kind as well in which things which cannot be or are not true are assumed as if they were so, so that by demonstration it may become evident what would follow if they were. This kind also had a role when leaving geometry they transferred the method of demonstration to related sciences.) When they came to the process of demonstration they used to call both kinds,[34] and other things which I omit for the sake of brevity, 'hypotheses' or 'suppositions' with respect, that is, to what they wanted to demonstrate. For evident things were laid down and less evident things were built on them and revealed to the learner.

This was the origin of the word, and this was its usage amongst geometers. Those who contemplated things immediately discerned in geometrical figures and numbers, that is, in the business which is of all nature the clearest and most completely fitted to the human mind, that illumination of our mind which most especially thrives on geometrical figures, but also on all other things generally, and without which there would be nothing of which our mind could acquire knowledge. So they cultivated this method of demonstrating taken from geometry as the exemplar and reduced it to the form of that art, and indeed science,[35] that is called 'logic'. No one can deny that they frequently retained words and what we call 'terms' which they found in geometry. Thus the word 'hypothesis' is also very frequently used in Aristotle's teachings on demonstrations.[36] So since the jurisdiction of logic extends over all the sciences, the word 'hypothesis' was introduced from logic to astronomy as well; though the word was used even without derivation from logic, because of the presence of particular geometrical demonstrations in astronomy, and thus in the authentic sense.[37]

We, however, call 'a hypothesis' generically whatever is set out as certain and demonstrated for the purpose of any demonstration whatsoever. Thus

geometer's assumption that the line he draws is a foot long or that it is straight'. On this obscure passage see A. Gomez-Lobo, 'Aristotle's hypotheses and the Euclidean postulates', *Review of Metaphysics*, 30 (1977), 430–9.

[34] The two 'kinds' referred to here are axioms and postulates, *not* the two varieties of postulates introduced in the interpolated lines above.

[35] An echo of the traditional definition of 'logic' as *ars artium*, *scientia scientiarum* (see, e.g., Peter of Spain, *Summulae logicales*, Tract. I).

[36] I.e., in his *Posterior Analytics*, or perhaps more specifically Bk I of that work, which was often entitled *De demonstratione* in the period. Useful accounts of Aristotle's usage are to be found in H. Vaihinger, *Die Philosophie des Als-Ob* (Berlin, 1911), Ch. XXIX; and H. D. P. Lee, 'Geometrical method and Aristotle's account of first principles', *Classical Quarterly*, 29 (1935), 113–24. (Vaihinger's account is mostly omitted in the English translation of his work [London, 1924].)

[37] Kepler's claim that the term was extrapolated from geometry to other disciplines may have its ultimate source in Proclus' *Commentary on Plato's Timaeus*, I, 226 and 236–7.

in every syllogism the hypotheses are what we otherwise call 'propositions' or 'premisses'. But in a longer demonstration, which includes many subordinate syllogisms, the premisses of the initial syllogisms are called 'hypotheses'. Thus in astronomy suppose we demonstrate with the help of numbers and figures some fact about a star we have previously observed, from things we have seen when carefully and meticulously examining the heavens. Then in the demonstration we have set up, the above-mentioned observation constitutes a hypothesis upon which that demonstration chiefly rests. This is an authentic meaning of the word 'hypothesis'. Specifically, however, when we speak in the plural of 'astronomical hypotheses', we do so in the manner of present-day learned discourses. We thereby designate a certain totality of the views of some notable practitioner,[38] from which totality he demonstrates the entire basis of the heavenly motions. All the premisses, both physical and geometrical, that are adopted in the entire work undertaken by that astronomer are included in that totality. They are included if the practitioner has for his convenience borrowed them from elsewhere. And they are likewise included if he has already demonstrated them from observations, and now, in the reverse manner, requires that what he has demonstrated should be conceded to him by the learner as hypotheses: from which hypotheses he promises to demonstrate with syllogistic necessity both those observed positions of the stars (which had in the first place been used by him as hypotheses) and also, so he hopes, those which are about to appear in the future. I have said what a hypothesis is. I shall also say of what kind a legitimate hypothesis must be.[39]

As in every discipline, so in astronomy also, the things which we teach the reader by drawing conclusions we teach altogether seriously, not in jest. So we hold whatever there is in our conclusions to have been established as true. Besides, for the truth to be legitimately inferred the premisses of a syllogism, that is, the hypotheses, must be true. For only when both hypotheses are true in all respects and have been made to yield the conclusion by the rule of the syllogism shall we achieve our end – to reveal the truth to the reader. And if an error has crept into one or other of the premissed hypotheses, even though a true conclusion may occasionally be obtained, nonetheless, as I have already said in the first chapter of my *Mysterium*

---

[38] In rendering *artifex* by 'practitioner' I follow Duncan, *Johannes Kepler*, 13.

[39] In accordance with the dialectical rules Kepler has treated first the definition of *hypothesis*, then its genus and species. He now goes on to consider the *proprium* of hypotheses, their proper function or distinctive nature. The whole of the rest of the *confirmatio* is a riposte to Ursus' claim that it is the *proprium* of hypotheses 'to inquire into, hunt for and elicit the truth sought from feigned or false suppositions' (*Tractatus*, sig. Biv, v).

cosmographicum, this happens only by chance and not always, but only when the error in the one proposition meets another proposition, whether true or false, appropriate for eliciting the truth.[40] For example, Copernicus postulated too small a latitude for the moon. At one time the moon was seen to occult the Heart of the Scorpion. On the basis of this false hypothesis about the latitude Copernicus nevertheless inferred that at that time the star ought to have been occulted by the moon. But another error occurred, for the latitude he assumed for the star was too small by exactly the same amount as was the latitude he had postulated for the moon.[41] And just as in the proverb liars are cautioned to remember what they have said,[42] so here false hypotheses, which together yield the truth once by chance, do not in the course of a demonstration in which they have been combined with many others retain this habit of yielding the truth, but betray themselves. Thus in the end it happens that because of the linking of syllogisms in demonstrations, given one mistake an infinite number follow. So since, as I have said earlier, none of those whom we honour as authors of hypotheses would wish to run the risk of error in his conclusions, it follows that none of them would knowingly admit amongst his hypotheses anything liable to error. Indeed they worry not so much about the outcome and conclusions of demonstrations, but often more about the hypotheses they have adopted: thus almost all notable authors to date assess them on both geometrical and physical grounds and want them to be confirmed in all respects.

Why then, you may ask, given that they all demonstrate the same motions of the heavens, is there nevertheless so great a diversity of hypotheses? Certainly one should not suppose that it happens because what is true customarily follows from what is false. For, as I said before, in the long and tortuous course of demonstrations through diverse syllogisms of the kind which customarily occur in astronomy, it can scarcely ever happen, and indeed no example occurs to me, that starting out from a posited unsound

---

[40] See the passage from his *Mysterium cosmographicum* quoted in Ch. 6, p. 215.

[41] *De revolutionibus*, IV, 27, where Copernicus appeals to his observation of a lunar occultation of Aldebaran to confirm his account of lunar parallaxes. I am indebted to Professor E. Rosen for identification of this reference. As Professor Rosen pointed out, Kepler confuses the star with 26° in Scorpio, which Copernicus reports as rising at an angle of $59\frac{1}{2}°$ at the time of observation. Part of the passage in which Copernicus reports this, his first recorded observation, is underlined in Kepler's copy, *Nicolaus Copernicus de revolutionibus orbium coelestium: Facsimile* (Leipzig and New York, 1965), f. 128r–v.

[42] Ursus had quoted the proverb both in his attack on Rothmann in the preface to the *Tractatus* (sig. Aiii, v) and in his notes on the letter in which Tycho had charged him with theft and plagiarism (sig. Fii, r). It has several classical sources, including Quintilian, *Institutio oratoria*, IV, 2, 91, and is in Erasmus' *Adagia*.

hypothesis there should follow a result altogether sound and fitting to the motions of the heavens or a thing of the kind the demonstrator wants. But the same result is not always in fact obtained from different hypotheses whenever someone relatively inexperienced thinks it is. The results which follow from Copernicus' hypotheses were demonstrated with respect to the numbers by Maginus from different hypotheses, which were to be as far as possible in agreement with the Ptolemaic ones.[43] Do both of them really prove the same result? Far from it! Copernicus wanted also to demonstrate both the cause and the necessity of the greater proximity of the superior planets to the earth when they are in opposition to the sun, as well as wanting to represent their future motions by numbers. Maginus left out that which Copernicus found wanting in the Ptolemaic hypotheses and to which he attached no little importance in his own suppositions, and only imitated Copernicus' numbers. In Copernicus it follows that Mars suffers a greater parallax than the sun. In Maginus this does not follow at all. I here forbear to mention many further matters. And though some disparate astronomical hypotheses may provide exactly the same results in astronomy, as Rothmann claimed in his letters to Lord Tycho of his own mutation of the Copernican system, nevertheless there is often a difference between the conclusions because of some physical consideration.[44] Thus even were Tycho to have elicited from his hypotheses exactly the same numbers as Copernicus has (which are, however, defective), there would still be this difference in intention between Tycho's and Copernicus' demonstrations: that as well as wanting to predict the future motions, Tycho wants also to avoid postulating the immensity of the fixed stars and some other things that Copernicus admitted into his hypothesis. So when the conclusion is altered different hypotheses must be set up. But a thoughtless man who pays attention only to the numbers will think that the same result follows from different hypotheses and indeed that the truth can follow from falsehoods.

If a man assesses everything according to this precept, I doubt indeed whether he will come across any hypothesis, whether simple or complex, which will not turn out to have a conclusion peculiar to it and separate and different from all the others. Even if the conclusions of two hypotheses coincide in the geometrical realm, each hypothesis will have its own peculiar

---

[43] By 'numbers' Kepler means here, as elsewhere, 'apparent celestial coordinates'. The reference is to Giovanni Magini, *Novae coelestium orbium theoricae congruentes cum observationibus N. Copernici* (Venice, 1589), which presents a geocentric system in which the individual planetary models (*theoricae*) are designed to yield results in accordance with the Prutenic Tables.

[44] Rothmann to Tycho, October 1587 and October 1588 (*Epist. astron.*, 81–91 and 120–32 [*T.B.o.o.*, VI, 110–19 and 149–61]).

corollary in the physical realm. But practitioners are not always in the habit of taking account of that diversity in physical matters, and they themselves very often confine their own thinking within the bounds of geometry or astronomy and tackle the question of equipollence of hypotheses within one particular science, ignoring the diverse outcomes which dissolve and destroy the vaunted equipollence when one takes account of related sciences. Given that this is so, it is proper that we too should adapt our argument and response to their manner of speaking.[45]

Well then, isn't it necessary for one of the two hypotheses about the primary motion (to take an example) to be false – either the one which says the earth is moved within the heavens, or the one which holds that the heavens are turned about the earth? Certainly if contradictory propositions cannot both be true at once, these two will not both be true at once: rather one of them will be altogether false. But is not the same conclusion about the primary motion demonstrated by both means? Do not the same emergences of the signs of the zodiac follow, the same days, the same risings and settings of the stars, the same features of the nights? Does what is true follow equally from what is false and what is true then? Far from it! For the occurrences listed above, and a thousand others, happen neither because of the motion of the heavens, nor because of the motion of the earth, insofar as it is a motion of the heaven or of the earth. Rather, they happen insofar as there occurs a degree of separation between the earth and the heaven along a path which is regularly curved with respect to the path of the sun, by whichever of the two bodies that separation is brought about.[46] So the above-mentioned things are demonstrated from two hypotheses insofar as they fall under a single genus, not insofar as they differ. Since, therefore, they are one for the purpose of the demonstration, for the purpose of the demonstration they certainly are not contradictory propositions. And even though a physical contradiction inheres in them, that is still entirely irrelevant to the demonstration. So this example certainly does not show that what is true can follow both from what is true and from what is false. But perhaps the best-known example, that of a concentric with an epicycle which is equivalent to an eccentric, will press us harder. Is it not true that if a planet, or whatever it is that propels a planet, looks to the centre of the eccentric in its motion, it will not at the same time look to the centre of the epicycle? So that if

---

[45] Throughout the following paragraph Kepler argues *ex concessione*, as if it were true that the sole purpose of hypotheses was to save the phenomena.

[46] Cf. the passage from *Mysterium cosmographicum* quoted in Ch. 6, p. 216, in which the two hypotheses are said to have in common the assumption that there is 'a separation of motions between the earth and the heaven'.

one astronomer affirms the former and another the latter, is it not necessary that one or the other utters a falsehood, despite the fact that both demonstrate the same thing, namely, that although the motion of the planet remains steady, it spends a longer time in one half of the circle? So doesn't what is true follow accordingly from what is false as well as from what is true? Far from it! I offer the same in reply as above. That from which it is inferred that the delays are longer in one of the two halves of the circle is said by both, both by the one who posits the eccentric and by the one who posits the concentric with epicycle. For each of the two introduces this generic proposition: that the larger portion of the planetary circle lies in that half of the circle [of the zodiac]. Since this genuine and commensurate middle term of the demonstration inheres generically in each hypothesis, those things which each hypothesis says specifically, whether or not they conflict, will remain quite irrelevant to the demonstration.[47] And we certainly do not ascribe eyes and the human exercise of discernment to the planets, so that they can mark a point here or there with compasses;[48] and practitioners introduce those specific views I have mentioned as conceits of their own rather than for the sake of explaining nature. Therefore neither the former nor the latter supposition is worthy of the title 'astronomical hypothesis', but rather what both have in common, namely, that it is assumed and posited that there is a definite and measured part of the circle that the planet traverses, which lies in one half of the circle of the zodiac. This, I say finally, is a proper hypothesis from which the length of the time spent in that half of the circle may be demonstrated. Therefore, to conclude, each consequence in astronomy is derived only from one single middle term and presupposes a hypothesis which is of single form even if it is differentiated insofar as it is considered apart from that demonstration. And conversely, every hypothesis whatsoever, if we examine it minutely, yields some consequence which is entirely its own and is not shared with any other hypothesis. Nor can it happen in astronomy that what was originally founded on a false hypothesis should be true in every respect. So this is the distinctive nature of hypotheses (if we are to depict the form of legitimate hypotheses), to be true in every respect. And it is not right for an astronomer knowingly to assume false or ingeniously contrived hypotheses in order to demonstrate from them the celestial motions.

Now let us consider briefly the absurd opinions of Ursus. [*Refutatio*]

Firstly, he declares [a body of] astronomical hypotheses to be a fictitious

---

[47] For an explanation of the Aristotelian terminology used here see Ch. 6, p. 217; and N. Jardine, 'The forging of modern realism: Clavius and Kepler against the sceptics', *Studies in the History and Philosophy of Science*, 10 (1979), Appendix.

[48] On the meaning of this see Ch. 6, p. 219.

portrayal of an imaginary, not a true and authentic, form of the world-system.[49] In these words he openly denies that there exists a hypothesis which is not false. He confirms this opinion of his a little later when he says that hypotheses are nothing but certain 'fabrications'; and further on he says that they would not be hypotheses if they were true; and again, that 'it is the mark of hypotheses to yield knowledge of what is true from what is false'.[50] So on this view the earth will neither be moved nor stand still. For Ursus will acknowledge both as hypotheses. And there are lots of such examples. But this travesty of a definition has already been sufficiently refuted and it is not worth wasting words on it. The whole string of errors arises from a perverse understanding of the original sense of *hupotithesthai*, which seems to Ursus to be the same as 'to feign'.[51] But we said above what it meant amongst geometers.

Next, he says that hypotheses are contrived for the purpose of observing the celestial motions.[52] This view is diametrically opposed to the truth. We can observe the celestial motions even if we hold absolutely no opinion about any disposition of the heavens. The fact is that observation of the celestial motions guides astronomers to the formation of hypotheses in the right way, and not the other way round. In all acquisition of knowledge it happens that, starting out from those things which impinge on the senses, we are carried by the operation of the mind to higher things which cannot be grasped by any sharpness of the senses. The same thing happens also in the business of astronomy, in which we first of all perceive with our eyes the various positions of the planets at different times, and reasoning then imposes itself on these observations and leads the mind to recognition of the form of the universe. And the portrayal of this form of the universe thus derived from observations is afterwards called '[a body of] astronomical hypotheses'.

Thirdly, he would have astronomers work at least on this: that from the hypotheses they have adopted they should both calculate the celestial motions and tell us what sort of motions will appear at future times.[53] But what he says here is not without qualification true. For even though what he mentions is the primary task of an astronomer, the astronomer ought not to be excluded from the community of philosphers who inquire into the nature of things.

[49] *Tractatus*, sig. Biv, v, 1–7. [50] *Tractatus*, sig. Biv, v, 9–10, 18–19, 23–4.

[51] *Etumon*, here rendered 'original sense', has the connotations both of 'true sense' and of 'original or pristine sense': H. G. Liddell and R. Scott, *A Greek–English Lexicon*, 9th edn (Oxford, 1940), 704.

[52] *Tractatus*, sig. Biv, v, 4–6. The objection depends on the ambiguity of *observare*, which can mean both 'observe' and 'keep track of'. Kepler is unfair in adopting the former reading: see Ch. 2, fn. 37.

[53] *Tractatus*, sig. Biv, v, 16–18; also *Tractatus*, sig. Ci, r, 11–14.

One who predicts as accurately as possible the movements and positions of the stars performs the task of the astronomer well. But one who, in addition to this, also employs true opinions about the form of the universe performs it better and is held worthy of greater praise. The former, indeed, draws conclusions that are true as far as what is observed is concerned; the latter not only does justice in his conclusions to what is seen, but also, as was explained above, in drawing conclusions embraces the inmost form of nature. And I am here not just describing some ideal of a good astronomer; there are plenty of examples. To predict the motions of the planets Ptolemy did not have to consider the order of the planetary spheres, and yet he certainly did so diligently.[54] To predict and expound the heavenly motions by a method of calculation Copernicus and Lord Tycho after him did not have to ask why it is that the planets at their evening risings become nearest to the earth.[55] For they could have produced the same results even by using the Ptolemaic form of the heavens with the dimensions corrected. But love of finding out about nature made astronomers take up the exploration of this part of physics on astronomical grounds; nor are they wrong to rejoice in the conformity of their hypotheses to the nature of things. What about Copernicus? He censured a certain non-uniformity of the motion of the epicycle in Ptolemy, not on the grounds that this motion conflicts with what is seen and with our experience or observation of the stars, but because it is in conflict with the nature of things; so from this, he declared, he derived his motive for parting company with Ptolemy.[56] And O wretched Aristotle! Wretched is that philosophy of his in the part said to be most excellent,[57] if astronomers are such faithless witnesses on philosophical questions. For when he had reached the point at which he had to expound the particular number and order of the heavenly intelligences, he referred us to the astronomers Eudoxus and Calippus. And although he was aware that they disagreed in certain respects, he wanted us to revere and admire both of them because of the careful study they had carried out and the teaching

---

[54] *Almagest*, IX, 7.

[55] Cf. Rheticus, *Narratio prima* (Danzig, 1540); transl. by E. Rosen, *Three Copernican Treatises*, 3rd edn, revised (New York, 1971), 137. *Planetae acronuctioi*, 'planets rising in the evening', are planets in opposition to the sun.

[56] In his preface to *De revolutionibus* this is one of the reasons Copernicus gives for abandoning 'the traditional opinion of the astronomers'.

[57] Astronomy was variously said in the period to be preeminent amongst the branches of natural philosophy by virtue of the perfection and divinity of its subject matter, the heavenly bodies, and by virtue of the certainty it derives from its mathematical character (cf. Aristotle, *Metaphysics*, XII, 8, where astronomy is treated as the branch of mathematics closest to philosophy by virtue of the eternal character of its objects; and Ptolemy, *Almagest*, I, 1, where the excellence of astronomy is related to the divinity of the heavenly bodies and its certainty to its mathematical character).

about the motions which they had imparted. But he wanted us to learn from whoever should expound the matter most accurately.[58] So Aristotle's opinion of the work of those ancient astronomers and their most excellent undertaking (and those who followed after them were no less praiseworthy) was different from that of Ursus.

Fourthly, he appears to offer in support of his opinion the argument that it is not necessary for hypotheses to be true, for the most crass absurdities are present in all hypotheses to date.[59] Truly he proves what is in question by means of something else far more questionable.[60] To be sure, if we have to consider as false and absurd whatever appears to some single group of men to be so, there will be nothing in the whole of natural philosophy that will not have to be regarded as the most crass absurdity. For diverse are the opinions of men and the notions of the learned. Copernicus did not think it absurd that the earth is moved and that the immensity of the fixed stars is so great. Tycho does not think it absurd that the planets follow a moving sun wherever it goes as their centre and yet do not depart from their own motions (on these matters see the dispute between Tycho and Rothmann in the *Astronomical Letters*).[61] To the followers of Ptolemy that inestimable swiftness of the outermost heavenly spheres is not absurd. These things are relatively remote both from our senses and likewise from the grasp of our minds, and are disputed about on various grounds. But those things which cannot be true at the same time are most crassly absurd; from which it follows that not more than one form of hypothesis can be true. And a philosopher certainly has the capacity to judge which of them is true and which false. Yet it is certain that not all are crassly absurd, as it seems to Ursus. To me, indeed, it seems that Copernicus said nothing false, and William Gilbert the Englishman appears to have made good what was lacking in my arguments on Copernicus' behalf through his admirable skill and his industry in collecting observations in the study of magnets.[62]

58 *Metaphysics*, XII, 8, 1073b, 9 *et seq.*   59 *Tractatus*, sig. Biv, v, 15–16.

60 Ursus is here convicted of a form of *petitio principii*; cf. *Prior Analytics*, II, 16, 65a, 13–15: 'What is as uncertain as the question to be answered cannot be a principle of demonstration.'

61 Rothmann to Tycho, October 1588: 'Who would ever believe that the centre of the major epicycle of the sun is endowed with such an efficacy that it can drag after it all the planets and, moreover, draw them out of their spheres and return them to them again' (*Epist. astron.*, 129 [*T.B.o.o.*, VI, 158]). Kepler had himself protested at the 'monstrous motions' of the planets required by Tycho's hypotheses in his letter of February 1599 to Maestlin (*K.g.W.*, XIII, 291).

62 This is the first mention in Kepler's writings of Gilbert's *De magnete* (London, 1600) and his first allusion to a specifically magnetic account of the motive power of the sun.

Fifthly, he despairs of true hypotheses in astronomy, holding everything to be uncertain in the Pyrrhonian manner. No one denies that there are still some flaws in even the best-constructed astronomy, and hence in the hypotheses also. That is why even today there is so much labour expended on repair. Let no one say that therefore all the rest is uncertain. Ptolemy discovered by means of astronomy that the sphere of the fixed stars is furthest away, Saturn, Jupiter and Mars follow in order, the sun is nearer than they, and the moon nearest of all. These things are certainly true and consonant with the form of the universe. He also postulated too great an ascent and descent of the moon. Regiomontanus and Copernicus discovered that from this hypothesis there follows something false and not in accord with the apparent magnitude of the moon.[63] So they said that the moon does not make such great jumps up and down. And who doubts that this is the way things are in the heavens? And who, I ask, but a melancholic and despairing person would have doubts about the ratios of the diameters of the earth, sun and moon? Thus today there is practically no one who would doubt what is common to the Copernican and Tychonic hypotheses, namely, that the sun is the centre of motion of the five planets, and that this is the way things are in the heavens themselves – though in the meantime there is doubt from all sides about the motion or stability of the sun. Given that with the help of astronomy so many things have already been established in the realm of physical knowledge, things which deserve our trust from now on and which are truly so, Ursus' despair is groundless.

Sixthly, he says that numerous forms of hypotheses can be fabricated. The construction of hypotheses is such light work for him![64] But it is laborious and troublesome, as those who have tried it will be able to testify. This opinion has some support from the fact that Ursus holds hypotheses to be only fictitious, without need of truth. But in fact it would not even be easy for Ursus to construct many kinds of the sort of hypotheses he describes. For though he does not care whether they are true or not, nonetheless he would have them yield a method of calculation of the motions. And this does not happen with no matter what fictitious hypothesis one comes upon.

---

[63] Regiomontanus, *Epytoma in Almagestum Ptolemei* (Venice, 1496), prop. 22; Copernicus, *De revolutionibus*, IV, 2. Cf. Rheticus, *Narratio prima* (transl. by Rosen, *Three Copernican Treatises*, 133–4).

[64] Cf.:

...because I now hold, grasp and comprehend with my mind better hypotheses [than Tycho's] and indeed various species and forms of hypotheses; and moreover, it is not to be denied or gainsaid that daily my faculty of thinking things up with my small wit yields new and different ones besides. So I have begun to hold my own like the other ones almost in contempt.

Tractatus, sig. Aiv, r, 10–14; see also *Tractatus*, sig. Biv, v, 8.

Seventhly, he says that it suffices for hypotheses to agree with a method of calculation of the celestial motions, even if not with the motions themselves. What an amazing doctrine – if Ursus' meaning can really be judged from these words.[65] What then, Ursus, is a method of calculation? If a method of calculation, that is, a method of reckoning the positions of the stars, is sound, then clearly it is in agreement with the motions and is based on a hypothesis which is likewise in agreement with the motions. But if some method of calculation does not represent the motions, the hypotheses from which it was derived are likewise found wanting. So if we do not care whether a method of calculation agrees with the motions, clearly there is nothing to stop us from concocting numerous hypotheses (which Ursus kept claiming was easy for him) and setting up just as many kinds of method of calculation that are, to be sure, defective and not in agreement with the heavens. Congratulations, Ursus, on your prolific inventiveness! But perhaps you would make the point that you mean only the hypotheses employed in astronomical expositions, and that the use of these consists solely in the teacher showing the pupil through a hypothetical portrayal the reasons for observing particular rules in calculating the celestial motions. This use, whilst it is not foreign to hypotheses, is secondary and of less importance. For we first depict the nature of things in hypotheses, then we construct out of them a method of calculation – that is, we demonstrate the motions. Finally, retracing our steps we expound to the learner the true rules of the method of calculation.

Eighthly, he thinks nothing of using hypotheses which contradict the Sacred Scriptures. But he imagines that he has demonstrated from the Scriptures the hypotheses he himself throws up.[66] But if being false is the mark of hypotheses, he himself will have no hypotheses, because he maintains those he himself throws up to be supported by the infallible authority of the Scriptures. Thus he is forgetful of his own words, and, indeed, even more so in that though he has made falsity the mark of hypotheses, he later says more reasonably that astronomers are at liberty to contrive either true or false hypotheses.[67]

Ninthly, he claims that what he has said about the falsity of astronomical hypotheses is made use of in most other branches of learning.[68] It is indeed

---

[65] *Tractatus*, sig. Biv, v, 16–18. Kepler is unfair in his reading, for Ursus clearly has in mind calculation of the *apparent* celestial motions. *Calculus* is a difficult word to render: Ursus often uses it in the sense of 'calculation' or 'reckoning', but Kepler appears to use it consistently to mean 'method of calculation or reckoning'.

[66] *Tractatus*, sig. Biv, v, 20–4; contrasted with *Tractatus*, sig. Div, v, 9–12.

[67] *Tractatus*, sig. Biv, v, 18–19 and 24–6.

[68] *Tractatus*, sig. Biv, v, 28–9.

made use of, but erroneously and by the inexperienced. Nor does what is good and true follow in due course, save very rarely and by chance, but rather what is false and pernicious. Thus in medicine suppose that someone is suffering from a quartan fever and an inexperienced doctor forms the false hypothesis that the cause of the fever is sluggishness of the blood. He would no more succeed in making the sick man better by venesection than Ursus would derive the motions of the heavens from a false hypothesis.[69] But let us follow Ursus through his examples. For instance, he cites as an example the rules of algebra in which the number sought is posited as equal to a unity cossically designated, although in fact it is not unity;[70] yet from this false posit one attains knowledge of the truth. But, Ursus, the example is neither appropriate nor properly explained. When the number we seek is unknown, we give it an unknown name, and with that name we follow the prescribed rules until that name is revealed to us by the procedure. So in positing that a unity is designated we say nothing false. For by the very process of giving a name to a unity there is indicated a definite number which is always equal to the number sought. So here there is no false posit. But nothing similar happens in astronomical hypotheses. For whatever hypothesis one maintains, one maintains it as known and established. In what is known as the 'rule of the false', which Ursus adduces as another example, he can rejoice in nothing but the name.[71] For the number posited is called 'false' not because the truth follows from it – if that were to happen it would not be so called – but because it leads to what is also false, as is readily apparent to one who considers the matter. A number is sought which when handled by the prescribed rules will come to some particular value. Whatever number comes to mind is chosen and handled by the rules laid down, and if it comes to the value hoped for it is the number sought; but if it comes to less it is false. The same thing is tried out on another [number] and the two are compared, as are their deviations, and from the inspection of these things the true [number] is eventually elicited. I shall now say, using the same form

---

[69] A matter of some concern to Kepler, who was suffering from intermittent fever at the time. Venesection is rejected as a treatment by Fernel, whom Kepler cites later: Jean Fernel, *Universa medicina* (Paris, 1567), 364–5.

[70] 'I' was used for the unknown in an equation; on this passage see Ch. 6, p. 221.

[71] *Regula falsi* is the method for solution of equations by successive approximation. An account of the use of the method in the period is given by A. J. E. M. Smeur, 'The *rule of the false* applied to the quadratic equation in three sixteenth century arithmetics', *Archives internationales d'histoire des sciences*, 28 (1976), 66–101. Ursus' example may derive from Ramus: 'For just as Aristotle taught that in logic what is true can be derived from false premisses, so in arithmetic the discovery of this arithmetical rule [the *regula falsi*] reveals how what is true may be derived from what is false' (*Scholae mathematicae* [Basel, 1569], 145).

of words, what in astronomical hypotheses is similar to this. Hypotheses are sought which will correspond to the motions of the heavens. The Alphonsine hypotheses are found to err, likewise the Copernican. But the skilful practitioner, having made a comparison of the two and having removed the sources of error, establishes some third [hypothesis] which avoids all error in the prediction of the motions of the heavens and in that way corrects both hypotheses. What is there here to support Ursus? In his example we are offered one way, and that both indirect and artless,[72] in which astronomers are in the habit of searching for hypotheses. But he wanted to show how what is true and fitting to the motions of the heavens is demonstrated from an already established false hypothesis. In the 'rule of the false', on the other hand, a false number leads to something likewise false. Ursus insists that what is true is obtained from false astronomical hypotheses.

Tenthly, after Ursus has brought his opinion to our attention, he defends it with a laughable testimony. For why should someone unknown sway me? For that reason Ursus himself endorses that testimony with his own testimony, saying that the author is very learned.[73] I will help the floundering Ursus. In case you don't know, the author of the preface was Andreas Osiander, as testifies the hand of Hieronymus Schreiber of Nuremberg to be seen in my copy (that Schreiber to whom several prefaces of Schöner are addressed).[74] Osiander, however, though he does very closely support Ursus' view, is both more reasonable in the matter and expresses himself more clearly. It is clear what had motivated him. He was master of Nuremberg when Copernicus' work was being published there.[75] So since he admired

---

[72] As opposed to *entechnos*, 'artful' or 'skilled'; *atechnos* also has in the context of oratorical presentation the connotation of 'unmethodical' or 'unsystematic'. Cf. C. Chevalley's note on Kepler's use of the term in *Paralipomènes à Vitellion (1604)* (Paris, 1980), 469.

[73] Ursus had cited the author as 'extremely learned, but unknown' (*Tractatus*, sig. Ci, r).

[74] A copy of *De revolutionibus*, presented by its printer Johannes Petrejus to Schreiber, in which 'Andreas Osiander' is written in above 'Ad Lectorem', is in the library of the Karl Marx Universität, Leipzig. It was identified as Kepler's copy by L. A. Birkenmajer, *Mikołaj Kopernik* (Cracow, 1900), 649. For details of Schreiber's life and his connections with Osiander and Rheticus see List, 'Marginalien zum Handexemplar', 443–60. The point of the reference to Schöner's prefaces may well be to establish the credentials as a witness of the little-known Schreiber: Johannes Schöner, 'The German Ptolemy', to whom Rheticus' *Narratio prima* was addressed, was well known as mathematician, astronomer, geographer and belle-lettrist. E. Zinner, *Entstehung und Ausbreitung der coppernicanische Lehre* (Erlangen, 1943), Appendix E, lists over a dozen copies of the first edition of *De revolutionibus* in which a sixteenth-century hand has identified Osiander as author of the preface.

[75] As Preacher at the Church of St Lorenz Osiander had greatly influenced the governing City Council.

mathematical learning, he could not but admire so magnificent a work. But when he saw the most paradoxical dogma about the motion of the earth he decided that the reader's desire for novelties ought to be restrained in this way. But, Osiander, what desperation drove you to go so far as to say that nothing certain about the true form of the universe can be derived from astronomy? Most experienced as you were in these matters, how could you still be in doubt about the most evident things? Were you still uncertain about the ratio of the sizes of the earth and the sun reported by astronomers and a thousand things of this kind? If this art knows absolutely nothing of the causes of the heavenly motions, because you believe only what you see, what is to become of medicine in which no doctor ever perceived the inwardly hidden cause of a disease except from the external bodily signs and symptoms which impinge on the senses, just as from the visible positions of the stars the astronomer infers the form of their motion? But let us return to Ursus, to whom I say this about his witness: he is not exempt from all contrary testimony. He means to persuade the reader in that preface that Copernicus was not of the opinion that the earth really is moved. Copernicus himself refutes him right at the outset. Osiander advised Funck the chronologer, if that man is to be trusted, that in the Grace of Hezekiah the sun did not move back, but only the shadow of the pointer of the sundial – as if these two things could be separated from each other, when it is agreed from the reading of Sacred Scripture that the miracle was observed universally.[76] More serious absurdities that also took possession of this clever man I leave for others to discover.[77] Small wonder, then, if he went astray in mathematical science

[76] In the second edition of his *Chronologia* (Wittenberg, 1570), *Commentaria*, sig. B6, r, Johann Funck, Osiander's son-in-law, had written:

Moreover, I certainly ought not to omit to say, lest the pious reader recall it, that in the first edition I asserted that the whole machinery of the heavens was moved backwards by God in the pledge of health to the prostrate Hezekiah. And indeed I then vehemently and plausibly elaborated the point. For, by I know not what oracle, I thought this to be quite certain. But some years later, since there was an author of the same opinion as me who was about to discuss this point publicly in the School at Königsberg, I was reproved by that most excellent man Master A. Osiander. And I saw that I had not inspected the passage in the Sacred Scriptures carefully, nor examined it as thoroughly as it deserved. For the text clearly speaks of the shadow on the sundial of Ahaz, not of the sun in the sky (Kings 4: 20; Isaiah 38). So if anyone has followed that opinion of mine, I would advise him to give it up in favour of a saner interpretation of the Sacred Scriptures.

[77] This innuendo may refer to the unorthodoxy of Osiander's Lutheran faith, which culminated in his quarrel with Luther and Melanchthon and his retreat to East Prussia; other possible targets are his cabalistic methods of biblical exegesis and his Paracelsian sympathies. On Osiander's intellectual career see E. Hirsch, *Die Theologie des Andreas Osiander und ihre geschichtlichen Voraussetzungen* (Göttingen, 1919); B. Wrightsman, 'Andreas Osiander's contribution to the Copernican achievement', in R. S. Westman, ed., *The Copernican Achievement* (Berkeley, 1975), 213–43; and K. Goldammer, 'Paracelsus, Osiander and theological Paracelsism in

also, which he admired but did not practise. I cite Aristotle against him, whose far sounder judgement of astronomers I mentioned above. It should not be passed over and should be remarked on that Osiander seems to have written insincerely and not from personal conviction, employing, indeed, the Ciceronian strategies of *De republica*.[78] For since he feared for Copernicus at the hands of the herd of philosophers, in case readers should be scared away from his most excellent work by the absurdity of the hypotheses, he wanted to put these things of his in a preface in order, so to speak, to sugar the pill of Copernicus' work. In fact, Copernicus, steadfast in his Stoic resolution, believed that he should declare his opinions openly even to the detriment of this science. Osiander, following rather the dictates of expediency, preferred in his preface to conceal Copernicus' most true and earnest opinion. Osiander's plan has clearly succeeded for the past sixty years. At last the time has come to expose this dissimulation, as it certainly seems to me, from Osiander's private correspondence.[79] After Copernicus had written to Osiander on 1 July 1540, Osiander replied to him on 20 April 1541, saying, amongst other things, the following.

I have always been of the opinion that hypotheses are not articles of faith, but bases for calculation, so that even if they are false it does not matter provided they yield the phenomena of the motions exactly. For who could make us surer that the unequal motion of the sun is due to an epicycle than that it is due to an eccentric, if we follow Ptolemy's hypotheses, since it could happen in either way. So it would seem to be a good idea for you to say something on this matter in the preface. For thus you would pacify the peripatetics and the theologians whom you fear to be about to raise objections.

the middle of the sixteenth century', in A. G. Debus, ed., *Science, Medicine and Society in the Renaissance: Essays to Honour Walter Pagel*, 1 (London, 1972), 105–20.

[78] The *In Somnium Scipionis*, preserved with Macrobius' commentary, was the only substantial part of Cicero's *De republica* known at the time. Kepler's reference to the precedent set by Cicero may derive from Macrobius' account of the role of Scipio's dream:

He [Plato in the *Republic*] realised that in order to implant this fondness for justice in an individual nothing was quite so efficient as the assurance that one's enjoyments did not terminate with death...Cicero proved to be equally judicious and clever in following this method of treatment: after giving the palm of justice in all matters concerning the welfare of the state, he revealed, at the very end of his work, the sacred abodes of immortal souls and the secrets of the heavens and pointed out the place to which the souls of those who had served the republic prudently, justly, courageously and temperately must proceed, or rather, must return.

Transl. by W. H. Stahl, *Macrobius' Commentary on the Dream of Scipio* (New York, 1952).

[79] These letters are known only from Kepler's quotations, which may well be excerpts.

On the same day, in fact, Osiander wrote to Rheticus as follows.

The peripatetics and theologians will be easily placated if they hear that there can be diverse hypotheses about the same apparent motion and that they are not advanced as being certainly so, but rather as governing the calculation of apparent and composite motion as expediently as possible; that it could happen that someone else should think up other hypotheses; that one man may think up appropriate constructions and another more appropriate ones, both giving rise to the same appearance of motion; and that anyone is free and, moreover, is to be congratulated if he thinks up more expedient ones. Thus, called away from severity in condemnation and summoned to the pleasures of inquiry, they will at first be more reasonable and then, seeking in vain, will go over to the author's opinion.[80]

Besides the dissimulation employed in the preface, which I have mentioned, there appears in these words of the author a clear equivocation on the word 'hypothesis'. For some hypotheses are said above to be like small change, scarcely worthy of the name, others to be proper and truly astronomical hypotheses. As when, in Osiander's example, we determine and set out the part of the planetary circle which lies in one half of the circle of the zodiac, it is worthy of the name 'hypothesis' and cannot be changed or altogether false. But when we set up a method of calculating the ascent and descent of a planet in those unequal parts, it can often be achieved in several ways, and so, indeed, we set up further hypotheses for the purpose of implementing that prior and astronomical hypothesis, one man by placing the centre of the planetary circle away from the centre of the world, another by inserting an epicycle into the concentric. But these are not in themselves astronomical, but rather geometrical hypotheses. Thus if some astronomer says that the path of the moon describes an oval shape, it is an astronomical hypothesis. But when he shows by what circles a drawing of this sort of oval can be constructed, he uses geometrical hypotheses. And when Ptolemy said that the motions of the planets slow down at the apogee and are speeded up at the perigee, he set up an astronomical hypothesis; but when he introduced the equant, he did so as a geometer for the sake of calculation, so that it could be worked out how much that motion of the planet slows down at any given moment in time. No one is to be taken seriously who does not acknowledge and grasp this diversity in hypotheses. Indeed,

---

[80] In translating the letter to Rheticus I have been guided by the versions of K. H. Burmeister, *Georg Joachim Rhetikus, 1514–1574: eine Bio-Bibliographie*, III (Wiesbaden, 1968), 25–6; and H. Hugonnard-Roche *et al.*, '*Georgii Joachimi Rhetici Narratio Prima*: édition critique, traduction française et commentaire', *Studia Copernicana*, XX (1982), 208.

Osiander makes a mockery of Copernicus, or rather of his readers, by his equivocation, transferring those things that are true of geometrical hypotheses to astronomical hypotheses, even though the natures of the two sorts are very different.

Altogether there are three things in astronomy: geometrical hypotheses; astronomical hypotheses; and the apparent motions of the stars themselves. Accordingly, there are two distinct tasks for an astronomer: one, which truly pertains to astronomy, is to set up astronomical hypotheses such that the apparent motions will follow from them; the other, which pertains to geometry, is to set up geometrical hypotheses of whatever kind (for there can often be various kinds in geometry) such that from them those prior astronomical hypotheses, that is, the true motions of the planets unadulterated by the distortion of the sense of sight, both follow and can be worked out.

Here the very tenor of my discussion invites me to add Patricius to Osiander and to say something in reply to his attempt to deride all astronomers and reject all astronomical hypotheses.[81]

He is infuriated with astronomers for attempting to construct the apparent motions of the planets from various circles and solid orbs and to impute to the nature of things those circles, those hypotheses, figments of their own minds. He himself asserts this about the planets. They move amongst the fixed stars in the liquid ether exactly as they appear to, free from the fetters of solid orbs, which do not exist. And exactly as appears to our eyes they truly describe with non-uniform motions spirals and lines variously contorted back and forth, never exactly repeating themselves. Nor ought we to be surprised by this diversity, because the planets are in truth animals with the faculty of reason – a view he supports with the authority of pagan philosophy – and it would not have been impossible for the divine omnipotence to create creatures with enough wisdom to perform those ordained motions until the end of the world.[82]

---

[81] In his letter to Kepler of December 1599 Tycho had protested at the misrepresentation of his views in Patrizi's *Nova de universis philosophia* (first publn, Ferrara, 1591; Venice, 1593):

He, moreover, explicitly attributes [my hypothesis] to me, but he so perverts it, and through ignorance ill comprehends it and, assuming and postulating things of his own contrary to mine, so pollutes and abuses it that many of the things there set out in his account I do not recognise as my own. And I shall protest about this in print.

K.g.W., XIV, 92. This is the first known published reference to Tycho's system.

[82] This account is based on *Pancosmia*, f. 105v–107r. On the peculiar pagination of the work see *Catalogue générale des livres imprimés de la bibliothèque nationale*, Vol. 131 (Paris, 1935), 393.

First, I shall readily concede this to him: there are no solid orbs. On this point I both happily subscribe to Tycho's reasons and have my own besides. Patricius, indeed, does Tycho an absurd injustice when he says that, enthralled by the view that there are solid orbs, he contrived a new form of the universe.[83] Tycho rejects this view in all his writings and by the very nature of his hypotheses. Nor do I deny that the planetary circuits are ruled by the highest reason. But I shall say this as well. If God the Creator had wanted the planets to execute spirals and irregular circuits of the sort that appear to us when we look, he could easily have brought it about. In fact, on this point Patricius is already at odds with astronomers who philosophise properly, when asked the following question. What would God have preferred: that the planets should fly around in composite, ever changing and irregular curved motions of this kind; or that each should describe a circle, uniform and as regular as possible, clearly distinct from the sort of confused motion which meets our eyes? Reasons can certainly be found for our deluded sight perceiving those regular motions otherwise than they have in reality been ordained. On this point there is not one of those who philosophise soberly who would not affirm the latter opinion and altogether reject the former, and who would not, moreover, at the same time most heartily thank astronomers should they discover the reason for the deception and distinguish the true motions of the planets from those which are accidental and derived from the phantasms of the sense of sight, and claim for their circuits simplicity and well-ordered regularity. Patricius, indeed, philosophises in such a way that one who paid heed to him could not move a foot without granting that a miracle occurred. For where, walking through the fields, he encounters hedges and things near to his path, he would believe, on the testimony of the sense of sight, that distant mountains are really following him. He would not use the judgement of reason, nor even consult the common sense.[84] Instead he would trust the sense of sight and say that the mountains are

---

[83] *Pancosmia*, f. 106r, col. 2, and *Pancosmia*, f. 92v, col. 1, where Copernicus' and Tycho's hypotheses are said to have originated from the assumption that stars are fixed into the heaven like knots or nails in a plank.

[84] Not common sense in its modern connotation, but the common sense of Aristotelian faculty psychology. Here is a typical account of it:

Aristotle did not indeed postulate a sixth sense but rather a common sense, the judge of composite forms to which the individual senses send their images. It comprehends motion and change of shape; is midway between partaking and not partaking of reason; participates in memory and the mind; and is attributed to beasts as well insofar as they have a spark of intelligence. Objects of the common sense are shape, evident to sight and touch, also distance between things, evident to sight and hearing, and, moreover, the motion, size and number of all things.

Stobaeus, *Eclogarum libri duo* (Antwerp, 1575), 151.

moving no less than he himself. Likewise, if he sets out in a carriage from Strasbourg or Vienna, he will swear by Jove that the towers, which in those cities are very high, are collapsing on him. Once a certain man of my acquaintance saw through a glass window some oxen grazing in the meadows by the town. And because a spider was hanging in front of the window, he, confusing it with the oxen as a result of his line of sight, and not perceiving the intervening space, was the butt of our ridicule, exclaiming: 'A miracle, a many-legged ox!' Clearly Patricius' wisdom on this point is no greater when he maintains that the planets run their courses amongst the fixed stars, just as it appears to us.

Let this, then, be said to Patricius. To set down in books the apparent paths of the planets and the record of their motions is especially the task of the practical and mechanical part of astronomy; to discover their true and genuine paths is (despite Osiander's and Patricius' futile protests) the task of contemplative astronomy; while to say by what circles and lines correct images of those true motions may be depicted on paper is the concern of the inferior tribunal of geometers. He who has learned to distinguish these things will easily disengage himself from the deluded seekers after abstract forms who quite heedlessly despise matter (the one and only thing after God),[85] and from their importunate sophisms. And I have wanted to say this in answer to Patricius because he supports Ursus and Osiander in their errors on certain points. I return to Ursus.

Eleventhly, in order utterly to discredit the astronomy of our age, Ursus praises that of the ancients in the manner of querulous old men or whatever other hopeless and melancholic type of persons there may be.[86] But, indeed, he uses an argument which is, in my opinion, of the best form.[87] Most of the arts, he declares, were once more perfect, medicine, music, oratory – why not astronomy as well? There are indeed those who imagine Thessalian medicine to have been something more excellent than is apparent from the surviving records.[88] There are, on the other hand, authors of no less note, Fernelius and others, who with greater plausibility credit our age with more.[89] Ursus bases his judgement about music on an effect which is not in the power

---

[85] On the significance of this see Ch. 7, p. 236.

[86] *Tractatus*, sig. Ci, v, 25–7.   [87] I.e., an induction.

[88] Cf. *Tractatus*, sig. Ci, v, 27–31.

[89] 'But indeed, whether because the art itself has now shone forth more clearly or because there is a certain change in all things, anyone who considers every point and weighs it up according to its importance will adhere to a more reasonable and serious judgement, and settling for the more plausible position will think that scarcely anything of the ancients can suffice for this age' (prefatory letter to Henry II of France, *Universa medicina*, sig. Ei, r; first publ. in *De naturali parte medicinae* [Paris, 1542]).

either of that art or of its practitioner.[90] I believe that many admire the music of the ancients simply because they do not know what it was like. If God grants me life, and I have time, I shall show them in my *Harmonics* how far the music of the ancients lagged behind ours.[91] And by the neighing of his ass, which he used to prefer to the best singing, that Scythian, equally an ancient, readily refutes Ursus' insistence that music today no longer has as much power over men's minds.[92] Good singers were quite rarely heard; as in comedies and tragedies, so in every kind of singing the tone was modulated; the sense of the words was adapted to the occasion, the characters and the place with the delivery clear, articulate and colloquial; the Greeks were a highly susceptible people; and the performers on the lute and lyre and Lesbian Sappho would have moved even a stone. All of which we can imitate today, together with anything else of the ancients that was admirable; and we sometimes do so. For our skill was quite unknown to the ancients. But more of this on another occasion. As for oratory, I will not dispute the point.[93] The main reason lies, however, in the nature and habits of the [Greek] race, which, being most susceptible, yielded easily to any emotion; and being most avid for glory, did not hesitate to do that which, whilst to them it was praiseworthy, ought today to be detested by everyone, committing suicide, for example. Concerning astronomy the matter is indeed clear unlike anything above. Let all the skilled mathematicians of Europe come forward: Ursus will not find even one who will concede to him that the astronomy of the ancients was more perfect. So if anyone should offer what Ursus has promised, namely, the restoration of the ancient astronomy, I leave it to the thoughtful reader to judge how valuable it would be.[94] These were the main

[90] Cf. *Tractatus*, sig. Ci, v, 31–8.

[91] In Bk 3, Ch. 15 of the *Harmonice* Kepler presents an account of the way in which music moves the emotions, a theory which implies a rejection of the traditional view to which Ursus subscribes according to which the various modes *per se* have specific emotive effects. According to Kepler only polyphonic music has the full range of affective powers, and in Ch. 16 he argues that the ancients did not have polyphony. In his letter to Herwart of September 1599 Kepler relates this difference between ancient and modern music to the ancient adoption of Pythagorean intonation, in which thirds and sixths are dissonant, as opposed to the just intonation of the moderns, in which they are consonant. He goes on to express his wonder at Ursus' claim of superiority for ancient music (*K.g.W.*, XIV, 72). On these aspects of Kepler's music theory see M. Dickreiter, *Der Musiktheoretiker Johannes Kepler* (Berne, 1973), 148–87; and D. P. Walker, *Studies in Musical Science in the Late Renaissance* (Leiden, 1978), 36–7.

[92] Kepler's direct source may well be Rheticus' *Narratio prima*: 'Surely the Scythian had no such soul who preferred hearing a horse's neighing to a talented musician' (transl. by Rosen, *Three Copernican Treatises*, 196). A classical source is Plutarch, *Moralia*, 174F, 334F and 1095 F.

[93] Cf. *Tractatus*, sig. Ci, v, 38–Cii, r, 2.   [94] Cf. *Tractatus*, sig. Cii, r, 2–8.

points that were to be dealt with in this first chapter on the nature of hypotheses.

CHAPTER 2

[*Propositio*]   On the history of hypotheses

[*Narratio*] Ursus employs a clever device in composing a history of hypotheses. For starting out with genuine accounts, he gains the reader's confidence in his knowledge of historical matters. Having obtained this, he slips into his lies without at all breaking the thread of his narrative, and foists them on the reader as historical facts.

[*Confirmatio*]   Thales was the first of this series of astronomers in Greece (though Diogenes Laërtius records Linus and Musaeus as the most ancient of all, the latter as the discoverer of the sphere, the former as the discoverer of the motions of the sun and moon).[95] And if the things Pliny and Herodotus say about his prediction of an eclipse are true, then quite certainly he must have had accurate knowledge in this branch of astronomy.[96] And Laërtius, indeed, asserts that he was the first to have explored the secrets of astrology, to have ascertained the course of the sun from solstice to solstice, to have established the length of the year as three hundred and sixty-five days, to have written about solstices and equinoxes, to have measured the seasons, to have compared the magnitudes of the sun and moon and to have predicted an eclipse of the sun.[97] But he cannot have obtained a method for predicting eclipses on his own, since it is not the work of one lifetime to trace enough anomalies of the moon for prediction of an eclipse of the sun. Shortly before that very eclipse and the war recorded by Herodotus that was waged at that time, there was certainly peace between the Lydians and the Babylonians; and Thales in fact lived at Miletus, close to Lydia. So it is reasonable to believe that Thales learned the art of computing eclipses from the Babylonians. Besides, Laërtius tells us that he went to Egypt.[98] And according to the same author Thales wrote of himself to Pherecydes as follows: 'I have travelled to Egypt there to confer with the priests and astronomers.'[99] Laërtius intimates who these astronomers were in his [life of] *Pythagoras*, of whom he says that he travelled to that same country, Egypt, to converse with the Chaldaeans.[100] For Egypt was under the yoke of the Babylonians. Indeed,

[95] Diogenes Laërtius, *De clarorum philosophorum vitis*, I, 3–4.
[96] Pliny, *Naturalis historia*, II, 53; Herodotus, *Historiae*, I, 74.
[97] Laërtius, *De clarorum philosophorum vitis*, I, 23–4.
[98] *De clarorum philosophorum vitis*, I, 27.
[99] *De clarorum philosophorum vitis*, I, 43.
[100] *De clarorum philosophorum vitis*, VIII, 43.

at the end of the above-mentioned letter Thales added: 'But I, who write nothing, travel around all over Greece and Asia.'[101] So whether he went to Chaldaea by land or to Egypt by sea, in either place he would have met plenty of Chaldaeans and could have introduced astronomy from there into Greece. Moreover, the fact that almost all the Greek states used the lunar year suggests that the Babylonians used to be consulted by the Greeks about the motion of the moon. On this point see Aristotle's words (*De caelo*, Book 2, Chapter 12): 'A similar account of the other planets as well is given by the Egyptians and Babylonians' (where and when he concealed from us) 'who have observed their behaviour over a period of very many years, and from whom we learn many reliable things about each of them.'[102] On this point it pleased me to compare these conjectures of mine with those of Ursus, for I am not one to carp at what is well said. But now Ursus immediately goes wrong, supposing that the hypotheses which he says belong to physics were only established shortly after the time of Aristotle, and that there is no way of knowing of what sort were the hypotheses of astronomers in the time of Thales and Pythagoras. On this point he does not appear to have read Pliny, Book 2, Chapter 22. For these currently used hypotheses of physics, as Ursus calls them, that is, that ordering of the spheres which Ptolemy adopted, were familiar to Pythagoras.[103] Listen, reader, to Pliny's words: 'But occasionally Pythagoras, drawing on musical doctrine, calls the distance between the earth and the moon a whole tone, between the moon and Mercury half that interval, between Mercury and Venus almost the same, between Venus and the sun a tone and a half, between the sun and Mars a tone, between Mars and Jupiter half a tone, between Jupiter and Saturn half a tone, and from there to the zodiac a tone and a half.'[104] Thus, reader, you learn the order of the spheres from Pliny. So either Pliny is an unreliable witness or Pythagoras held the same opinion as Ptolemy about the order of the spheres, and did so two hundred years before the time which Ursus assigns to the origin of these hypotheses. And this is what Ptolemy, Book 9, Chapter 1, intimates when he says that the orbs of Venus and Mercury were located beneath that of the sun by earlier authors (by Pythagoras), but above that of the sun by certain more recent authors (by Heraclitus, Plato, Eudoxus and

---

[101] *De clarorum philosophorum vitis*, I, 44.   [102] *De caelo*, 292a, 7–9.

[103] Cf. *Tractatus*, sig. Cii, v, 7–37. At first sight Kepler misunderstands Ursus, who definitely identifies the 'physical' hypotheses as a system of concentric orbs and says that they were superseded by the hypotheses of Aristarchus, Apollonius and Ptolemy. Kepler, however, tacitly assumes that Ptolemy's planetary models were embedded in a system of concentric solid orbs as was customary in the *Theorica* literature of his period (see Ch. 7, pp. 230–1).

[104] Pliny, *Naturalis historia*, II, 84.

Aristotle). And finally he follows the more ancient opinion as the more probable, denying that anything certain is to be had from astronomy on this point. And Patricius attributes this same opinion to Zoroaster the Persian, who came many generations before Pythagoras, on the strength of these words of his: 'Six he formed them' (or else, 'He set up six circles'), 'hurling the seventh fire of the sun into the middle.'[105] If the sun is in the middle of the planets, then Venus and Mercury will be beneath the sun (for the same Zoroaster surely located the earth in that middle which is the centre). And since Pythagoras derived this wisdom from Zabracus the Assyrian, a fellow-countryman of Zoroaster (so Patricius relates, from Proclus I think),[106] it is likely that he derived this order from the same source. Nevertheless those who were later called 'Pythagoreans' abandoned their master's ordering. For, as will shortly be shown, long before the time of Aristotle they held the same opinion about the form of the world as did Aristarchus of Samos fifty years after Aristotle (under Ptolemy Philadelphus), and in our time Copernicus. And by his discoveries Copernicus made it possible for us to understand these reports of Aristotle and Plutarch about the Pythagoreans and of Archimedes about Aristarchus.[107] Here another of Ursus' errors comes to light, for he cites Aristarchus as the originator of the Copernican hypotheses, whose origin is far more ancient.[108]

Philolaus the Pythagorean lived halfway between the time of Pythagoras and that of Aristotle, and was the first to make public the Pythagorean doctrines, since before that they were not known to any but the Pythagoreans themselves. Not only does Laërtius say obscurely of him that he thought that the earth 'is moved beside the first circle' (interpret as 'under the zodiac'), but Plutarch also [attributes this view to him] much more clearly.[109] And since Copernicus himself had cited Plutarch's words in the preface of his work,[110] it is a great wonder that Ursus did not read them, Ursus who boasts so much and so often of his reading of Copernicus and unjustly reproaches others for not having done so.[111] According to Plutarch, Philolaus

---

105 Patrizi, *Nova de universis philosophia*. The first wording is in *Pancosmia*, f. 76r, col. 2, 31–2; *Pancosmia*, f. 105v, col. 2, 20–1; *Zoroastris oracula*, f. 9v, col. 1, 32–3. The second wording is in *Panaugia*, f. 16v, col. 1, 47–8; *Pancosmia*, f. 97r, col. 2, 12–13; *Pancosmia*, f. 108r, col. 1, 16–17.

106 *Pancosmia*, f. 90[91]v, col. 2; *Pancosmia*, f. 112r, col. 2. Patrizi's source is Porphyry, *Vita Pythagorae*, 24–5, not Proclus.

107 I.e., the reports which Kepler is about to analyse.

108 *Tractatus*, sig. Cii, v, 25–7.

109 Laërtius, *De clarorum philosophorum vitis*, VIII, 85; Pseudo-Plutarch (Aëtius), *Placita philosophorum*, III, 1.

110 Copernicus cites the passage in Greek in the version of *Plutarchi opuscula* (Venice, 1509).   111 Cf. *Tractatus*, sig. Ci, r, 15–16.

held that 'the earth is moved in a circle about the fire' (interpret as 'the sun') 'under the zodiac in the manner of the sun and moon'. Moreover, in Book 2 of *De caelo*, Chapter 13, Aristotle [says] of the Pythagoreans generally: 'Those who live in Italy and are called "Pythagoreans" declare that there is fire in the middle.'[112] (Here interpret [the fire] as the sun, for both the Ionian Thales and the Italian Pythagoras, the founders of philosophy, believed that the mysteries of philosophy should be concealed from ordinary people.) 'The earth, indeed, which is one of the stars' (planets) 'creates day and night by its motion in which it is carried about the middle.' Here Aristotle did not properly understand their opinion, or else, indeed, there is a deficiency [in the text] which I restore as follows: 'The earth creates the year by the motion in which it is carried about the centre, and it creates days and nights by the motion in which it is carried around on its axis.' He continues: 'Besides they imagine another contrary earth which they call "counterearth".' They designated the moon by this term, because it is carried around with the earth in the same orb and because of the many properties it shares with the earth, as students of optics tell us. Aristotle, indeed, who did not realise this, having prejudged the absurdity of their opinion, declares: 'They do not seek reasons and causes, so as to derive and produce the appearances, but attempt violently to accommodate and reconcile those phenomena to certain preconceived opinions and reasonings of their own.' But it is clear from their arguments, which Aristotle expounds as follows, that for the Pythagoreans the fire is the same as the sun. 'It would appear to many others also that the earth should not be assigned the middle place, if they wanted to derive some probability from rational considerations rather than from appearances. For that place

---

[112] In the long quotation which follows (*De caelo*, II, 13, 293a, 20–293b, 16; *De caelo*, II, 13, 293b, 18–21, 26–30) Kepler originally copied the elegant but inaccurate translation of Joachim Périon, *Tripartitae philosophiae opera omnia absolutissima...* (Basel, 1563), III, col. 179. He then deleted the material from 'They do not seek reasons' to the end and substituted a more accurate translation. This translation does not match, even approximately, any of those listed in F. E. Cranz, *A Bibliography of Aristotle Editions, 1501–1600* (Baden-Baden, 1971). Since Cranz' list often omits texts to be found in commentaries this is inconclusive. Kepler's translation also fails to match the text in any of the commentaries listed by C. H. Lohr, 'Renaissance Latin Aristotle commentaries', *Studies in the Renaissance*, 21 (1974), and then in successive instalments in *Renaissance Quarterly*. That this is Kepler's own translation is strongly suggested by his having made a German translation of this chapter of *De caelo* in 1613: see F. Rossmann, *Nikolaus Kopernicus, erster Entwurf seines Weltsystem sowie eine Auseinandersetzung Johannes Keplers mit Aristoteles* (Munich, 1948), 56–77. On Kepler's Greek scholarship see E. Rosen, 'Kepler's mastery of Greek', in E. P. Mahoney, ed., *Philosophy and Humanism: Renaissance Essays in Honour of Paul Oskar Kristeller* (Leiden, 1976), 310–19.

should be assigned to what is most noble and excellent, and fire is indeed more noble than earth.' (Here, if you note that there is no body in the whole universe nobler than the sun itself, you will increase the force of the argument remarkably and will understand what that fire is.) 'Limits are, indeed, nobler than what lies between them.' (For they are, as it were, certain bounds which other quantities do not contain; a bound is surely more noble than that of which it is a bound.) 'The outer boundary' (of the universe) 'and the middle' (the centre) 'are limits' (of all places that lie in between, in which the planets and the sun are commonly located; so the centre and the outermost surface are the most excellent places in the universe). 'Arguing from these premisses, they think that it is not the earth but the fire' (the sun) 'that is located at the middle of the sphere. The Pythagoreans think, for this very same reason, that the most important thing in the universe should be guarded above all others; the middle (which, indeed, they call Jupiter's watch-tower) is such a point; so the fire' (the sun) 'should occupy that region' (the centre), 'as if what is in the strict sense called "the middle" is at once the middle of magnitude and of matter and of nature.' Aristotle tries to refute these arguments in the following words: 'But as in animals the middle of the body and of the animal are not identical, so the same thing ought all the more to be realised about the heavens as a whole. So there is no need for them to fear for the universe or to summon a guard to its centre. They should rather inquire what kind of thing that middle is, and where it is. For that precious middle is a starting point. But the middle of place seems more similar to an end than a starting point. A middle, indeed, is bounded, whereas an end bounds. For that which encompasses and bounds is nobler than that which is bounded, for the latter is constituted by matter, the former by essence.' There is an equivocation on the word 'middle', which for the Pythagoreans is the centre, but is taken by Aristotle in some other sense. And if this equivocation is eliminated, Aristotle's argument coincides with the one opposed to it, and supports the Pythagoreans.[113] For the middle of which they speak plays the role of an end here in Aristotle. These are the points he makes against the Pythagoreans, and it is not difficult to answer them, though this is not the place. But from what he says their opinion that the place

---

[113] Kepler evidently has in mind the following refutation of Aristotle. Concede, with Aristotle, that what is nobler should be assigned to the sun, that 'starting points' (*principia*) are nobler than 'ends' (*termini*), and that 'the middle' is a starting point. Only by refusing to take 'the middle' of the world at face value as the centre of place can Aristotle avoid the conclusion that the sun is at the centre of place. This is confirmed by Kepler's refutation of Aristotle's argument in his *Epitome astronomiae Copernicanae*, where he denies that the centre of place and 'the heart' (*cor*) can be distinguished in the universe as they are in animals (*K.g.W.*, VII, 262).

of the earth is away from the centre is made especially clear. And now let us hear them also on the subject of its motion. 'But those who deny that it rests located at the centre of the universe assert that it revolves about the centre; and not only it, but also that which is opposite it' (quite clearly the moon), 'as we just said above.' And after he has interpolated the opinions of others: 'For because the earth is not the centre' (a point) 'but is distant from it by a whole hemisphere' (interpret this too by concession to those who set the earth in the centre of the universe), 'they see nothing to stop the phenomena appearing to us who live away from the centre' (not just by one semidiameter of the earth, but by as much as the distance between the earth and the sun) 'just as if the earth were in the middle. For nothing noticeable now results when we are half a diameter from the centre' (that is, nothing if the earth is left in the middle). This is what Aristotle says about the Pythagoreans, and it is altogether more and more explicit than what Archimedes says about Aristarchus.[114] Had Ursus read this, he would have attributed the invention of the Copernican hypotheses not to Aristarchus, who came after Aristotle, but to Philolaus and the Pythagoreans long before Aristotle.

However, not all the subsequent Pythagoreans held Philolaus' conception of the universe. For, according to Aristotle, Eudoxus of Cnidus, a contemporary of Plato and Aristotle and himself a Pythagorean, followed the view of his master Pythagoras more closely, in that like him he held that the sun is moved.[115] Ursus takes the liberty of saying that it is not known what hypotheses Eudoxus and Calippus, who corrected him, employed. And yet after dealing with some other matters, he says, with the same rashness in making assertions, that concentrics were derived by them from the Pythagoreans and combined with eccentrics by Hipparchus.[116] But we shall see from Aristotle what can be gathered about Eudoxus' opinion. And it seems especially worthwhile to assess this passage very carefully, since it has been either ignored or badly handled by interpreters.

However, we are faced with a double difficulty in that Aristotle too, a man quite skilled in mathematical matters as is apparent in many passages, either misunderstood Eudoxus' meaning or expressed it somewhat obscurely; and the numbers [in the text] are clearly inconsistent with the words. So let us do what we can, let us use conjectures to strip away the masks from the words and by correction of the errors in the numbers bring the words into

[114] Archimedes, *Arenarius (De arenae numero)*, I, 4–7.
[115] *Metaphysics*, XII, 8, 1073b, 17–18, quoted by Kepler below. Kepler presumably considers Eudoxus as a Pythagorean because he is placed in the *successio* of followers of Pythagoras in Laërtius, *De clarorum philosophorum vitis*, VIII.
[116] *Tractatus*, sig. Cii, r, 25–8.

agreement with them. So then, in Book 11 of the *Metaphysics* Aristotle turns to the contemplation of simple substances devoid of matter, and he is raised to that height of sublimity by his careful disquisition on motion and the prime cause of motions. When he has inferred the prime mover from the simple, uniform and eternal motion which we call 'the prime motion', he undertakes in Chapter 8 to investigate the number of the secondary intelligences subordinate to the prime intelligence. And he especially insists that individual intelligences govern individual motions and that nothing should be introduced into the universe that lacks a function or is superfluous. Indeed, he bases the number of movable celestial bodies on the variety of the motions which appear in the heavens. But he seeks to find out from astronomers what the motions themselves are and how many of them are perceived; hence his words:[117] 'And it is evident even to those who are moderately conversant with this subject that there are, indeed, many motions of each of them that is carried. For the individual planets are carried by more than one motion. On the question of their number, we shall now in the interest of knowledge relate what certain mathematicians say, so that we can grasp some definite number. As for the rest, in part it should be investigated by us and in part it should be asked of those who study this subject whether they, who deal with these matters, believe anything different from what we have said. And then each of them should be admired and thanks should be given them, but we should only have confidence in those who have given the more accurate and careful account.' Let Osiander and Ursus, and whoever else would belittle the ideas astronomers have about the nature of things, take heed of this preface.[118] This is what follows: 'Eudoxus assigned the movement of the sun and of the moon to three spheres each, of which the first is that of the fixed stars.' In textbooks of astronomy this [first sphere] is commonly portrayed as embracing the whole system of each planet, and it forms a concentric; it is moved on the poles of the equator from east to west and revolved together with [the system of the planet] in twenty-four hours.

[117] In the series of quotations from *Metaphysics*, XII, 8, 1073b, 8–1074a, 14, which follows, Kepler originally copied Périon's translation, *Aristotelis Stagiritae opera omnia* (Basel, 1563), III, col. 1142 (in which, as in most editions of the period, Book Λ is given as Book XII, Books A and α being lumped together as Book I). He then deleted all but the first quotation, substituting the version to be found in Fracastoro, *Homocentrica sive de stellis* (Venice, 1538), f. 61v–63r. Below Kepler implicitly attributes this translation to Fracastoro by suggesting, incorrectly, that Fracastoro had amended the text. Though Fracastoro does not claim the translation as his own, the attribution may well be correct, for the version does not match any of the translations listed in Cranz, *A Bibliography of Aristotle Editions*.

[118] Cf. *Apologia*, pp. 145–6, where this passage is cited against Ursus. Kepler evidently takes Aristotle to mean by 'the more accurate account' 'the one which better portrays the true nature of the universe'.

Sometimes another function is assigned to it as well, that of carrying the nodes, whence it gets its usual name; but this is not in accord with Eudoxus' views. 'The second [sphere] is carried in that circle which runs through the middle of the zodiac.' Eudoxus seems to have so disposed this orb in the case of the moon that its poles are fixed in the first orb, the orb of daily revolution, in the solstitial colure inclined to its poles by the amount of the obliquity of the ecliptic. Its motion, therefore, takes place beneath the first [orb] from east to west, that is, in antecedence, and it returns to the same point of the first [orb] – that is, the same point of the zodiac – every nineteen years. In that period it carries the nodes round. I shall go on to say, in [my account of] the third [circle], of what sort the motion of this [second] circle is in the model for the sun, insofar as it can be determined.[119] 'The third sphere is carried in a circle inclined to the latitude of the zodiac.' I interpret this as follows. The third [orb] is inclined to the equator of the second. So it is this one which carries the moon herself. Its poles are fixed in the second, being inclined to them by the amount of the maximum lunar latitude. It is moved in consequence, that is, from west to east, and is returned to the same point of the first [orb] – the same point of the zodiac – in about a month. So it comes about that the orb which is posited because of the motion in longitude provides for the latitude, and that which should produce the latitude brings about the motion in longitude.

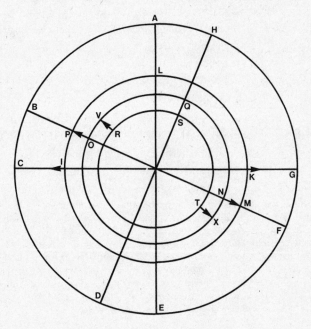

119 My translation of this sentence is conjectural.

Let ADG be the sphere of the fixed stars, ADG itself being its solstitial colure, AE its equator, and CG its axis and poles. Let HD be the zodiac and BF its axis and poles. IKL will be the first orb in the model for the moon according to Eudoxus, with I and K as its poles. NOQ will be the second [orb], with its poles N and O fixed in the first at points P and M. RST will be the third [orb], with its poles R and T fixed in the second at points V and X. Here, Ursus, you have the concentrics of Eudoxus in the model for the moon.[120] Now go and extol the ancient astronomy! This model is consistent only with concentrics, clearly because Eudoxus did not know of the remaining inequalities of the moon.

But I have long asked myself why Eudoxus had introduced the circle of latitude in his model for the sun as well. Admittedly Lord Tycho Brahe, who detects some change of latitude in the fixed stars, does this.[121] But this most subtle matter can be detected only after many ages. And even if it could have been detected in Eudoxus' time it would not have been known to Eudoxus, who overlooked many more evident matters. Fracastoro's interpretation is surely out of place. 'So this [sphere]', he says, 'was turned through the zodiacal longitude by the second [sphere]. Moreover, by its own motion it brought it about that the heavenly body did not remain always at the same maximum declination, but rather the moon appeared at one time in one sign of the zodiac and at another in another, and the sun declined more or less from the celestial equator.'[122] But what need is there to introduce a new circle to vary the declination of the sun, since this function is proper to the third [circle] which lies under the zodiac? And his words are not worthy of an astronomer. 'The moon', he says, 'was perceived now in one sign of the zodiac and now in another, according to the point of maximum declination from the equator.' But, Fracastoro, the moon never declines further from the equator than when in Cancer and Capricorn, just like the sun. So some time ago I almost came to the conclusion that Eudoxus had perhaps wanted to demonstrate the motion of the fixed stars using the second circle NOQ, rotated in another manner and on other poles. For the following words occur: 'But the [circle] in which the moon is carried is deflected to a greater latitude than that in which the sun is moved.' All difference between the models for the sun and moon is removed by these words with this one

---

[120] Cf. *Tractatus*, sig. Cii, r, 21–3.

[121] In his *Progymnasmata* (Prague, 1602), in whose preparation for publication Kepler was at the time involved, Tycho infers a small variation in the plane of the ecliptic from collation of his own observations with those of Timocharis, Hipparchus and Ptolemy (*T.B.o.o.*, II, 233–47); cf. Kepler's account of the matter in *Epitome astronomiae Copernicanae* (*K.g.W.*, VII, 410 and 527–30).

[122] *Homocentrica*, f. 61r, 37–61v, 2.

exception, that OV (or NX) is smaller in the case of the sun than in the case of the moon. But finally it occurred to me that Eudoxus' intention was not so much astronomical as philosophical. For, indeed, the astronomer Eudoxus is accounted by Laërtius as one of the philosophers and, moreover, one of the Pythagoreans.[123] So Eudoxus' plan was as follows. Since the sun and moon appear to be very different from the others both in their brightness and in their motions, he wanted to establish equality in all respects between the two luminaries and to attribute latitude to the sun on the pattern of the moon, but undetectable by the senses, or rather, undetectably variable. And I think this is implied by those words according to which he attributed a greater latitude to the moon than to the sun. For this equality was perhaps altogether in conformity with his philosophical designs. And it was no doubt for that reason, amongst others, that he also placed the sun next above the moon, the lowest. And Aristotle derived this, like other things, from him. Now we have settled the point, let the diagram above become the model for the sun. The position and motion of the first circle ILK remains the same; the second NOQ almost completely lacks motion; the third RST carries the sun and is returned in the space of a year to the same point of the first – or rather, of the zodiac. Nevertheless, I would not deny the possibility that in the case of the sun, when Eudoxus portrayed the solstitial colure (in which he showed that the sun has its maximum declination), Aristotle took that colure to be a mobile sphere. Nor would I reject that conjecture which is provided by Fracastoro's words: 'In fact, at its maximum declination the sun appeared to be sometimes further away from and sometimes nearer to the equator.'[124] For Eudoxus could either have supposed there to be or really have perceived some variation in the solstitial declination, either as a result of error and variety in his solstitial observations or through collation of what was handed down from the ancients concerning the obliquity of the zodiac (for they were diligent in studying this because of the beginnings of the Olympiads),[125] and could have set the sun in a circle of latitude to bring this about. As for Iphitus, it is generally known that the basis of his institution of the Olympiads was astronomical. And Iphitus' use of this kind of period makes it quite clear that the nature of the solstices and likewise of the maximum declination of the sun was known to him, since he started the Olympiad from the full moon that was closest to the solstice. It is, indeed,

[123] Cf. fn. 115.

[124] *Homocentrica*, f. 60r, 30–1.

[125] The Olympiads, four-year periods dated from the first Olympic games said to have been instituted in 776 BC by Iphitus of Elea, provide the chronological framework for the later Greek and Roman historians.

four hundred years from him to Eudoxus. See also what we retailed above
from Laërtius about Thales, who was midway between Iphitus and Eudoxus,
and about Linus, the most ancient of all.

There follows: '[He assigned the movement] of each of the other planets
to four spheres, of which the first and second are the same, both that of the
fixed stars carrying them all, and the one placed beneath it whose motion,
common to all, is through the middle of the zodiac.'[126] [That is to say,] all
the planets have one orb which corresponds to the sphere of the fixed stars
in motion and position, and another which moves under the zodiac in the
opposite direction to the first and obliquely. This [orb] was NOQ, above,
moving on poles N and O. But here his purpose is different. For it is in
agreement with Eudoxus' intentions that this [orb] should not now carry the
nodes, but is turned in the case of Saturn in thirty years, in the case of Jupiter
in twelve years, in the case of Mars in twenty-three months and in the case
of Venus and Mercury in twelve months.

Let ABC again be a part of the sphere of the fixed stars and of the solstitial

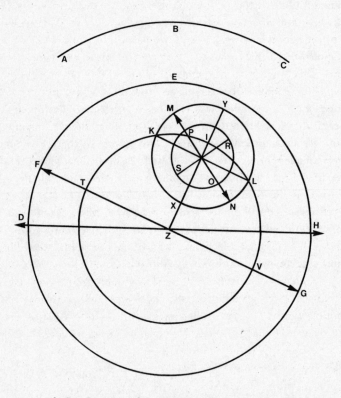

---

[126] Continuation of the quotation from *Metaphysics*, XII, 8.

colure in that [sphere]. To this corresponds DEG, the first Eudoxan orb which the five planets have. It is turned on the poles and axis DH coinciding with the axis of the prime motion. Let TIV be the second orb, whose poles T and V are fixed in the points F and G of the first; and let DF be the obliquity of the ecliptic. On these poles the planet is moved in consequence, in the case of Saturn returning to the same point of the first [orb] – that is, of the zodiac – in thirty years. Let KML be the third orb, whose poles K and L are in the surface of the second TIV, which goes through the middle of the signs. It is moved on these poles in consequence in the upper part along the path of the great circle YIX, and thus is moved as the nodes require. But if YXZ, the part of the line representing the ecliptic, were set out as a circle, the colure TIV would be a line in the diagram as would KYL, and the poles K and L would coincide with the centre [of KML]. Let PRO be the fourth, whose poles P and O are fixed in the third at points M and N, on which it is likewise moved in consequence along the path of its great circle RS. The planet is carried around this path in the time required for its anomaly of commutation. This is the philosophical opinion about the five planets of Eudoxus the philosopher, who did not know of the other inequality with regard to the orb of a planet.

'The poles of the third sphere of all [the planets] are in that circle which runs through the middle of the signs; and the turning of the fourth is in a circle which is inclined to the middle of this circle. Besides, the poles of the third sphere, which are different in each of the others, are the same in Venus and Mercury.' To tackle this truly Gordian knot by a reliable method and procedure we must proceed as follows. In the case of the moon Eudoxus perceived three motions: first, the diurnal; second, the menstrual, in the opposite direction to the first and obliquely under the zodiac; and third, the one which changes its digression from the ecliptic. Corresponding to this number of motions he postulated as many orbs. In the case of the other planets he observed four quite evident motions: first, diurnal, under the equator in antecedence; second, under the zodiac in consequence; third, in latitude from the ecliptic; and fourth, that according to which the planets alone among the luminaries have the peculiarity of performing retrogradations. Aristotle's own words bear sufficient witness to the fact that the first and second motions are common to all. And though it does not follow from Aristotle's words, reason persuades us that Eudoxus was not unaware that the third too, [the motion] of latitude, is common to the moon and the five planets. For even if he did not employ more care in observation than sufficed for detection of the retrogradations, he could not have failed to notice several times a year the conjunctions of the planets, in which one is either further north or further south than the other, and thus differs from the other in latitude and is not carried precisely through the middle of the signs. And since the same words appear in the description of the fourth [orb] as in the description of the third, above – 'the middle circle of the fourth is oblique to the middle [circle] of the third' – there is all the less reason why we should doubt that one of these [orbs] is assigned to the latitude of a planet, the other to its retrogradation. So on this point we should pay no attention to Fracastoro, who asserts that the latitudes of the planets were unknown to Eudoxus and that retrogradation alone was produced by both these circles, forcibly accommodating Eudoxus to his own *Homocentrica* and to the kind of motions clearly invented by himself most ingeniously. He does, in fact, achieve his intention of making the planets retrograde and moving and stationary by postulating two orbs having the earth at their centre, of which one has its poles in the ecliptic and its motion in the direction of the poles of the ecliptic, and the other has its poles carried round by the former at a distance from its poles sufficient for the retrogradation.[127] And on these poles this interior orb is carried round in the same time in the opposite direction to the exterior one. But if this were so, a planet would remain always in the ecliptic, or always at the same latitude outside the ecliptic, running from side to side of the sun with a motion of libration; or, if the period of revolution were inconstant, the planet would sometimes have a latitude equal in amount to its retrogradation. And even if we concede that Eudoxus was unaware of the latitude of the planets, nevertheless it is improbable that he employed hypotheses of so recondite a sort for the purpose of bringing about retrogradation, since the notion of an epicycle offers itself spontaneously, so to speak, to one who considers retrogradation, and in establishing disciplines we follow above all whatever is simplest and easiest, depending on necessity and not, as Fracastoro does, on ingenuity. And it is not only implausible but clearly altogether unreasonable to suppose that Eudoxus thought that this motion of retrogradation takes place in a concentric through a motion of libration, so that the planet maintains the same distance from the earth. For the huge magnitude of Mars at its evening rising is evident even to the eyes of ordinary people, but when approaching the sun it is surpassed by stars of the third magnitude, from which the great difference in its distances from the earth on the two occasions is apparent. And even if Eudoxus had set up this libration, he would have been censured by Calippus, and so the number and order of Eudoxus' orbs would have been changed. But Aristotle affirms the contrary. So since the most diligent Calippus did not hold that this point in Eudoxus should be

---

[127] Kepler here gives a précis of *Homocentrica*, f. 61v.

censured, we should not attribute such lack of knowledge to Eudoxus. But let us return to Eudoxus. We have declared that one of the third and fourth orbs should be responsible for the latitude, the other for the retrogradation. This we maintain as the first point. Secondly, though Aristotle definitely says that the first and second orbs are common to all the planets, he does not say that either of the third and fourth [orbs] of the five planets is the same as the third [orb] of the luminaries. From this it is, I think, right to conclude that even if one of these has the same function as the third orb of the luminaries, that of carrying the planets in latitude, nevertheless they differ very greatly in position. Thirdly, we must consider what follows from the fact that the poles of the third sphere are the same in the case of Venus and Mercury. Well then, since nothing is common to these two (that is, in the accepted view of the universe) except the line of mean motion which is projected through the centre of the epicycle – for their latitudes are different, and their times and extents of retrogradation likewise different – it follows accordingly that the poles of the third orb are attached in the line of mean motion. And further, since it is certain that the latitudes of Mercury and Venus culminate at different points in the zodiac and these points are not always the same distance apart, though the midpoints of retrogradation of the planets are always in conjunction, it follows accordingly that the third orb is responsible for the retrogradation, the fourth for the latitude. This is, indeed, clearly indicated by Aristotle's words. It is true, however, as it was in the case of the moon, that the third orb carries the nodes, but the fourth, thanks to the third, can bring about both the retrogradation and the latitude together. Fourthly, and lastly, let us consider Aristotle's assertion that the poles of the third orb are in a circle through the middle of the zodiac. One can bring that about in many ways: either the opposite poles might be set apart in opposite points of the zodiac – that is, when the earth is in the centre of this third [orb]; or both poles might be at the same point of the zodiac – that is, when the earth is placed outside this orb, so that the orb is an epicycle and the line from the earth through the epicycle's centre coincides with its axis; or, finally, the placing of the axis of this epicycle along the longitude of the ecliptic would locate both poles at neighbouring points of the zodiac. The first of these satisfies Fracastoro, but for the reasons given above it cannot be correct. For an orb of this kind is not placed in such a way as to be able to bring about the retrogradation. The second also cannot be correct, if we stick precisely to the words. For thus the planet would indeed be made retrograde, but at the same time a latitude equal to the retrogradation would be brought about. Nor can the third be correct. For the planet would be carried from side to side of the ecliptic. In these straits I cannot but think that one of the following happened: either when Aristotle looked at the drawing of Eudoxus, who projected a line from the earth through the centres of the epicycles, he thought it to be the axis of the sphere or epicycle and likewise the points at which the line cut the epicycle to be [its] poles; or, indeed, given that for optical reasons such an episphere cannot be portrayed in the correct position, projected onto the plane of the ecliptic perpendicular to the [line of] sight, unless the poles and centre and, moreover, the entire axis are superimposed at a single point,[128] Aristotle himself confused the centre and the poles when he saw the line from the earth to the centre of the epicycle produced to the ecliptic. Thus it came about that he thought that the motion of the sphere (or as he calls it, episphere)[129] was about that line. And so driven by necessity we shall suspect this of Aristotle and shall explicate Eudoxus as follows. The centre and midplane of the third sphere (which was an epicycle, or, to conform to Aristotle, episphere) coincided with the plane of the ecliptic in such a way that the lines produced from the earth through the centre of this epicycle were different in the case of different planets, but in the case of Mercury and Venus they coincided and formed a single line. In this third [sphere], indeed, another episphere was set up, whose poles declined from those of the third by the amount which sufficed for the latitudes. And, moreover, by the motion in consequence of the fourth the planet is made both to digress in latitude and to become retrograde in the lowest part of the epicycle; and by the motion of the third it comes about that the planet is at its maximum latitude sometimes when in conjunction with the sun, sometimes when at its stations, etc.

In Aristotle there follows: 'Moreover, Calippus posited both the same disposition of the spheres and the same intervals as Eudoxus. Further, to both Saturn and Jupiter he assigned the same number of spheres.' He changed almost nothing in the case of Jupiter and Saturn, for because of the length of their periods not even he had been able to detect their inequalities, though he was by profession an astronomer. 'He thought that two further spheres should be added to the sun and moon, if one were to demonstrate the appearances; and, indeed, one each for the other planets.' Calippus was not attached to any sect of philosophers (for he is not listed by Laërtius among the philosophers),[130] but was, as I have said, an astronomer by profession, and he dealt only with astronomical matters and did so in a more accurate

---

[128] I.e., if Eudoxus projected the third orb, KML in the above diagram (*Apologia*, p. 168), into the plane of the ecliptic, its centre and its poles, K and L, would coincide, as Kepler has already pointed out (*Apologia*, p. 169).

[129] See fn. 10 above.

[130] He does not figure at all in *De clarorum philosophorum vitis*.

manner than Eudoxus, seeking elegant philosophical arrangements. Accordingly, through diligent observation he noticed that there are in the sun, moon, Mars, Mercury and Venus more inequalities than the hypotheses of Eudoxus allow; in particular, there is that [inequality] relative to the parts of the zodiac which we call 'eccentricity'. So he added one orb to each, but whether it was an epicycle or an eccentric deferent cannot be ascertained from these words (it could have been either, but more likely the former). Here again Fracastoro offers an incongruous interpretation. Firstly, he says that in the case of the sun and moon the [orbs] added by Calippus – which are one each – are two each, and that they were arranged in such a way as both to induce in these luminaries the slowing down and speeding up which is apparent in each of them (and which we bring about by postulation of eccentricity). Moreover, he disposes the spheres in the manner of his own *Homocentrica*,[131] as if the ancients had once employed such awkward devices, but it had been shrouded in silence by later authors. It is true that these orbs exist for the sake of the eccentricity, but single ones suffice, a point on which Fracastoro appears to despair – single ones, namely, epicycles, or little circles about the centre if Eudoxus wanted to bring about also the progression of the apogees. Nor do Aristotle's words proclaim twos and threes. Next, Fracastoro thinks that that [orb] was added in the case of the others because of their declination from the equator, on the model of the sun and moon,[132] as if, perhaps, Eudoxus had not already been aware of the very evident digression from the equator both of Mars, Venus and Mercury and of the sun and moon; and as if Eudoxus had not placed an individual orb in the zodiacal circle for each of the planets to bring about the variation in its declination; and as if not even Calippus had noticed that Saturn and Jupiter are at different distances from the equator at different times just like the other five. Thus Fracastoro has nothing in his favour except for the resulting numbers, and they are variously corrupted in the versions [of Aristotle's text].[133] So I think the matter is settled as follows. Two orbs only, not four, were added to the sun and moon by Calippus; that is, one to each. Further, the one added to some of the five planets was not to bring about latitude or declination, but to bring about eccentricity. (I leave it to others to decide whether the agreement with the numbers is enough for us to conclude that the double inequality of the moon was known to Calippus and that because of the similarity of its motions he established a double in-

[131] *Homocentrica*, f. 62r, 18–19 and 25–7.
[132] *Homocentrica*, f. 62r, 34–62v, 1.
[133] See fn. 138.

equality for the sun as well.)[134] So on this reckoning there remain four orbs each for Jupiter and Saturn; five orbs each fall to Mars, Mercury and Venus; and likewise four each fall to the sun and the moon, the innermost (in the Aristotelian order of the spheres). Hence light is cast on an obscure passage in Aristotle, Chapter 12 of Book 2 of *De caelo*. Aristotle says there: 'Those planets which lie in the middle are moved by more motions.'[135] That is, Mars, Mercury and Venus, the ones in the middle according to Aristotle, are moved by five motions; Jupiter and Saturn, the uppermost, and the sun and moon, the lowest, by only four. And our interpretation is indeed vindicated against Fracastoro's by this passage alone.

Up to this point, then, Aristotle has carried out one part of his plan, as earlier he had proposed that those who investigate these matters should be questioned. The other part now follows, for he declared that 'in part he should inquire for himself'. So he added: 'But if all the [spheres] together posited are to satisfy the appearances, for each of the planets there must be other spheres, one less in number, which revolve and always restore the first sphere of the star next in order below to the same place; for only on this assumption will it come about that all the apparent motions of the planets are demonstrated.'[136] Here Fracastoro proceeds in the opposite way to me. But I shall yet again disagree with him on certain points, of which the first is that, without due consideration, he attributes to Calippus as well these matters which Aristotle appends here, which are philosophical rather than astronomical.[137] Thus Eudoxus and Calippus posited as many orbs in each of the individual planetary spheres as the kinds of motion that were observed in the star. Aristotle the philosopher employed a philosophical axiom, namely, that the motion of an outer sphere is communicated to the inner spheres, and that this is through a certain impulse, so that inferior bodies are set in motion by each and every one of the motions by which each superior body is set in motion, unless they are restored by some special and innate principle of motion. So because, for example, the innermost orb of Saturn goes round in thirty years, it is going to impel the first [orb] of Jupiter so that it cannot comply with the motion of the fixed stars. For this reason the first orb of Jupiter ought to proceed in the opposite direction so that it may

---

[134] Kepler here raises and, by implication, rejects the conjecture that Calippus knew of the lunar evection as well as the lunar eccentricity, and so added two orbs to Eudoxus' model for the moon and, by analogy, two to his model for the sun as well.

[135] *De caelo*, II, 10, 291b, 30–1.

[136] Continuation of *Metaphysics*, XII, 8.

[137] Cf. *Homocentrica*, f. 62v, 14–20, where Fracastoro attributes Aristotle's counteracting spheres to Calippus.

run its course in thirty years, always in opposition to the lowest [orb] of Saturn, and thus remain in the same place. But the first [orb] of Jupiter is unable to perform this, for it has another duty, that of following the fixed stars and going round the earth in twenty-four hours. So a restoring sphere is needed here in place of the first Jovian [sphere], to bring about what we have said to be necessary. Further, when it is freed by this restoring sphere from the single motion of the innermost Saturnian sphere, it is still subject to the other Saturnian spheres. For this reason there will be needed for Jupiter, which is immediately below, and for its first orb, as many restoring spheres as the total number of orbs by which Saturn is carried, less one. For when the first sphere of Jupiter has been freed from the other motions of Saturn by restoring spheres, it remains subject to the first [sphere] of Saturn; but there is no need for it to be freed from its motion, since the motion of both is the same and in common with the sphere of the fixed stars. Therefore, just as the first [sphere] of Saturn does not bring about any motion through an innate active principle, but is impelled by the fixed stars in twenty-four hours, so the first [sphere] of Jupiter is driven by the first sphere of Saturn, and thus in succession the first [sphere] of each superior body impels the first [sphere] of the one inferior to it as far as the moon.

Besides, I think that the things Fracastoro says by way of interpretation on this point condemn themselves if they are compared with this interpretation. Truly, this basis of Aristotle's for arranging the heavenly orbs is not only not evident in astronomy, but is also not altogether appropriate on natural grounds: firstly, in that he asserts without discrimination that motions of superior [orbs] are communicated to inferior [orbs], both [motions] of those which embrace inferior bodies and [motions] of those which do not embrace them, namely, epicycles; and next, in that Aristotle argues from the total number of orbs to an equal number of motive intelligences, despite the fact that seven of that number, namely, the ones designated as the first [orbs] of the heavenly bodies, are devoid of motion. What need then do they have for motive intelligences – except perhaps for passive ones, through whose force they show obedience to the impulsion of the one above and follow its command spontaneously? Thirdly, since, as said above, seven of the number lack a motion, why does he not instead of a restoring sphere supply each with some motion or other and thus free its sphere by its very nature from one motion of the one above it? Perhaps [he does not do so] in order to preserve equality between the first [sphere] of Saturn and the first [spheres] of the remaining planets, so that because it is endowed with this leisure, they too may rejoice in the same privilege.

But let us continue with Aristotle's account. 'So since there are eight moving spheres for one lot, twenty-four for the other' (in other versions 'twenty-five' is found). The eight are those of Saturn and Jupiter, four each. The number of the rest comes to twenty-five if one grants, in accordance with Fracastoro's view, that the sun and moon have five. So I fear that this emendation of twenty-four to twenty-five is Fracastoro's.[138] If the sun and moon have only four each the sum of the latter will be twenty-three, not twenty-five. Consider whether it ought not to be interpreted as follows: 'eight for one lot', that is, for the sun and moon, 'twenty-three for the other', that is, for the five others, so that the reason for this division into eight and twenty-three may be more apparent.

'And of these the only one for which it is not necessary to have a restoring sphere is that which bears the lowest star.' All the six higher [spheres] which carry stars, each one the lowest in the arrangement [of spheres for that star], have beneath them counteracting spheres, because they also have beneath them another star. Only the lowest of all, the one bearing the moon, the last of the stars, does not have a restoring sphere beneath it, for it has no star beneath it whose first [sphere] is to be restored. 'Moreover, the restoring spheres of the first two will be six.' For Calippus had given four to Saturn and the same number to Jupiter, the first of the planets. In each case Aristotle added that same number less one, as he has told us above. So he added twice three, that is, six. 'Sixteen, indeed, for the remaining four.' For Eudoxus gave Mars, Mercury, Venus and the sun, respectively, five, five, five and four. Aristotle added the same numbers, in each case less one, that is, four, four, four and three, which makes fifteen altogether. If, however, the sun too has five carrying spheres, as Fracastoro holds, it will have four restoring spheres as well, and the total will clearly be sixteen. 'So the total number of carrying and restoring spheres will be fifty-five' – just so, if you grant Fracastoro in the former case eight and twenty-five and in the latter case six and sixteen. But if we add together eight and twenty-three, and to that six and fifteen, we obtain a total of fifty-two. 'But if, indeed, one does not add all the motions we have mentioned to the sun and moon, the total number of spheres will be forty-nine.' If, on the one hand, Calippus added two each, we lose first four carrying spheres, and then two restoring spheres of the sun, because two of its carrying spheres have been lost; if you subtract six from fifty-five, forty-nine remain. Some versions, however, have forty-seven erroneously.[139]

---

[138] In transcribing Fracastoro's translation above Kepler has substituted '24' for '25'. Of the sixteenth-century published translations listed by Cranz all but that of Périon have '25' here. Kepler's suggestion that Fracastoro may have emended the text suggests that Périon's was the only other translation he had consulted.

[139] Cf. Fracastoro's assertion that '49' occurs in the better texts: *Homocentrica*, f. 63r, 18–19.

If, on the other hand, one each was added by Calippus, there will be lost first two carrying spheres, one of the moon's and one of the sun's, and consequently one of the sun's restoring spheres; when three are taken away from fifty-two, there are left, again, forty-nine. These remarkable numbers, if they really are genuine, reveal I know not what kind of portent. For eight, the number of carrying spheres of Saturn and Jupiter, is two cubed. And four, the number of stars after those two which have restoring spheres beneath them, is two squared. Six, the number of restoring spheres of Saturn and Jupiter, is the first perfect number. Sixteen, the number of the restoring spheres for the last four [stars], is the square of four. The number of carrying spheres for the last five [stars] is the square of five. And, finally, forty-nine, the sum of them all, when the additions of Calippus in the case of the sun and moon have been subtracted, is the square of seven. But there is no evident reason, not even a hint, why Aristotle thought this addition superfluous in the case of the sun and moon, but not likewise in the case of the others, provided that Aristotle is taken simply to have sought the number [of the spheres]. But for the fact that Aristotle had so often poured scorn on the Pythagorean numbers,[140] I would believe that he sought the square of seven deliberately on contrived grounds, perhaps because he was persuaded by some mystery of religion that it was the number of the gods. Now with justice I doubt whether Aristotle, if free from the taint of superstition, would have so written these things. To summarise, his plan was an excellent one, to seek out the hidden nature of things through the eyes and observations of astronomers. He is, indeed, to be censured because he mixed with astronomical observations his philosophical reasonings, which were altogether disparate in kind. For what do the thirty-one orbs and as many motions of Calippus have in common with the twenty-one [orbs] of Aristotle which bring about no new motions? And this is the nature of that famous passage in Aristotle. From its explication it is clear, as I said earlier, that Ursus writes heedlessly in his history of hypotheses when he declares that Eudoxus and Calippus employed concentrics, something which he could in no way have inferred from the text. Fracastoro is not, indeed, a reliable witness on this point, for when he wrote his work on concentrics he tried to assimilate the ancients to himself. On the remaining points, [Ursus] confesses that he knows nothing about them except from an obscure passage of Aristotle.[141]

Indeed, in saying that the Pythagoreans after Plato adopted eccentrics after [*Refutatio*]

---

[140] Notably in *Metaphysics*, I, 5; particularly apposite is the passage 986a, 9–13, where Aristotle charges the Pythagoreans with having invented the *antichthon* so as to bring the number of heavenly spheres up to the perfect number 10.

[141] *Tractatus*, sig. Cii, r, 25–8.

spurning the inadequate concentrics of Eudoxus and Calippus, he betrays the impudence of his account on three counts.[142] In the first place, Calippus himself did this in correcting the hypotheses of Eudoxus, in that he censured their defects and added one orb, which can be held with the greatest plausibility to have brought about eccentricity. There was, therefore, no need for this to happen to Calippus' emendation after Plato, except perhaps in the case of Saturn and Jupiter, whose motions he had not thoroughly investigated. Next, when he says that Eudoxus was spurned by the Pythagoreans, he does not properly accommodate his account to the historical evidence, though by his tone he intimates that he has done so. For on the authority of Laërtius, Eudoxus himself was also one of the Pythagoreans.[143] But who would gather these things from Ursus' account? Finally, I would like him to tell me who they are whom he calls 'Pythagoreans'. If he regards all who embrace some doctrine of Pythagoras as Pythagoreans, then he certainly does not speak in accordance with the usage of history, knowledge of which he arrogates to himself. But if, as the histories do, he calls 'Pythagoreans' those devotees of this sect who modelled their disputations on Pythagoras' maxims and their customs on his regimen of life, let him learn from Laërtius that the sect died out with Epicurus. But Epicurus was born at precisely the time that Aristotle was becoming famous, Aristotle who had been taught by Plato at almost the same time as Eudoxus. Note, moreover, that much later, when Epicurus was a grown man, the now famous philosopher Aristotle made use of Eudoxus' opinions and retailed them in his book *De caelo*. But what had Epicurus, the last of the Pythagoreans, who constructed the world out of atoms, written about astronomy?[144] So this account too goes up in smoke.[145]

But did Hipparchus, perhaps, substitute eccentrics for the concentrics of Calippus? As for the myth of Calippus' concentrics, we have already dealt with it above. And his readers know of no eccentrics of Hipparchus mentioned by Ptolemy (except in the demonstration of the single apogee of the sun, where its convenience for purposes of demonstration commends

---

[142] *Tractatus*, sig. Cii, r, 21–4.   [143] *De clarorum philosophorum vitis*, VIII, 86–9.

[144] Eudoxus, not Epicurus, is treated by Laërtius as the last of the Pythagoreans. Possible sources of Kepler's mistake would be a misreading of *De clarorum philosophorum vitis*, VIII, 50, where Laërtius promises to deal first with the Pythagoreans and then with those who belonged to no particular sect, ending with Epicurus; or a misunderstanding of x, 11, where Epicurus' rejection of one of the Pythagorean maxims is retailed, which might be mistaken to imply that he was an aberrant disciple.

[145] The implication of these remarks is presumably that Ursus must be wrong in supposing that the Pythagoreans rejected Eudoxus' concentrics because the only known Pythagorean later than Eudoxus himself was Epicurus, who wrote nothing on astronomy.

the eccentric scheme).[146] And Cardano gives as warrant for this assertion its author, Fracastoro.[147] But in Ptolemy, Book 9, Chapter 1, they will rather find that Hipparchus in no way turned his hand to correcting errors, but in fact only left tables of observations. So Hipparchus attempted less than either Eudoxus or Calippus in astronomy and confined his thoughts and cares within narrower bounds.[148] What then is this so magnificently announced transformation of hypotheses that Hipparchus applied to those of Calippus?

We have come to the time of Hipparchus, who lived after Aristotle. Ursus says that at this time physical hypotheses were introduced, that is, that order of the spheres which is still commonly accepted today.[149] I have, however, shown above that this is all a mere dream when I declared that Pythagoras held the same view about the order of the spheres as afterwards did Ptolemy. And what more can I say?[150] In whatever way these three bodies of the sun, of Venus and of Mercury are placed between Mars and the moon (that is, in the customary hypotheses), it always yields the same result in astronomy. For once embarked on this course they leave behind observations. The plausibilities of physical reasons alone lead authors in different directions as each especially follows this or that reason, though they agree on the remaining points. Since this is so, whether the commonly held present-day order of the spheres was discussed before or after Eudoxus, [the views] of those men and of Eudoxus should not be thought to be different astronomical hypotheses. For although from the time of Pythagoras to the present day the majority of astronomers have agreed on the main point, this doubt about the order of these three bodies has always persisted, nor has it yet been resolved by students of physics. Heraclitus made the sun second nearest, using the same argument as Aristotle, that the remaining stars give less heat.[151] Plato, mixing Pythagorean, Socratic and Heraclitean philosophy – this on the

---

[146] *Almagest*, III, 4.

[147] *De rerum varietate*, II, 11, 55; 'We accept what Hipparchus said according to Fracastoro against Eudoxus and Calippus, who provided homocentrics.' The source is *Homocentrica*, f. 1, v, 6–7.

[148] Ptolemy relates that Hipparchus contented himself with showing the inconsistencies between observations and the hypotheses of others, but provided no hypotheses of his own.

[149] *Tractatus*, sig. Cii, v, 9–11.

[150] The purpose of the following rather disorganised material is not immediately evident. I take it that Kepler's purpose is to discredit Ursus' tacit assumption that the difference between the ordering of the planetary spheres adopted by Plato and Aristotle and that adopted by Ptolemy marks a substantial development in the history of astronomy.

[151] On Heraclitus cf. Laërtius, *De clarorum philosophorum vitis*, IX, 10. Aristotle's argument is in *Meteorologica*, I, 3.

authority of Laërtius – embellished this view with an optical inference.[152] If Venus and Mercury were beneath the sun, they would sometimes be seen with diminished faces, like the moon. Ptolemy added also in accordance with their opinion the argument that they never darken the sun.[153] The moderns, rejecting these arguments, hold these bodies to be transparent, unlike the moon, and to give way to the sun when it approaches, as Ptolemy held, or even to pass beneath it, as the moderns maintain on the basis of doubtful passages from Averroes and Proclus.[154] But they have not yet avoided the interpenetration of the orbs of the sun, Venus, Mercury and the moon, so that Alpetragius was on firmer ground when he interposed only one of them between the sun and the moon.[155] And on conjectural physical grounds they hold that Mercury, which Crates held to be above, is beneath Venus, because it is moved with more motions and there is a very close resemblance between its motions and the moon's.[156] And, indeed, they place one beneath the sun,

[152] *De clarorum philosophorum vitis*, III, 9: 'He [Plato] mixed together the doctrines of Heraclitus, the Pythagoreans and Socrates.' Since Laërtius goes on to say that Plato derived his doctrine of *sensibilia* from Heraclitus, his doctrine of *intelligibilia* from the Pythagoreans and his political philosophy from Socrates, it is hard to discern the relevance of the citation. The argument attributed to Plato is not in Plato; from the deleted draft of this passage, given in fn. 14 above, it is clear that Kepler's source is Copernicus, *De revolutionibus*, I, 10, where the argument is attributed to Plato's followers.   [153] *Almagest*, IX, I.

[154] The reference to Averroes is derived from *De revolutionibus*, I, 10, and is underlined in Kepler's copy of the work (*Nicolaus Copernicus de revolutionibus*, f. 8r); for details of the ultimate source see Rosen's note in J. Dobrzycki, ed., *Nicholas Copernicus on the Revolutions*, translation and commentary by E. Rosen (Warsaw and London, 1978), 356–7. The reference to Proclus is probably drawn from Maestlin's preface to the edition of Rheticus' *Narratio prima* appended to the *Mysterium cosmographicum* (*K.g.W.*, I, 83–4). Kepler had not read Proclus' *Hypotyposis* at the time he wrote the *Apologia* (*K.g.W.*, XIV, 334).

[155] Sources for the reference to Alpetragius include Regiomontanus, *Epytoma in Almagestum*, IX, I; Copernicus, *De revolutionibus*, I, 10; and *Tractatus*, sig. Cii, v, 7–8 (derived from Copernicus).

[156] The meaning of the interpolated reference to Crates (omitted in Frisch's edition) is obscure, as is that of the partially deleted marginal note which refers to Crates and Metrodorus (see fn. 13 above). Pseudo-Plutarch (Aëtius), *Placita philosophorum*, reports that Anaximander, Metrodorus of Chios and Crates placed the sun highest of all, then the moon, then the planets and fixed stars. In Clavius, *In Sphaeram Ioannis de Sacro Bosco commentarius* (first publn, 1571, 3rd edn, Rome, 1585), 70, this view is attributed to Metrodorus and Crates; and in *Heraclidis Pontici . . . Allegoriae in Homeri fabulas de diis* (Basel, 1544), 37, it is attributed to Crates alone. But I can find no attribution to Crates of an opinion on the ordering of the planets. It seems to me possible that Kepler has here misremembered material from Clavius, whose reference to Crates and Metrodorus immediately follows an argument for placing Mercury beneath Venus because of the greater irregularity of its motions (*In Sphaeram Ioannis de Sacro Bosco*, 69–70). Kepler's misinterpretations of Laërtius, above, suggest that he did not check his sources very carefully in this section.

so that it can be in the middle and the space between the sun and the moon be filled. So this dispute is physical, not astronomical. But I return to Ursus. Ursus, indeed, speaks everywhere as if there were astronomical schools in those times, as there were philosophical schools; as if there was a continuous succession of practitioners, as there was of philosophers;[157] and as if there was either maintenance or public and formal change of astronomical dogmas. This is far from being so. It has always been the fate of astronomy for there to be few who devoted themselves to it. And even though there are cases where one man learned from another in matters of astronomy, nevertheless at that time the business of astronomy took place more in private, and no famous astronomical school frequented by students is known from the histories. Likewise not all astronomers, even when they were contemporaries, knew of one another; nor did all leave written books, owing to the lack of printing, I imagine. For we said above what Philolaus thought. And if his more perfect astronomical views were known to Eudoxus, whose life overlapped his, what need was there for Eudoxus with due care to construct a less perfect astronomy? For it is known from Archimedes that Eudoxus determined the ratio of the diameter of the sun to that of the moon as nine to one and that Aristarchus, who adhered to the same principles as Philolaus, set it at a different value, namely, approximately twenty-one to one.[158] The former is far removed from the truth, the latter very close to it. So each man adhered privately to that type of hypotheses that he took on trust from his teacher, or thought fitting on physical grounds, or found to be most appropriate for his demonstrations. Certainly no kind of hypotheses 'prevailed' generally or was formally 'inaugurated', as Ursus holds.

Now, at last, the historical sequence brings us to the famous Aristarchus of Samos. We are already sufficiently agreed that the hypotheses of his that Archimedes retails were also those of Philolaus and the Pythagoreans before Plato. He certainly fitted his discourse to his demonstrations and said that the circle by which the earth is carried round has the same ratio to the fixed stars as has the centre to the circumference. And Archimedes, expounding this, uses almost the same words as those in which, according to Aristotle as cited above, the Pythagoreans had expounded their opinion, as anyone who consults both passages will see.[159] This all the more confirms what has

---

[157] Diogenes Laërtius' work arranges the philosophers into schools, and within schools into master–disciple successions.   [158] Archimedes, *Arenarius*, I, 9.

[159] *Arenarius*, I, 4–7:

Aristarchus of Samos set out certain posits from which it follows that the universe is many times greater than the universe as mentioned above. For he supposed that the fixed stars and the sun remain immobile; *that the earth itself is carried along the circumference of a circle around the sun which is set in the middle of the circuit*; and, moreover, that the sphere of the fixed stars,

been often reiterated above, that the hypotheses of Aristarchus and the Pythagoreans were the same. Nevertheless, Archimedes distorts the sense of Aristarchus' words somewhat by a conjecture. Archimedes holds that Aristarchus should be interpreted as follows: the ratio of the earth to the sphere of the sun is the same as that of the sphere of the sun (or earth) to that of the fixed stars. But this ratio still does not suffice for Aristarchus' demonstrations. For the ratio of the earth to the sphere of the sun is detectable by the senses, in that it produces a solar parallax of some minutes. So a parallax of the same number of minutes would occur between the fixed stars and the sphere by which the earth is carried round. But Aristarchus plainly excludes the senses altogether by introducing a proportion which is infinite as far as the senses are concerned. But I digress. Yet Ursus' doubt about the period in which Aristarchus flourished should not be left uncensured, in case those who know that Ursus did indeed study the histories should think without reason, as he does, that it is doubtful. Ursus, I declare, raises a doubt whether Aristarchus lived during the reign of Ptolemy Philadelphus or of Ptolemy Philometor, an intolerable uncertainty given that a hundred years elapsed between these two kings.[160] But it is a simple task to resolve this doubt. The Roman consul Marcellus stormed Syracuse in the ninth year of Ptolemy Philopator, and Archimedes was killed in that storming. So he wrote his little

placed about the same centre as the sun, is of so great a size that the circle in which he supposed the earth to be carried has the same ratio to the distance of the fixed stars as the centre of a sphere has to its surface. This, indeed, clearly cannot be so; for the centre of a sphere has no magnitude, so it cannot be supposed to have a ratio to the surface of the sphere. So it is probable that Aristarchus is to be understood as follows. Since we suppose the earth to be, as it were, placed at the centre of the universe, the ratio which the earth has to the universe as we describe it is the same as that which the sphere containing the circle in which he supposes the earth to be carried has to the sphere of the fixed stars. For he adapts the demonstrations of the appearances to a posit of this kind, and in particular he appears to assume that the sphere in which he holds that the earth is moved is equal to 'the universe', as we use the term.

This is translated from the version probably used by Kepler, *Archimedis opera non nulla a Federico Commandino Urbinate nuper in Latinum conversum* (Venice, 1568), f. 49v–50r. The italicised passage is vaguely reminiscent of part of Aristotle's description of the Pythagorean cosmology, but nowhere in the Greek or in the Latin of Kepler's translation of Aristotle and Commandino's translation of Archimedes is the wording 'almost the same', as Kepler claims. I suspect that Kepler has in mind a more subtle similarity. Aristotle says of the Pythagoreans that 'they see nothing to stop the phenomena appearing to us who live away from the centre just as if the earth were in the middle'. On Kepler's interpretation of the Pythagorean cosmology this would imply that the distance of the earth from the centre is negligible compared with the distance of the fixed stars. This is precisely the interpretation Kepler is about to offer (against Archimedes) of Aristarchus' opinion about the size of the sphere of the fixed stars. But it must be admitted that if this is the similarity Kepler has in mind his presentation is elliptical to the point of downright obscurity.   [160] *Tractatus*, sig. Cii, v, 23–5.

book on the number of grains of sand, in which he sets out Aristarchus' hypotheses, before the ninth year of Philopator and perhaps towards the end of the reign of Euergetes, whom Philadelphus immediately preceded. But Epiphanes succeeded Philopator, and, finally, Philometor succeeded him. Thus it is established that Aristarchus, whose books Archimedes read, flourished under Philadelphus, and there is no reason to suppose [he flourished under] Philometor. The historian has historical confirmation; let the astronomer hear astronomical confirmation. According to Hipparchus, as related in Ptolemy, Chapter 1 of Book 3, Aristarchus observed the summer solstice in the fiftieth year of the first Calippic period. That was the sixth year of Philadelphus.[161]

As well as going astray on so many points in his history, Ursus wrongly interprets Aristarchus and Copernicus, and clearly reveals in himself that dimness in understanding Copernicus that he attributes to others. For he interprets the 'inclination' which Copernicus established between the axis of the earth (which corresponds to the equinoctial axis) and the plane of the ecliptic as a 'nutation amongst the planets, like the pitching of a ship at sea, now onto its bow, now onto its stern'.[162] But he completely misses Copernicus' meaning. For although at the solstices the axis of the earth is inclined to a line produced from the sun to the centre of the earth, and at the equinoxes it is at right angles to it, this is not brought about by nodding toward or nodding away, but by translation or *phora* – being carried – of the centre of the earth through a quadrant of its circle, with the axis of the earth always maintaining the same orientation in different positions and not nodding at all. And the simple assertion that this angle varies is not true. For the inclination of the axis of the earth to the plane of the ecliptic always remains the same, whether at the equinoxes or at the solstices, but the earth does not always face the sun. (I am not now considering the very small change which occurs over many ages.) For the earth is carried round with its own inclination which always points in the same directions of the universe. So this inclination, while remaining the same with respect to the plane [of the ecliptic] itself, nevertheless forms different angles with different lines in the plane, as is known from solid geometry. Look in Euclid.[163]

So one who would aptly portray for himself the first and third motions of the earth according to Copernicus (though the third is not, in itself, a

---

161 The source here is perhaps J. J. Scaliger, *De emendatione temporum* (first publn, 1583; Frankfurt, 1593), 226 and 230, which dates the start of the first Calippic year in the Julian year 4384 and the accession of Ptolemy Philadelphus in 4429; this duly places Aristarchus' observation in the sixth year of his reign.
162 *Tractatus*, sig. Cii, v, 31–4.
163 A marginal note omitted by Frisch. The remark is a corollary of *Elements*, XI, 4.

motion) should cut a cylinder obliquely at an angle of sixty-six and a half degrees in two parallel planes. When this is done the corresponding sections, or faces, form ellipses, by corollary 16 of Book 1 of Serenus' *De cylindri sectione*.[164] Let him imagine, for the time being, that they are perfect circles. So, as it is carried around at an angle of inclination, the axis of the earth describes the same kind of surface as that between the circumferences of the faces, sometimes at right angles, sometimes inclined.

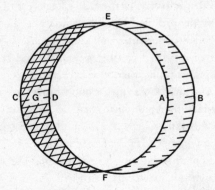

AC and BD are the ends of the cylinder, the former the lower, the latter the upper; CD is a surface of height equal to the axis of the earth, here as at AB inclined, upright at E and F. So CD and AB are the solstitial points, E and F the equinoctial.

Nor has Ursus happily rendered the Greek words *kulisis* – 'rolling' – and *dinesis* – 'whirling' – and by this alone he succeeds in making Copernicus obscure to one versed in the Greek language.[165] For you, Ursus, the earth whirls, as you maintain in the hypotheses you throw up. In Copernicus the orb which carries the earth, if it were something palpable, would whirl. In the customary hypotheses each of the orbs, and in particular the entire universe and the outermost sphere, whirls. On the lathe the disc whirls, with its poles at rest, and remains in its place as it moves.[166] In Copernicus the earth simply *peripheretai* – is carried around – provided we consider only the annual motion, distinguishing it in our minds from the diurnal. But if you say that in Copernicus the earth is whirled, I shall take you to speak not of the annual motion but of the diurnal, for we distinguish the annual motion from the diurnal in our minds. If, however, you say it rolls, now you have

---

[164] Commandino's translation of this together with Apollonius of Perga's *Conics* was published in Bologna in 1556.   [165] Cf. *Tractatus*, sig. Ciii, r, 3–12.

[166] Rheticus uses this image in his *Narratio prima* (transl. by Rosen, *Three Copernican Treatises*, 148).

combined both the diurnal and annual motions. For rolling, in case you do not realise it, is a motion compounded out of being carried along and whirling. And that bowl used by players, with which they aim to knock down the skittles, rolls. And, by Hercules, there could not be a better illustration of the composite motion of the earth than to imagine the globe of the earth as like the players' bowl, but with an oblique axis, rolling along in a directed line, but one which is directed not in a plane but rather in an orb. In the above diagram it is as if the earth were to roll forward in the surface EAFC along the path EGF; and thus the three motions of Copernicus are present in this one rolling-along.

Ursus takes his history of hypotheses up to the time of Christ with a measure of success, in that wherever he goes wrong his error can be regarded as a venial sin. For all of us slip up through our manifold ignorance. But now, when he comes to Apollonius of Perga, he is so thick-headed and consciously and deliberately retails such an altogether worthless story that one does not know whether to laugh at the fellow or to be angry with him. He furnishes his account with lots of details, seeing that he has ventured to swear to anyone who is not already informed about the business that it is no mean matter of which he speaks. Since it is on this matter that he builds up the main support for his case and, moreover, perpetrates his main fraud, it seems right to devote a special chapter to it. At the beginning of the chapter we shall declare the things known to us about Apollonius from trustworthy sources; then we shall compare Ursus' triflings with them.

CHAPTER 3

No hypotheses of Apollonius of Perga are extant   [*Propositio*]

In architecture it is not usual for everything, mortar, stones, keys, bolts and [*Narratio*] windows, to be obtained from one workshop; nor is one and the same person architect, stonemason, carpenter, smith and cabinet-maker. The same holds in the business of astronomy. For I maintain that to the architect there corresponds the man who devotes all his thought to setting up certain hypotheses representing the form of the universe, derived by inference from things which happen at different times in the heavens. His task, however, is so great that it is impossible for him to seek out everything by himself. So he takes some or all of his observations from other men. There are those who construct for him the tables needed for everyday use. Another man, renowned for precepts of arithmetic, has in his possession something which the practitioner appropriates for his own use. Another devises compendia for working with numbers. Another sets out the doctrine of triangles most skilfully.[167] Another shows us how to measure triangles more easily. And lastly there are those who, by solving geometrical problems by demonstration, considerably alleviate the labour of the chief practitioner. However, the man who cuts keys, puts together windows or cuts stones into the required shape is not to be taken as the architect, even though the architect uses his work; and the construction of the house is to be ascribed not to him, but to the architect. Likewise, in astronomy a man who is a good arithmetician or geometer is not at once an astronomer just because one cannot be an astronomer without arithmetic or geometry, the two wings [of astronomy].[168] Nor if one who has set out the doctrine of triangles greatly helps the practitioner in constructing astronomical tables, is he on that account the author of tables or hypotheses.

Since these things are so, I am at the outset greatly amazed that Ursus attributes astronomical hypotheses of his own to Apollonius, who as far as astronomy is concerned left nothing at all in writing except a geometrical problem,[169] which Ptolemy used for the purpose of demonstrating the stations of the planets, that is, as part of the embellishment of his edifice.

[Partitio]   Well then, so as to make it quite clear what Apollonius demonstrated and what Ursus most recklessly fabricates, in these pages we shall derive and set out the whole of Apollonius' design from [Ursus'] source, namely, Ptolemy. And I am greatly amazed that Ursus passes over Ptolemy and cites as his authority a derivative source, namely, Copernicus. Indeed, the fact that Ursus makes no mention of Ptolemy, who expounded Apollonius' opinion most clearly, and rests the main defence of his case on Copernicus, who expresses the matter less clearly, is such as to arouse great suspicion that Ursus never read Ptolemy, but rather that there was some teacher who told Ursus that

167 Perhaps a barbed compliment intended to turn the tables on Ursus, who had claimed on the strength of his skill as a mathematician to be a true astronomer where Tycho was a mere 'mechanic' concerned with instruments and observations (*Tractatus*, sig. Biv, r).

168 The image of arithmetic and geometry as the Platonic wings of astronomy is commonplace in the period. I cannot find it in Plato, though in the *Republic*, VII, and more explicitly in *Laws*, 817 *et seq.* and in *Epinomis*, 990 (formerly attributed to Plato), geometry and arithmetic are presented as prerequisites for the proper study of astronomy; cf. also Geminus in Simplicius, *Commentary on Aristotle's Physics*, 291. Johannes Stadius, *Tabulae Bergenses* (Cologne, 1560), attributes the image to Plato's *Timaeus*, but it is to be found neither there, nor in the commentaries on *Timaeus* of Proclus and Chalcidius, nor in passages of Cicero, Macrobius and Martianus Capella inspired by the *Timaeus*.

169 Kepler appears to use the term *problema* in the technical sense derived from Euclid and Proclus in which *problemata*, unlike *theoremata*, concern the possibility of specific geometrical constructions: see T. L. Heath, *The Thirteen Books of Euclid's Elements*, 2nd edn (Cambridge, 1926), 124–9. I have rendered it throughout as 'problem', though on occasion 'construction' might be more appropriate.

Apollonius' problem was reminiscent of something like the matter contained in Tycho's hypotheses, and, indeed, that the problem had two parts. For when he wrote these things Ursus had grasped only this much, and nothing more or more definite. But [let us turn] to Ptolemy.

In that age geometrical demonstrations were greatly prized, and both [*Confirmatio*] mechanical descriptions and related calculations not involving the skill of demonstrations were little valued, as can easily be gathered from Ptolemy's lengthy justification at the beginning of Book 9.[170]. Accordingly, the way in which the particular point at which a planet appears to be at rest could be designated geometrically was widely discussed, for it seemed that this matter could only be settled approximately and established by calculation. Pergaeus the geometer certainly excelled the others. He applied geometrical demonstration to the problem, and he assumed those things that would, he hoped, once granted, facilitate the demonstration. And, indeed, that demonstration, carried out with admirable skill and clever inventiveness, is clearly reminiscent of the *Conics* of that author.[171]

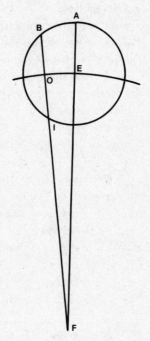

Ptolemy says of him that having posited at least one inequality of motion, that which is relative to the sun, he set up two forms of the problem.[172] First, suppose this inequality to occur in an epicycle, in such a way that the centre of the epicycle is moved in a concentric in consequence through the signs and the star in the epicycle is likewise moved in consequence in the upper part.

---

[170] *Almagest*, IX, 2, where Ptolemy cites Hipparchus as having called for hypotheses sufficient to demonstrate all the planetary inequalities of motion as opposed to the inadequate methods then used.

[171] Perhaps Kepler refers to the similarity with *Conics*, III, 37, noted by G. J. Toomer, 'Apollonius of Perga', in C. C. Gillispie, ed., *Dictionary of Scientific Biography*, I (New York, 1970), 179–93.

[172] *Almagest*, XII, I. The reconstruction of Apollonius' proofs which follows is quite close to that of Reinhold, *Theoricae novae planetarum Georgii Purbachii...ab Reinholdo Salveldensi pluribus figuris auctae, et illustratae scholiis* (first publn, Wittenberg, 1542; revised edn, Wittenberg, 1553). Kepler had referred to Reinhold's treatment in his first assessment of the *Tractatus* (see Ch. 3, p. 62). J. L. E. Dreyer, *A History of Astronomy from Thales to Kepler*, 2nd edn (New York, 1953), 152–6, provides a clear exposition of Apollonius' demonstrations.

That is, F is the earth, and around it is a concentric OE, and in that an epicycle ABI about the centre, the sun being always in the direction FEA. Having posited these things, Apollonius draws a line FIOB, such that the ratio of OI, half the part [of the line] within the epicycle, to IF, the rest [of the line] from the eye F to the convex part of the epicycle I, is the same as that of the velocity of the epicycle from E to O to the velocity of the star from A to B. And he demonstrates that if the planet is at the point I, it will appear to be stationary. If, however, one prefers to bring about this inequality with respect to the sun by means of an eccentric, that method will be applicable only in the case of the three superior planets, which are fashioned in all respects in conformity with the sun. The centre of this eccentric should be moved about the centre of the zodiac in consequence, in the same way as the sun. Moreover, the star in the eccentric should be moved around its centre in precedence, in the same way as the inequality of motion.

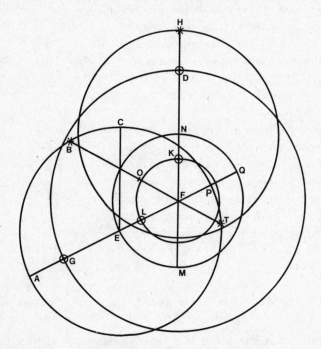

F is the earth; N and E the centre of the eccentric revolved about F in an annual motion along the path NEQ; and A is the planet in the eccentric performing a motion of commutation (as Copernicus calls it)[173] along BCT.

---

[173] Cf. *De revolutionibus*, V, I, where Copernicus calls the retardations and retrogressions which reflect the earth's annual motion *motus commutationis* by analogy with parallax, *commutatio*.

Having posited these things, he again draws through F, the earth, a line such that the ratio of OT, half the line produced from B to T, to FT, the smaller part of the line, is the same as the ratio of the velocity of the eccentric from E through Q to N to the velocity of the star from B. And he demonstrates that when the star is at point T, nearer to the earth, it appears to stand still. This is the whole of the design and undertaking of the famous Apollonius. And it is clear that except for this nothing that he wrote in astronomy is anywhere to be found. See in Copernicus himself, in the passages cited by Ursus, how Copernicus adapted the things demonstrated [by Apollonius] to his own hypotheses.[174] For Copernicus says nothing more about Apollonius than does Ptolemy, from whom he derived these things in his work. Besides, this invention, in itself most ingenious, was not even of much use to practitioners [of astronomy], because in order to obtain a conclusion to his demonstration Apollonius left out the other inequality in the motions of the planets, which is relative to the parts of the zodiac. So Ptolemy and Copernicus were not able to use that demonstration, except when adapted to special cases of the other inequality. So as yet we have no astronomical hypotheses of Apollonius really worthy of the name and deserving inclusion among the other types. For what he posits in his demonstration he posits not as if he thinks that is how matters really are (as practitioners generally do), but so as to make it clear what would happen in the case of the stations of the planets if that were how things really were. He uses two kinds of principle for his demonstration, because he does not see any difference, as would a practitioner, an astronomer, but only deals with this kind of motion as a geometer.

Now see how smoothly Ursus lies about this invention of Apollonius. [*Refutatio*] Firstly, he says that Apollonius transmuted the form of the universe of the Aristarchan [hypotheses], as if he had been an astronomer like Aristarchus and the others. Ptolemy calls him a mathematician, and certainly nowhere is the title of astronomer applied to him. Then, since there hovers before him a memory of the double form of the Apollonian demonstration, but a feeble memory like a thing seen in a dream (for I conjectured above that Ursus did not himself read these things in Ptolemy, but was advised by another who had perhaps read them), he himself goes on to misascribe a double transformation of the Aristarchan hypotheses to Apollonius. He says that the first transformation was undertaken so as to remove that immensity of the fixed stars which, according to Archimedes, Aristarchus introduced; and likewise, that the transformation was so carried out that the centre of

---

[174] *De revolutionibus*, v, 35, quoted by Ursus in *Tractatus*, sig. Civ, v.

the fixed stars, together with the earth, was turned around with an annual motion.[175] But it will be readily apparent to anyone who cares to compare this interpretation of Ursus' with Ptolemy's words, cited above, about the first form of the Apollonian demonstration, that all this is a mere figment, Ursus' monstrous brain-child. There is, indeed, more resemblance between black and white than between Ursus and Apollonius. And if anyone attends to the spirit of the Apollonian posits, it will not be in any way plausible to him that Apollonius was so greatly concerned about that great gap and that immensity of the fixed stars. Ursus pretends that the other transformation was carried out by Apollonius in such a way that whereas before the earth (the centre of the fixed stars) was carried round with an annual motion about the immobile sun, the centre of the planets, now the sun and the centre of the planets go round the earth and the centre of the fixed stars, which are at rest, in the same period of a year.[176] And, indeed, this form of hypotheses was invented and established by Tycho Brahe, for reasons widely expounded by that author and from motives which are not to be belittled (and Ursus concocted these things in order to harass him). This assertion has, to be sure, some plausibility if one compares it with the Apollonian posit. Let F in the above diagram be in Apollonius' opinion the earth, and let NE be the circle around it which the centre of the eccentric describes in an annual motion, so that the line produced from F, the earth, through N, E, the centre of the eccentric, is always the same as that from F, the earth, through the sun, whether [the sun] is moved in the larger circle DG or in the smaller circle KL. And let H be the planet at apogee; so it will be in line with the sun. Let the apogee and its line FH, and in particular the centre of the eccentric N, be carried from that place to E, and let the position of apogee be at A, again [in line] with G or L, the sun. The planet, indeed, is moved by a motion of commutation from the line of the apogee in the contrary direction, namely, from A to B. It is clear that once every year, approximately, it lies at the perigee, though it does not move away to the opposite parts of the zodiac. For this reason it happens that with the rising of the sun it is borne aloft and with the setting sun it too subsides. In fact the hypotheses of Tycho also assert the same things. Thus where Apollonius says that D (or K), G (or L), the sun, and N, E, the centre of the eccentric, are in the same line, Tycho says that N, E, the centre of the eccentric, is the sun itself. This opinion of Apollonius cannot hold good unless the ratio of FE, the eccentricity, to EA, the radius of the eccentric, is right (otherwise the correct extent of the retrogradations will not be obtained). So it is definitely determined in the

---

[175] *Tractatus*, sig. Civ, r, 7–11.   [176] *Tractatus*, sig. Civ, r, 26–35.

case of the planet Mars that the eccentric EP intersects the circle FQ described by the centre of the eccentric. That, indeed, follows conclusively from Apollonius' premisses, given above. For since BT is drawn as a straight line outside the centre E, but AP, on the other hand, is drawn through it, by [proposition 8] of Book 3 of Euclid, AP will be longer than BT. But AP has its midpoint at the centre E, and BT, by hypothesis, at O; so too, by the same token, the half line EP is longer than the half line OT. Further, two lines are produced from the point F, which is away from the centre of the circle CP, both incident on its circumference PT, one, FP, coming from its centre, the other, FT, not so. It follows by [proposition 8] of Book 3 of Euclid that FT will be greater than FP. Now Apollonius has said, above, that OT is to TF as the velocity of the centre of the eccentric (of Mars), N, E, is to AB, the velocity or motion of commutation of the star; that is, approximately 59 to 28. But since by [proposition 10] Book 5 of Euclid the ratio of a whole to the larger of two parts is less than its ratio to the smaller, the ratio of OT to FP will be greater than 59 to 28. And, again, since by [proposition 10] of Book 5 of Euclid of two larger parts the one which is greater has a greater ratio to a smaller part than the one which is smaller, the ratio of EP to PF will be greater than the ratio of OT to PF. Hence, in the first place, OT to PF is certainly greater than 59 to 28; and so the ratio of EP to PF is much greater than 59 to 28. And since the ratio 59 to 28 is more than two, the ratio EP to PF will be, on three counts, more than two. If EP is more than twice PF, then PF is less than half EP, the semidiameter of the eccentric. So when PF, which is less than half [EP], is subtracted, FE (or FQ) remains by the same amount larger than half [EP]. So FP, the distance of the perigee of Mars from the earth, is less than FQ, the semidiameter of the circle which carries the centre of the eccentric.[177] But Tycho says the same thing about the circle of Mars intersecting the sun's circle. From these considerations the resemblance that I have mentioned between Tycho's hypotheses and Apollonius' geometrical problem is evident. But the point at issue, that Apollonius published hypotheses that can correctly and properly be called 'astronomical', does not yet follow from these considerations; much less, indeed, does it follow, as Ursus asserts with no hesitation, that by carrying out a second alteration of the Aristarchan hypotheses Apollonius expounded the form of the system of the world that today is proposed by Tycho. For to start with Apollonius was not an

---

[177] In translating this passage I have been helped by Christine Schofield's exegesis of Kepler's argument: C. J. Schofield, 'The geoheliocentric planetary system: its development and influence in the late sixteenth and seventeenth centuries' (Ph.D. thesis, Cambridge, 1964; New York, 1981), 130–2.

astronomer by profession, but a geometer. And he did not himself, as the office of astronomer requires, apply in practice what he demonstrated from a problem derived from astronomy in order, having adopted this hypothesis, to infer and demonstrate from observations the motion of some planet. Rather, he handed over to astronomers a merely geometrical demonstration, as a retailer hands over keys or an axe to an architect in case someone should need these things for his work. Secondly, like Ursus, we earlier defined [a body of] astronomical hypotheses as being a definite conception of the form of the universe and of the disposition of the celestial bodies.[178] But this demonstration of Apollonius' is clearly not redolent of any such thing. For, indeed, he leaves it open to astronomers whether they want to use a concentric with epicycle for all the five heavenly bodies which perform stations, or to use an eccentric for the three superior ones. And neither the reason for the disposition of the various orbs nor the comparison of their magnitudes is any concern of his. For his demonstration is concerned with one single planet, and that only from the point of view of the single phenomenon of its stations and retrogressions. So Apollonius' problem is not to be taken for a body of astronomical hypotheses. Moreover, he does not adopt hypotheses sufficient to demonstrate the motions of even one planet, for he deliberately passes over that inequality of motion which is returned to the same place in the zodiac.[179] And thus he makes it clear that he seeks nothing beyond the geometrical certainty of a conclusion deduced from assumptions, something which generally delights clever thinkers.

Now suppose that we concede that the matters demonstrated by Apollonius are to be taken as proper hypotheses, something which has already been refuted. Even so, who would believe Ursus that they were taken, or rather derived by transformation, from the Aristarchan hypotheses? For in fact it seems highly probable that Aristarchus' ideas were either unknown to, or not understood by, Apollonius. For it was because Apollonius knew that some of the astronomers adopted eccentrics and others concentrics with epicycles that he adapted his problem to both. Indeed, I believe that had he known about the Aristarchan revolution of the earth he would have adapted his problem to it as easily as did Copernicus after him. Finally, if he had wanted to imitate Aristarchus, what difficulty would he have had, I ask you, in asserting that the centre of the eccentric is moved in the same manner as the sun? How much greater the difficulty he would have had had he

---

[178] Cf. *Apologia*, p. 139; and *Tractatus*, sig. Biv, v, 1–4.
[179] Cf. *De revolutionibus*, v, 3, f. 141v, 2–8, a passage marked in Kepler's copy, where Copernicus points out that Apollonius' construction uses a concentric deferent which cannot account for the zodiacal anomaly.

asserted, as does Tycho today, that the centres of all the planets are in the sun. Now since he separates the centres of the planets from the sun and from each other, he imposes definite constraints in the construction of the form of the universe on those who would undertake it, constraints which are far removed from those of Tycho. For after the orbs of the moon there follows the system of Mercury with an epicycle, and above it Venus with a far larger epicycle, then the sun. And then E, the centre of the eccentric of Mars, will go round about the sun L (look at the previous figure), carried aloft at a great enough height for EC (which has a definite ratio to EF) to be longer than EP[180], lest on occasion the orb of the sun be compelled to offer hospitality to that of Mars, something which is bound to appear impossible to all those who believe with Aristotle in palpable and adamantine orbs. The following consideration establishes the very great difference between Apollonius' actual hypotheses and those which Ursus attributes to him falsely (since they are Tycho's). In both, indeed, as demonstrated above, the planet Mars in its eccentric passes within the limits of the centre of the eccentric at two points. For Tycho the limits are in fact the same, for that same circle is that of the sun itself; so for him Mars descends to a point far nearer to the earth than the sun. For Apollonius, on the other hand, this is not so at all, or certainly not of necessity. For Tycho this is established from observations, the diversity of [Mars'] appearance making this subtle matter clear.[181] For Apollonius there were no observations that could refute the opposite view. Thus it follows that in Apollonius Mars is never to be imagined lower than the sun. Moreover, just as it follows from the things demonstrated by Apollonius that the centre of the eccentric of Mars ought to be carried round above the sun, so there is nothing to stop Saturn's centre [from being carried round] below the sun, all three [superior planets] going round at the very greatest distances above the sun. So there will be no certain basis like that of Tycho and Aristarchus on which to find out the ratios of the orbs, except for the single one that is derived from the urge to avoid contact of the orbs and the collision of two planets. And on this basis nothing is shown beyond certain limits below which the planets do not descend, but how high above those limits

---

[180] A slip: Kepler has confused P in the above diagram (*Apologia*, p. 188), the point at which the line from E through F cuts Mars' eccentric, with the unlabelled point at which it cuts the sun's orb.

[181] In letters to Landgrave Wilhelm of Hesse and Rothmann, Tycho had reported that he had established by measurements of Mars' parallax that when in opposition it is nearer than the sun (*Epist. astron.* [*T.B.o.o.*, VI, 70 and 179]). On the mystery surrounding this claim and Kepler's later abortive attempts to make sense of it see J. L. E. Dreyer, *Tycho Brahe* (Edinburgh, 1890), 178–80; W. Norlind, 'Tycho Brahes Världssystem', *Cassiopeia*, 6 (1944), 57–75; Schofield, 'The geoheliocentric planetary system', 64–9.

they are carried along cannot be ascertained from it. Here, indeed, is a diagram of the whole of this kind of hypothesis, if anyone wants to set it out in accordance with the Apollonian demonstration.[182]

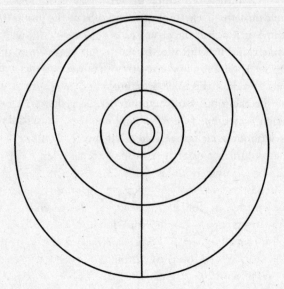

Already there is a great difference between the two hypotheses, in that one only moves the centres of the orbs with the sun, but the other in addition places them in the sun itself. Nor, as I have said above, should anyone maintain that in Apollonius the circuits of the planets are truly and by design subjected to the sun and change their position with it, as he would thereby put his trust in Ursus' declaration that Apollonius derived what he said from Aristarchus. But if you compare the two forms of the Apollonian problem with each other, you will see that the reason for thinking up the second of them arises not from the Aristarchan views, but from a transformation of the first form, which is Ptolemaic.

At the time it was generally agreed amongst astronomers and geometers that everything that is demonstrated by means of eccentrics can likewise be

---

[182] The meaning of this diagram, which Kepler leaves the reader to work out, is not obvious. Since there is a gap in the text at this point it is likely that he intended at least to letter the figure and provide a key or explanation. The diagram apparently shows the eccentrics of Jupiter and Saturn centred below the orb of the sun. From the preceding text one would expect it to show also Mars' eccentric centred above the sun. Perhaps Kepler realised that he had made Jupiter's and Saturn's eccentrics too small to allow Mars' eccentric to avoid both them and the sun's orb, and so left the diagram incomplete.

demonstrated by means of a concentric with epicycle.[183] Now in the first form of his problem Apollonius accepted from astronomers and assumed that the inequality of the five planets is brought about by an epicycle on a concentric. For in the case of the orbs of the three superior planets, Ptolemy's followers hold that there are three epicycles which are moved in the upper part in the same direction as the eccentric, in such a way that when the centre of the epicycle is in line with the sun the planet is always at the apogee of the epicycle. So Apollonius converted this form into another which is based on an eccentric. And for this to be done properly it was necessary completely to tie to the sun the motion of the centre of the eccentric, which stands in for the motion of the epicycle, because previously, indeed, the revolution of the epicycle about its centre was tied to the sun. This, and certainly nothing else, is the source of that hypothesis.

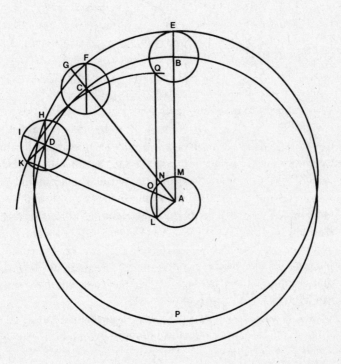

For the sake of demonstration, let the eccentric EFH be described with centre M and eccentricity AM. Since, therefore, the centre M is not displaced along

183 Kepler attributes knowledge of this equivalence to Apollonius and others on the strength of *Almagest*, XII, 1. For confirmation of this attribution see O. Neugebauer, 'The equivalence of eccentric and epicyclic motion according to Apollonius', *Scripta Mathematica*, 24 (1959), 5–21.

the circumference MN, a perfect circle will be described through EFH. From A the motion of the planet will appear slow about the apogee E and fast about the perigee P. The same will happen if a concentric BCD is described with centre A, and on it an epicycle whose radius BE (or CF or DH) is equal to the eccentricity AM in the first case. The motions are, in fact, so arranged that both the first orb ME and the second orb AB proceed equally and in the same directions about M and A, respectively; and the lines from the planet E, F and H produced through B, C and D, the centre of the epicycle, are always parallel to the line MA, the eccentricity, as it is carried around in the concentric BCD. Whence it comes about that in itself, and with respect to the whole universe, the epicycle is no less at rest than the point M. But geometers, attending to E, G and I, the apogee of the epicycle, say that the planet is not at rest in its epicycle, but is moved in the opposite direction to the eccentric, namely, from G to F, in an arc of the same [angular] extent as BC, the path of the centre of the epicycle. Now suppose that M, the centre of the eccentric, is moved, and moved, moreover, in a manner other than the eccentric. Let it be moved, then, in the same direction as is the eccentric, but faster than it, so that as the eccentric completes the angle QLK, the centre of the eccentric completes the angle MAL. Given this, it will come about that the planet appears to descend rapidly from E to K, and does not describe a perfect circle. The same thing is brought about by means of a concentric with epicycle, where the concentric takes over the motion of the above-mentioned eccentric, and the epicycle, or rather the planet K in it, takes over the motion of the centre of that eccentric, the motion being measured from the line DH, parallel to AM, so that HDK and MAL are equal angles. Hence the real consideration which had provided Apollonius with this kind of posit becomes clear. For with respect to the approaches and retreats of the planets relative to the cycle of the sun the Ptolemaic hypotheses yield us the same result. So there is, as yet, nothing on the basis of which Ursus can defend himself and maintain that Apollonius was the originator of the inverted Aristarchan hypotheses, which are due to Tycho.

Add to these considerations this one also. The inequality of motion which Apollonius decided to leave out of his demonstration cannot be ignored. Therefore the apogee and perigee ought to be set in the line HN, CE in the diagram three back, always in the same direction. From this it follows that we can no longer call AE 'the line of the apogee', and the whole nomenclature of Tycho differs from that of Apollonius. Therefore, unlike Apollonius, he makes the planet move not in precedence from A to B, but in consequence from C to B. Given such a difference in substance and terminology, Ursus labours in vain to attribute to Apollonius the form of hypotheses derived from Tycho.

CHAPTER 4

No mention is to be found anywhere of the hypotheses   [*Propositio*]
which Tycho claims as his own and Ursus falsely
attributed to Apollonius

Ursus, urged on by I know not what impulse, pursues this single end: in   [*Narratio*] open violation of truth to claim Tycho's ideas about the universe for antiquity. The first of the ancients he brought to our attention was Apollonius, as related above. If he had undertaken to maintain the view that there is a certain affinity between the Apollonian problem and Tycho's hypotheses, he could in part have been tolerated and for the rest excused. But as he clearly equates the two, it is evident, as shown in the last chapter, that Ursus was not fully acquainted with them, and, moreover, had not read about or understood Apollonius' problem at all. Moreover, that affinity between Tycho's hypotheses and Apollonius' problem, which I have both demonstrated above and already said may be conceded, is in fact perceived only now, after Tycho has set up and published his form of the universe motivated by entirely different considerations. But for this, I doubt if there would have been anyone sufficiently endowed with wit to have been able to come up with Tycho's ideas simply from having looked at Apollonius' problem, so obscure is this affinity and invisible the path.

Another of the ancients, Martianus Capella, is brought to our notice by Ursus as a rival to Tycho.[184] Ursus makes much more of a fool of himself about him than about Apollonius. For this author sought worthy recreation for his mind in scholarship (he was a politician and Governor of the Province of Africa). So he decided to put together a poetical encyclopaedia.[185] Having contrived a most eloquent fable about the marriage of Mercury and Philology, he dealt with and ran through both the Platonic philosophy and, specifically, the so-called 'seven liberal arts'. And he did so in such a way that he not only very briefly expounded the main points of each, but also earnestly conveyed in an ornate style whatever in each of them he found to be noteworthy, abstruse, admirable or especially appealing to him. In

---

[184] *Tractatus*, sig. Ciii, r, 8–12.
[185] For Kepler's sources in this introductory paragraph see the notes on his more detailed treatment in the subsequent paragraphs.

astronomy the Roman followed the Latin authors, especially Macrobius, Pliny and Vitruvius, so closely that he did not even set aside their errors, but rather retailed them in his own work. So, in accordance with their opinion, he made the sun the centre of the orbs of Venus and Mercury. But Ursus, who maintains that knowledge of Martianus Capella's doctrine accrued to Copernicus alone, makes no mention of the others, thinking Martianus to be the originator, to the great detriment of his cause. But how much he would have contributed to the idea that he eagerly sought for in the authors and in antiquity if he had set Macrobius before Martianus, Pliny before Macrobius and Vitruvius before Pliny. For Vitruvius yielded astronomical treasures to Pliny as Pliny did to Martianus. And, indeed, not even Vitruvius is the author of the astronomical doctrines which he recounts in his work. For what has an architect to do with astronomy? Vitruvius does not deliberately play the part of an astronomer, but taking the opportunity to make his work commendable and pleasant to read he intersperses it with both astronomical and other noteworthy matters gathered from elsewhere. Macrobius, indeed, and the ancient commentary on Bede attributed this opinion to Plato himself, despite the fact that his successors, the Platonists, had corrupted it like many other things. The same Macrobius, indeed, attributes the same ancient opinion to the Egyptians themselves, so that on his account it arose almost at the same time as astronomy itself. Ursus, however, overlooks this prize, because, having scarcely glanced at Martianus Capella's name in Copernicus, he focusses his attention on this alone, not even considering Copernicus' assessment carefully. For he might at least have tried to find out who those other Latins were whom Copernicus declared to have 'well understood' the same matter along with Martianus. And he might have asked to which of the Latin authors belonged the words *conversas absidas*, words which Copernicus quoted from Pliny.[186]

[*Confirmatio*]   On the question of this hypothesis about Venus and Mercury it is appropriate to quote the words, so that what I have already said may be clearly apparent. Capella's words, from Book 8, are explicit: 'Moreover, three of these travel round the earth with the sun and moon, but Venus and

---

[186] In my judgement we should not in the least disregard what was well understood by Martianus Capella, the author of an encyclopaedia, and by certain other Latin writers. For according to them Venus and Mercury revolve around the sun as their centre. This is the reason, in their opinion, why these planets diverge no further from the sun than is permitted by the curvature of their revolutions. For they do not encircle the earth like the other planets, but have *conversas absidas*.

De revolutionibus, I, 10; based on Rosen's translation in Dobrzycki, *Copernicus on the Revolutions*, 20.

Mercury do not go round the earth.'[187] And after interposing some further matters: 'Now although they rise and set daily, nevertheless their circles do not go round the earth at all, but are turned round the sun in a wider course. And, furthermore, they fix the centre of their circles in the sun, so that sometimes they are carried above it, and sometimes below it nearer to the earth.'[188] Further on, moreover, when he has first considered the sun and moon and their effects on one another and on the earth, he comes in due course to 'those which are turned about the sun in their wandering through the universe', Venus and Mercury, whose 'circles were known', he says, 'by our predecessors to be epicycles', that is, 'not to include the earthly globe in their own revolution, but somehow to be carried round on one side of the earth'.[189] And later, with reference to Venus, he tells us in these words how they are moved in that epicycle: 'Placed in her circle she moves about it with manifold diversity, in that sometimes she runs from one side of it to the other, or sometimes lags behind it and does not catch up with it, sometimes is carried above and sometimes lies below.'[190] It will now be made clear how these words and opinions are derived from others.

Pliny's opinion is in doubt, both from the obscurity of his words and in the judgement of interpreters of this difficult matter. It is attested by strong evidence from Chapter 8 of Book 2 that Capella's opinion was not anticipated by Pliny. For in that passage, when he explicitly arranges the spheres in order, he declares: 'Below the sun circles the huge star ascribed to Venus.' And a little later: 'Next to Venus the star of Mercury is carried in a lower circle.'[191]

These words prevent Collimitius, Zieglerus and Milichius, not unworthy interpreters of Pliny, from wanting to recognise Capella's view in Pliny when later on he speaks specifically of Mercury and Venus.[192] There are, however,

---

[187] Martianus Capella, *De nuptiis Philologiae et Mercuriae*, VIII, 854. The question of the editions from which Kepler drew this quotation, and the following quotations from Pliny, Vitruvius and Macrobius, is treated by B. Eastwood, 'Kepler as historian of science: precursors of Copernican heliocentrism according to *De revolutionibus*, I, 10', *Proceedings of the American Philosophical Society*, 126 (1982), 367–94. I am grateful to Professor Eastwood for having sent me a pre-publication copy of his paper. As Eastwood notes, the present quotation is derived from some edition based on the *editio princeps* of 1499 and not from the editions of Vulcanius (Basel, 1577) or of Grotius (Lyons, 1599), in which these passages are emended. The title page of the Basel edition of 1532 gives precisely the information about Capella that Kepler retails above.

[188] *De nuptiis Philologiae et Mercuriae*, VIII, 857.

[189] *De nuptiis Philologiae et Mercuriae*, VIII, 879.

[190] *De nuptiis Philologiae et Mercuriae*, VIII, 882.   [191] *Naturalis historia*, II, 8, 36–7.

[192] For the commentaries of Jacob Ziegler and Georg Tannstetter (Collimitius) see *Jacobi Ziegleri in C. Plinii de naturali historia librum secundum...Item G. Collimitii*

these words in Chapter 17: 'So first let us answer the question – why do these two never go very far away from the sun, and often go back to the sun? Both have contrary circles, being located beneath the sun, and as much of their circles is below [it] as there is of the aforesaid [circles] above it; and so they cannot go further away, because there the curvature of their circles does not have a greater extent.'[193] All declare that in these words Pliny is obscure. Collimitius, and following him Zieglerus, present one interpretation and simultaneously reject it as impossible. Zieglerus adds also another, supporting Milichius, which he himself denies to be derived from the Plinian words.[194] But Copernicus is accepted by both myself and Ursus as a witness in this difficult matter. And he, ignoring Chapter 8 and emending the text, as often turns out to be necessary in the case of this author, enables us to hear Capella's voice already in Pliny. For in Copernicus' opinion Pliny wrote as follows: 'Both have contrary circles, being located around the sun, and as for their circles as much is below as is above' – on the understanding that the word 'beneath' is corrupt and 'of the aforesaid' is really a later interpolation.[195] What especially led Copernicus to believe that Pliny wrote this was the fact that both Vitruvius, before Pliny, and Capella, Pliny's disciple, frequently appropriated one from the other, in turn, the same words, both those which come before and those which follow, as did Pliny, with the same dogmas about the planets, the same opinions and the same errors. On this point, where the concord of Mercury and Venus with the sun is dealt with, they express, moreover, the same view as that which Copernicus restored to Pliny. For since Pliny was master and guide to one of them and disciple of the other, it seems right to call on both as interpreters of Pliny's obscurity. And when I consider what Pliny said in Chapter 8, and what he must now be taken to have said in Chapter 17, it certainly seems to me that the man did not sufficiently understand the accounts of astronomers and the

scholia quaedam (Basel, 1531); Micyllus (Jacob Moltzer), poet, philologist and editor of Aratus' *Phenomena*, published no commentary on Pliny. As Professor Eastwood points out in 'Kepler as historian of science', Kepler's reference is in fact to Milichius (Jacob Milich), *Liber II C. Plinii de mundi historia, cum commentariis...Milichii...* (Frankfurt, 1563; Leipzig, 1573). Another such slip occurs later, where Kepler writes 'Vitruvius' for 'Macrobius'.

193 *Naturalis historia*, II, 14, 72.
194 For a detailed account of the interpretations offered by Tannstetter, Ziegler and Milich the reader is referred to Eastwood, 'Kepler as historian of science'.
195 On the strength of Copernicus' use of *conversas absidas* in the passage from *De revolutionibus*, I, 10, cited in fn. 186, Kepler identifies Pliny as one of the 'other Latin writers' Copernicus has in mind. He then fancifully reconstructs the emendations Copernicus would have had to make to the text to be justified in attributing Martianus Capella's view to Pliny.

words of Vitruvius. Rather, when he ran through all the authors like one caught in a torrent, he muddled everything up, and swollen with a surfeit of knowledge he produced many half-digested things which he himself did not fully understand, leading both himself and the reader astray by the obscurity of his words.[196] This clearly often happened to him on other occasions, when he dealt with subjects far removed from day-to-day life. So in the case of this author it is not absurd to suppose that he believed in that order of the spheres which is in Chapter 8, but nevertheless held in Chapter 17, like Vitruvius, that Venus and Mercury go round the sun.

But let us hear Vitruvius as well.[197] He expounded the prime motion, which is that of the fixed stars and signs of the zodiac, and then said in general terms how 'the moon, Mercury, Venus and the sun itself, as well as Mars, Jupiter and Saturn' – he lists the customary order of the planets from which Pliny greatly departs – pass through the signs in the opposite direction to the fixed stars. Then he went on to give an account of each of them in the order on which he had resolved. First he describes the motion and monthly period of the moon. Then he says of the stars of Mercury and Venus: 'But the stars of Mercury and Venus perform their retrogressions and retardations about the rays of the sun, wreathing about the sun itself as a centre in their travels.' Afterwards he comes to the stars of Mars, Jupiter and Saturn and expounds their motions. So the meaning of his words is clear.

The words of Plato himself are, indeed, obscure, but all the same they offer a hint of the opinion which Macrobius attributes to him: 'When God had made the bodies of each, he placed them in the circuits that the period of each would require, seven bodies in seven circuits: the moon in the first one around the earth; the sun in the second above the earth; the evening star and the star they say is consecrated to Mercury, in virtue of their velocity, in circuits of a kind that run on a par with the sun's orb, but are assigned a force contrary to it. For that reason the sun, Mercury and Venus catch up and are caught up by one another in the same places.'[198]

First, we see that the same orb is attributed to these three on the basis of their speed, but because of the evections of Venus and Mercury it is separated into three. That, however, comes about if the sun is in an eccentric and Venus

---

[196] Cf. Ziegler, who attributes Pliny's obscurity partly to corruption of the text and partly to Pliny's ignorance of astronomy (*In C. Plinii de naturali historia*, 12–13).

[197] *De architectura*, IX, I.

[198] *Timaeus*, 38C–D. The Latin translation, which appears to be Kepler's own, differs substantially from Cicero's (which Kepler had used in his *Mysterium cosmographicum* [*K.g.W.*, I, 23]), as it does from those of Chalcidius, Ficino and Serranus. As Eastwood notes ('Kepler as historian of science', 382), it is closer to the version of Cornarius (Basel, 1561).

and Mercury are in individual epicycles on it. Next, note that Plato is listing them as in Macrobius' arrangement in the order sun, Mercury, Venus, and not in the contrary order sun, Venus, Mercury; for on his account Mercury is said to be nearer the sun than Venus. Thirdly, Plato seems to have provided Capella with the use of the term 'catch up' for 'overtake'.[199] Fourthly, Pliny too is seen to have derived his expression 'contrary circles' from Plato's expression 'contrary force'. So both of them, Pliny and Capella, can be seen to have imitated Plato and to have drawn their opinion from him. This is clearest of all in the case of Macrobius, who lived in the fifth century after Christ, before the one and after the other. For he explicitly declares himself to be Plato's interpreter.[200] 'The arrangement of the Chaldaeans', he says, 'unites Archimedes with Cicero. Plato followed the Egyptians, the originators of all branches of philosophy, who thought the sun to be located between the moon and Mercury; accordingly they discovered and made known the reason why some believed that the sun was above Mercury and Venus. Nor do those who think so stray far from a semblance of the truth.' In a word, Macrobius says that it is in a sense true both that the sun is above Mercury and Venus and that it is below them. There follow these words: 'Reason, indeed, prompted belief in this kind of permutation.' The word 'permutation' appears to echo Plato's words 'they catch up each other and are caught up'. The reason for Plato's belief that the position of these two planets and the sun is sometimes permuted with its contrary was, he says, as follows: 'From the sphere of Saturn, which is the first of the seven, as far as the sphere of Jupiter, the second from the top, so great is the extent of the intervening space that the upper one completes its circuit of the zodiac in thirty years, the lower one in twelve. Again, the sphere of Mars is so far removed from Jupiter that it completes the same circuit in two years.' In these and the following words you have an interpretation of what Plato said,

---

[199] Kepler had earlier rendered καταλαμβάνουσι και καταλαμβάνονται by *comprehendunt et comprehenduntur*; καταλαμβάνω does indeed have the sense of 'catch up' or 'overtake' here, as does *comprehendo* in the last of the passages from Martianus Capella cited in fn. 190. Καταλαμβάνω and *comprehendo* are close equivalents and the connotation is in both cases unusual, so Kepler's argument is not entirely fanciful.

[200] The following quotations are from *In Somnium Scipionis*, I, 19, 1–7. This is an obscure passage. In rendering it I have made use of Eastwood's excellent literal translation and glosses ('Kepler as historian of science', 385–8); but on certain points, e.g., the rendering of *superiores vertices* by 'upper reaches', I have had to risk unfaithfulness to Macrobius' sense as understood by modern scholarship so as to avoid gross inconsistency with Kepler's reading. Eastwood shows, conclusively I think, that Macrobius' text does not (*pace* Stahl, *Macrobius' Commentary on the Dream of Scipio*, 252) allude to heliocentric orbits.

namely: 'God placed the bodies in those circuits which the period' – that is to say, the extent of the time of restitution – 'of each would require.' Macrobius continues: 'Venus, moreover, is so far below the region of Mars that one year is sufficient for it to run its course of the zodiac. Now the star of Mercury is, in fact, so near Venus, and the sun so near Mercury, that each of these three goes round its heaven' – he speaks of a heaven for each – 'in the same interval of time, that is, approximately a year. So Cicero, too, called those two courses "the sun's companions", for they never go far away from each other in their equal period.' Again, he has explicated a point in the text of Plato quoted above. There follows: 'The moon, however, lay so much further below these that what they completed in a year, she completed in twenty-eight days. Hence there was no disagreement amongst the ancients either about the order of the three superior planets, which the immense distance clearly and evidently distinguished, or about the location of the moon, which lay so much below the rest. But the proximity of these three close neighbours, the sun, Mercury and Venus, confounded their order, but only for others. For the following reason [for that confusion] did not escape the ingenuity of the Egyptians.' If the obscurity, of which there is some in this author too, leads anyone into doubt about his opinion, he should ponder well the words just quoted. For the following text has to be so interpreted that from it the basis of the order not only of the sun and its two companions, but also of the two themselves, is made clear. So we have to interpret a single one of the words which follow. 'The circle through which the sun travels' – interpret as epicycle on a concentric, by which we customarily save the eccentricity; unless you do so, you will not render Macrobius' words consistent – this, I say,[201] 'being inferior' – I would prefer 'interior' – 'is encircled by the circle of Mercury, and the higher circle of Venus also includes the latter. As a result, when these two planets run through the higher reaches of their circles they are understood to be located above the sun; but when they pass through the lower [reaches] of their circles the sun is thought to be above them' – supply in addition the words 'since they go around the heaven together', as he said earlier. Moreover he explicates those words of Plato: 'They catch up and are caught up by each other' (and are so in the same quarters, that is, before, behind, above and below).[202] 'So to those who said that their spheres were beneath the sun, this appeared to be so from the course of the stars, which, as we have said, sometimes seems to be below it, and which is indeed more noticeable, because then [the stars] are more

---

[201] 'This, I say', *hic inquam*, is Kepler's interpolation.

[202] The parenthetical remark is evidently Kepler's gloss of *secundum eadem* [*loca*], which he has used earlier to render the Greek *kata tauta*.

readily apparent. For when they are in the upper [reaches] they are more concealed by rays. And as a result the latter belief has prevailed and this arrangement has been accepted by almost everyone up to now. Nevertheless, more accurate observation reveals the better arrangement', etc. I have quoted these things from Macrobius for the sake of clarity, so that the meaning of Plato's words may be understood. The ancient commentary on Bede, which came soon after the time of Martianus Capella, clearly confirms this interpretation in Chapter 14,[203] when it locates the epicycle of Venus between the sun and the earth, and says that for Plato the course of its epicycle provided the reason why 'he had provided the planet Venus with a globe somewhat more elevated than that of the sun'.[204] To be sure, were I to consider the following words of that interpreter: 'Next to the sun she appears higher, next to the earth she appears lower', I would conclude that what I have called 'the epicycle of Venus' was located by him entirely below the sun.[205] But if you look at the other figure – the book was published at Cologne in 1537 – clearly it retails Capella's very opinion.[206] For the point representing the sun, that is, the letter κ, is placed in the embrace of the epicycle of Venus. Nor are the words just cited at odds [with this], provided they are backed up with a modicum of interpretation and 'next to the sun' is related to appearance, but 'next to the earth' to distance between the centres. Besides, the first figure is either defective or provided for the purpose of illustrating the opinion of those who placed Venus' epicycle above the sun.[207] The second figure, corrupted by the ignorance of the draughtsmen, I restore as follows.[208]

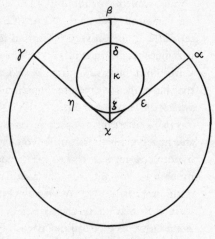

203 *Bedae presbyterii...opuscula* (Cologne, 1537). The *vetus commentarius* to which Kepler refers accompanies Chs. 13 and 14 of Bede's *De natura rerum*. The sources of the *vetus commentarius* are discussed by C. W. Jones, *Bedae Pseudepigrapha: Scientific Writing Falsely Attributed to Bede* (Ithaca, N.Y., 1939), 10; and Eastwood, 'Kepler as historian of science'. As Eastwood shows, neither the text nor the figures cited by Kepler in fact deal with the position of Venus' epicycle.

204 *Bedae opuscula*, f. 9v.   205 *Ibid.*

206 The right-hand one of the two figures on f. 10r.

207 The left-hand figure on f. 10r.

208 As Eastwood, 'Kepler as historian of science', points out, Kepler's restored figure is based on a differently labelled figure on f. 8r which forms part of the *scholia*

So, [Ursus,] compare this with the text of that figure, if you will. For, in passing, I wanted to help you.

But let us return to the matter in hand, and let us out of our generosity [*Refutatio*] despatch that as a reinforcement for Ursus' forces, which are in themselves quite pathetic. The view that the sun is the centre of the orbs of Venus and Mercury was not that of Capella alone, but also of the commentator on Bede, of Macrobius, Pliny, Vitruvius, Plato, and even of the Egyptians themselves. What follows from this? Does what Ursus maintains follow – that the Tychonic hypotheses were employed by those I have just mentioned long before Tycho? Not at all! For, in the first place, as I argued above in considering the Apollonian problem, this single point is but a part of the Tychonic world-system, so it cannot be held to be a proper and legitimate astronomical hypothesis that is distinct from the other ones. And should anyone suspect that Capella, Vitruvius, Pliny, Macrobius or someone else added the other parts of the Tychonic hypothesis to this idea, I appeal to their surviving works. In these they declare themselves in evident, clear and unmistakable terms to hold opinions contrary on all other points to Tycho's and identical to the hypothesis of students of physics (to use Ursus' terminology).[209] It would, indeed, take too long to spell out these passages; it suffices to indicate them. So Ursus' claim stands firm and immovable – verily, like a cart supported by one wheel! Only when he has convinced us that the part is equal to the whole, or two equal to twenty, will he establish what he has maintained!

Finally, as earlier I denied that Apollonius was by profession an astronomer and hence maintained that he was not to be regarded as the author of new hypotheses, so here I press the same point with an argument that is surely much more telling. For whereas Apollonius' problem concerned only the stations of the planets, the speculations of these authors range over the whole form of the fabric of the universe. He, moreover, is deservedly celebrated as the originator of his invention (insofar as he conceived it as a geometer), but they deliberately cobble together their conceptions of the world from elsewhere and embellish their works with them. And they do not arrogate these conceptions to themselves, with the exception of a few things in Pliny whose derivation from Vitruvius he conceals, not altogether to the latter's detriment since both of them were in error. Apollonius deserved praise for his cleverness and the admiration of all for his most profound and true

---

printed by Noviomagus, and not on the figure of the *vetus commentarius* to which Kepler refers.

[209] Cf. *Tractatus*, sig. Cii, r, 38–v, 25, where Ursus calls the Ptolemaic ordering of the planetary spheres 'physical hypotheses'.

speculation. But they abound in errors; and while they are indeed to be thanked for having handed on or preserved these speculations, neither they nor their masters thereby earned personal renown. We would be in a bad way if astronomy still remained in so great a state of uncertainty; if so much was still lacking in it; if our knowledge of latitudes and of the evections of Venus and Mercury was still no more certain; if we still thought that the path of the sun was bent and twisted;[210] and if we had no better grasp of the causes of astronomical phenomena than that we have derived from them.

Through the hypotheses which he justly claims as his own Tycho has provided us with a more perfect astronomy than they left behind. Had he not done so, even though he dealt with the whole universe, he would not object to his ideas being relegated to the figments of Pliny's mind and held unworthy to have the title 'astronomical hypotheses' applied to them.

But Ursus advances other authors of the Tychonic hypothesis as well. He presents not only architects, writers of natural history, poets, grammarians and men unversed in astronomy, whom he accepts despite the disparity of their professions. There follows an astronomer as well, that most famous of all astronomers after Ptolemy, renowned not only for theoretical knowledge but also for [its] outcome, [his] work on the celestial revolutions having been published, a work from which were derived tables; and those tables being accepted today in the service of almanacs, the tables of the ancients have been abandoned. Come on, Ursus, show your mettle — you are on the right track! Go on, tell me, was Copernicus the one who guided Tycho to his hypothesis? But when you tell me, remember that my concern will be with you and your words, not others. Now, as I have also intimated above, I do not deny that since the publication of Copernicus' hypotheses there are and have been many to whom contemplation of the Copernican hypothesis has indicated the same way of proceeding as Tycho's — Rothmann confesses himself one of their number.[211] And if any of them did not make use of Tycho as his instructor in this matter, praise should not, I think, be withheld from him. But, when Tycho sought his hypothesis, he did not, like the others, have in mind the transposition of the Copernican hypotheses to an immobile earth, but the approach of Mars to the earth when in opposition to the sun, the linking of epicycles to the sun and, in all, almost precisely the same points that had earlier been pointed out by Copernicus. But since the falsely

---

[210] Perhaps Kepler has in mind here Pliny's obscure claim that 'the sun travels unevenly in the middle of the zodiac between the two orbs with a serpentine course' (*Naturalis historia*, II, 13, 67).

[211] Rothmann had described his inverted Copernican system in letters to Tycho of October 1587 and October 1589 (*Epist. astron.*, 81–91 and 120–31 [*T.B.o.o.*, VI, 110–19 and 149–61]).

maintained solidity of the orbs had shut him off from the route to his hypothesis, it was fortunate that the comets intervened. By their passage they dissipated those clouds of solid orbs and removed the hindrances from the Tychonic hypothesis.[212] Besides, in the Tychonic [hypotheses] there is not only a mutation and transposition of the Copernican [hypotheses]; they contain many new things peculiar to them derived from the appearances themselves. And if Copernicus wanted to incorporate these things he would have to correct and amend his own hypothesis, something which I, indeed, who follow Copernicus on the main points, think should be done.[213] But you, Ursus, what do you say? What do you cite from Copernicus to confirm your claim that the hypotheses about the celestial motions propounded by Tycho are exactly described in the surviving works of Nicolaus Copernicus? If you had claimed that there is a similarity between the two and that it is an easy passage from Copernicus to Tycho, you have already seen my verdict given. Where, in fact, in Copernicus were they clearly described?[214]

[212] Cf. Tycho's own account of how at first his hypothesis was suspect to him because it involved interpenetration of the orbs of Mars and the sun, until study of the motions of comets shattered his belief in the solid spheres (*De mundi aetherei phaenomenis*, 190 [*T.B.o.o.*, IV, 159]).

[213] Cf. Kepler's account in his letter of May 1599 to Herwart of the way in which Tycho's innovations can be incorporated into the Copernican system (*K.g.W.*, XIII, 347–8).

[214] Had he continued, Kepler would surely have answered his rhetorical question by demolishing Ursus' claim that in *De revolutionibus*, III, 25, and *De revolutionibus*, V, 35, Copernicus explicitly states the Tychonic hypotheses. His earlier assessments of the *Tractatus* may well indicate the way in which he would have proceeded: see Ch. 3, pp. 61–2 and pp. 67–70.

III

The significance of the *Apologia*

6

Against the sceptics

The avowed intent of the first chapter of the *Apologia* is to establish the true nature of astronomical hypotheses and to demolish point by point Ursus' misrepresentation of their status. And Kepler makes it quite clear at the outset that it is above all Ursus' sceptical attitude to astronomy that he is out to refute. Later I shall consider the question of the nature and origins of Ursus' scepticism and of the prevalence of such scepticism in the astronomy of the period, questions that are important for an understanding of Kepler's innovative views on the place of astronomy in the scheme of human knowledge. But before venturing onto these large topics it seems appropriate to offer an exegesis of the arguments for scepticism in astronomy that Kepler attributes to Ursus and of his attempts to rebut them. This preliminary exegesis is frankly concerned to highlight certain arguments that are of obvious interest from the standpoint of modern philosophical concerns about the nature and status of scientific theories. At the end of the chapter I shall consider some of the dangers attendant upon this kind of reading with its inevitable imposition of categories alien to the period and the text. The subsequent chapters will move away from this 'Whiggish' style of interpretation, attempting more laboriously, and perhaps more rewardingly, to discern the significance of the *Apologia* in the context of sixteenth-century views on the status of astronomy and on the nature and history of the mathematical arts.

In the *Tractatus* Ursus never presents detailed arguments in support of his sceptical stance. His denunciation of the claims of astronomy to portray the true form of the world proceeds rather by innuendo and invective. Kepler, however, detects, shrewdly and accurately I think, four main arguments implicit in Ursus' account. Let us take them in the order in which Kepler tackles them.

1) *The argument from pristine usage.* Ursus equates *hypotheses* with *aitemata*.[1] This equation is to be found in Aristotle, who generally uses the term *aitema* in the sense of illegitimate or false postulate. Thus Ursus implies, or at least

---

[1] *Tractatus*, sig. Biv, v, 25; see also Ch. 2, fn. 38.

was reasonably taken by Kepler to have implied, that falsity was part of the *etumon*, the original connotation, of the term *hypothesis*.[2] This philological insinuation is abetted by the fact that the term *hypothesis*, like its Latin equivalent *suppositio*, sometimes has in the period connotations of spuriousness and inadequate warrant.

2) *The argument from evidential insufficiency.* The first step in the argument is to note that 'saving the phenomena', that is, accurate prediction and retrodiction of apparent celestial coordinates, does not guarantee the truth of astronomical hypotheses. For true conclusions can follow from false premises. To take saving of the phenomena as a demonstration of the truth of hypotheses is to commit the fallacy of affirming the consequent. This part of the argument is commonplace – following Averroes and Aquinas numerous authors, uneasily aware of the inconsistencies between Aristotle's cosmology and the planetary models derived from Ptolemy, had pointed out defensively that the truth of astronomical hypotheses is not demonstrated by their capacity to save the phenomena.[3] The argument can be strengthened by appealing, as does Osiander in the preface to Copernicus' *De revolutionibus*, from which Ursus quotes, to the existence of observationally equivalent planetary models. From the time of Apollonius of Perga it had been known that for suitable choice of parameters a planetary model using uniform motion on an eccentric is observationally equivalent to one using an epicycle carried on a concentric. If these devices are supposed to be real entities (extended, if not material or solid) the models are inconsistent; at least one must be false. So here, on the assumption that the models are predictively adequate, we are faced not merely with the logical possibility, but with the actual existence of a false hypothesis adequate to save the phenomena. With the advent of rival world-systems the force of the argument is further strengthened. For suitable choice of parameters Ptolemaic, Copernican and Tychonic world-systems can be made observationally equivalent to within the limits of instrumental accuracy required to detect stellar parallax. Not merely the details of individual planetary models but the general disposition and motions of all the heavenly bodies become open questions that cannot be resolved by observation of apparent celestial motions. On Ursus' assumption that predictive accuracy is the sole criterion of adequacy for hypotheses this implies that the most fundamental issues of astronomy are irresoluble.

3) *The argument by analogy with other disciplines.* Ursus claims that the derivation of true conclusions from false postulates is not peculiar to

---

[2] Cf. *Apologia*, p. 144.

[3] For references see P. Duhem, *ΣΩΖΕΙΝ ΤΑ ΦΑΙΝΟΜΕΝΑ* [*To Save the Phenomena*] (Paris, 1908); transl. by E. Doland and C. Maschler (Chicago, 1969), Chs. 3 and 4.

astronomy, but occurs in other disciplines as well. Thus in algebra the unknown in an equation is denoted by the symbol 'I', although its true value is rarely unity. Likewise in the solution of equations through successive approximation by the so-called *regula falsi*, we start out by choosing some arbitrary and generally incorrect value for the unknown. In both these cases, Ursus maintains, we derive a true conclusion from a false posit.[4]

4) *The argument by induction on the history of science.* Ursus claims that all hypotheses to date contain 'crass absurdities'.[5] His history of hypotheses, in which he mocks all the various world-systems that have been proposed, is evidently supposed to substantiate this claim. And though Ursus offers no explicit argument, Kepler is surely right to infer that he supposes it to follow from the absurdity of all past hypotheses that astronomers are unlikely ever to come up with true hypotheses.[6]

Whilst there appears to be no precedent for so sustained an attack on the pretensions of astronomy, there is little novelty in the materials Ursus deploys. His most obvious derivations are from Ramus. Though he takes issue with Ramus on the existence of a perfect *prisca astronomia* which employed no hypotheses,[7] Ursus' history of the absurd hypotheses put forward by astronomers is an elaboration of Ramus' brief history of astronomy.[8] His denunciation of hypotheses as 'fabrications' and 'lies' parallels Ramus' denunciations. And even his analogy between astronomical hypotheses and the use of false posits in algebra may be derived from Ramus.[9] His argument from observational equivalence is a standard argument, hinted at in Osiander's preface, which he quotes. And as we shall see in the next chapter, his crucial insistence on the irrelevance to astronomy of all criteria for the adequacy of hypotheses save predictive accuracy represents a stance widespread amongst professional astronomers in the period.

In considering Ursus' arguments (as generously reconstructed by Kepler) and Kepler's responses to them it is hard to avoid the impression that they are of very unequal weights. Thus the argument from insufficiency of evidence and the argument by induction on the history of science, arguments which have obvious counterparts in the modern onslaught on realist conceptions of the natural sciences, are apt to appear far weightier than the others. This appearance is potentially misleading. The authority of appeal to the *etumon* of a word, its original connotation, rests on a belief that the pristine

---

[4] *Tractatus*, sig. Biv, v, 28–33. [5] *Tractatus*, sig. Biv, v, 6–9.
[6] *Apologia*, p. 146. [7] *Tractatus*, sig. Ci, v.
[8] *Prooemium mathematicum* (Paris, 1567), 211–17, and in the second book of his *Scholarum mathematicarum libri XXXI* (Basel, 1569), 49–50, quoted in Ch. 8, pp. 266–8. [9] Cf. Ch. 5, fn. 71.

state of the arts and sciences is a repository of authentic meanings. Ursus openly subscribes to this proto-Heideggerian doctrine;[10] and Kepler, for all his insistence on the superiority of the 'modern' arts and sciences to those of the ancients, holds that the insights of the Pythagoreans into the harmonies of the world have a special privilege by virtue of their antiquity.[11] Nor should Ursus' apparently fatuous analogy with algebra be dismissed as merely foolish or opportunist. As Klein has shown, the status of the forms of reasoning employed in the 'magical' art of algebra remained deeply problematic throughout the period.[12] Instead of regarding the x's and y's of equations as designators of variables ranging over numbers, many sixteenth-century authors regarded them as names of the unknown quantity, as 'cossic denominators'. Though to us Ursus may appear inept – confused in the same way as the legendary pupil who complains: 'But, Sir, what if x is not the number of days it takes ten men to dig three ditches?' – in this conceptual framework his belief that all algebraic equations involve false postulates is perfectly apt.

With these provisos in mind, let us rehearse Kepler's counters to Ursus' cogent, if largely unoriginal, arguments.

In response to Ursus's equation of *hypothesis* with *aitema*, illegitimate or false postulate, Kepler concocts a somewhat fanciful history of usage, largely derived from passages in Aristotle and Proclus.[13] The pristine context was, he maintains, geometry. There the term was applied both to first principles of demonstration 'certain and acknowledged by all' and to other premisses, demonstrable but assumed without demonstration. It was the latter usage that led to extrapolation of the term, improper extrapolation Kepler implies, to premisses of doubtful legitimacy and evidence. From geometry the term was transferred to logic, and thence, because of the involvement of logic in all other disciplines, to astronomy, always preserving the original connotation of 'firm and secure foundation'. In this generic sense it is properly applied in astronomy both to observations and to general principles; but in the customary specific sense it applies to the latter alone. This response conforms to the rules of composition outlined in Chapter 4, above, by treating first

---

[10] *Tractatus*, sig. Ci, v.

[11] In the *Apologia* Kepler attributes a heliocentric cosmology to the Pythagoreans on the strength of Aristotle's account of their views in *De caelo*, and in the *Mysterium* and *Harmonice* he further speculates that they knew of the role of the Platonic solids as cosmographical archetypes (*K.g.W.*, I, 26–7; *K.g.W.*, VI, 17–18).

[12] See, e.g., J. Klein, *Greek Mathematical Thought and the Origin of Algebra*, transl. by E. Braun (Cambridge, Mass., 1968), Pt 2; and M. S. Mahoney, 'Die Anfänge der algebraischen Denkweise im 17 Jahrhundert', *Rete*, 1 (1971), transl. in S. Gaukroger, ed., *Descartes: Philosophy, Mathematics and Physics* (Brighton, Sussex, 1980), 141–55.

[13] *Apologia*, pp. 136–8.

the genus and then the species of hypotheses. And it is of considerable dialectical subtlety. Ursus' relation of *hypothesis* to *aitema* or illegitimate postulate is conceded, but shown to be a secondary usage, one which resulted from confusion. And Kepler makes clever use of the device of *coloratio*, introduction at the outset of an oration of material apt to prejudice the hearer in favour of positions to be argued for later. Thus he insinuates the impropriety of a sceptical attitude to astronomical hypotheses, intimating that in the transfer of the term from geometry *via* logic to astronomy the term preserved its original connotations of certainty and evidence.

The core of the *confirmatio* of the *Apologia* is Kepler's counter to the argument from insufficiency of evidence.[14] In the course of his rebuttal Kepler refers to an argument he had offered in his *Mysterium cosmographicum*, and this earlier argument does indeed cast considerable light on this complex and tortuous section of the *Apologia*.

At the beginning of the first chapter of the *Mysterium* Kepler announces:

I have never been able to agree with those who, relying on the example of an accidental demonstration, which with syllogistic necessity yields something true from false premises...used to maintain that it could be that the hypotheses which Copernicus adopted are false, but nevertheless the true phenomena follow from them as if from genuine principles.

Here is Kepler's opening gambit:

In fact the example is inappropriate. For this outcome of false premises is fortuitous, and that which is false by nature betrays itself as soon as it is applied to another related matter; unless you gratuitously allow him who argues to adopt infinitely many other false propositions and never, as he goes backwards and forwards [in his reasoning] [*nec unquam in progressu regressuque*], to stand his ground.[15]

I take Kepler's point in this rather cryptic passage to be as follows. A given set of false premises may, by a lucky accident, yield a conclusion known to be true. But when used for predictive or retrodictive purposes, that is, 'when applied to another related matter', it will fail. Repeated failure can be avoided only if one is prepared repeatedly to make *ad hoc* modifications, 'to adopt infinitely many other false propositions' and 'never to stand his ground'. As it stands this opening move is readily countered by appeal to the case of observationally equivalent but inconsistent world-systems. Kepler is aware of this, for he continues:

You might object as follows. It can be said with some truth today (and could have been said with some truth in the past) that the ancient tables and hypotheses satisfy

---

[14] *Apologia*, pp. 140–3.   [15] *K.g.W.*, I, 15.

the phenomena. Copernicus, nevertheless, rejects them as false. So, by the same token, it could be said to Copernicus that although he accounts excellently for the appearances, nevertheless he is in error in his hypotheses.[16]

Kepler's reply to this cogent objection has two parts. The first is brief:

I reply, to start with, that the ancient hypotheses clearly fail to account for certain important matters. For example, they do not comprehend the causes of the numbers, extents and durations of the retrogradations, and of their agreeing so well with the position and mean motion of the sun. Since in Copernicus' work a most beautiful regularity is revealed in all these things, the cause must likewise be contained therein.

We must be careful in interpreting this. Kepler is not denying the existence of observationally equivalent Copernican and Ptolemaic systems. Rather, he is claiming that where Ptolemaic astronomy merely predicts the phenomena Copernican astronomy reveals their cause. Thus whereas in the Ptolemaic hypotheses the 'numbers, extents and durations' of the planetary stations and retrogradations, and their relation to the mean sun, are predicted by independent assignments of parameters to each planetary model, in the Copernican hypotheses they can be seen to have a single cause, being reflections of the earth's revolution around the sun.

The second part of Kepler's reply is both much more substantial and much harder to follow:

Further, Copernicus denies none of the things in the [ancient] hypotheses which give the cause of appearances and which agree with observations, but rather includes and explains all of them. For though he appears to have changed many things in the accepted hypotheses, it is not in fact so. For it can, indeed, happen that the same [conclusion] results from two suppositions which differ in species, because the two are in the same genus and it is in virtue of the genus primarily that the result in question is produced. Thus Ptolemy did not demonstrate the risings and settings of the stars from this as a proximate and commensurate middle term: 'The earth is at rest in the centre.' Nor did Copernicus demonstrate the same things from this as a middle term: 'The earth revolves at a distance from the centre.' It sufficed for each of them to say (as indeed each did say) that these things happen as they do because there occurs a certain separation of motions between the earth and the heaven, and because the distance of the earth from the centre is not perceptible amongst the fixed stars [i.e., there is no detectable parallactic effect]. So whatever phenomena he demonstrated, Ptolemy did not demonstrate them from a false or accidental middle term. He merely sinned against the law of essential truth, because he thought that these things occur as they do because of the species when they occur because of the genus. Whence it is clear that from the fact that Ptolemy demonstrated from a false

[16] *K.g.W.*, I, 15.

disposition of the universe things that are nonetheless true and consonant with the heavens and with our observations – from this fact I repeat – we get no reason for suspecting something similar of the Copernican hypotheses.[17]

I take the central claim of this difficult passage to be that the Ptolemaic and Copernican hypotheses are observationally equivalent (within the limits of the accuracy of observation that would be needed to detect stellar parallax) precisely because they are kinematically equivalent – that is, for a suitable choice of parameters of the planetary models they ascribe the same relative motions to the celestial bodies. The account of relative motions shared by the two systems constitutes the 'genus' to which both sets of hypotheses belong and from which 'the appearances', the motions apparent to the earthbound observer, can be predicted. The 'specific' difference between the two sets of hypotheses, their different ascriptions of absolute motions, is irrelevant to the business of prediction. If this is the burden of the passage, it should occasion no surprise that Kepler appears to make heavy weather of the point. For, as Palter has emphasised, despite the fact that many sixteenth-century astronomers show a clear grasp of what from a later viewpoint may be regarded as special cases of the principle of kinematic relativity, the principle itself, and the associated terminology of absolute and relative motions, is not to be found until the mid-seventeenth century.[18] The main internal evidence for this reading is provided by Kepler's claim that to demonstrate the risings and settings of the stars it suffices to appeal to the generic hypothesis that there is a *motuum separatio* between the earth and the heaven. *Motuum separatio* must be translated as 'divergence', 'separation' or 'antithesis of motions', but it can surely be glossed as 'differential or relative motion'. The usage of Aristotelian terminology in this passage – 'proximate and commensurate middle term', 'in virtue of the genus primarily', 'law of essential truth' – diverges widely from that of its ultimate source, Aristotle's *Posterior Analytics*. For Aristotle is concerned with simple syllogisms whereas Kepler is concerned with complex geometrical deductions. And in Aristotle's account middle terms are predicates and causes are states of bodies, whereas Kepler calls hypotheses 'middle terms' and treats such complex states of affairs as systems of relative motions as causes. Elsewhere I have shown that Ramus anticipates Kepler's use of this terminology in his *Scholae mathematicae* of 1569.[19] In particular, Ramus had interpreted Aristotle's

[17] *K.g.W.*, I, 15–16.
[18] See R. Palter, 'Some episodes in the history of Copernicanism', in A. Beer and K. A, Strand, eds., *Copernicus: Yesterday and Today* (Oxford, 1975), 47–59.
[19] N. Jardine, 'The forging of modern realism: Clavius and Kepler against the sceptics', *Studies in the History and Philosophy of Science*, 10 (1979), Appendix.

requirement that the premisses of a demonstration be 'essentially true' as a stipulation that nothing irrelevant to the purpose in hand should be included in the demonstration or presentation of a subject. If Kepler is following Ramus, the sense of his claim that Ptolemy breaks the rule becomes evident: for the purpose of saving the phenomena it is otiose to appeal to a 'specific' hypothesis about absolute motions as Ptolemy did, for appeal to a 'generic' hypothesis about relative motions suffices.

With a modicum of reconstruction Kepler's overall contention is as follows. If one imagines, as the sceptic evidently does, that predictive success is the sole criterion of adequacy for hypotheses, then we do not have to choose between the rival world-systems; rather, we can rest content with what is common to them, an ascription of relative motions. If, however, we insist on more, on an account of the form of the universe which 'contains the causes' of the phenomena, then there are, Kepler intimates (and seeks to demonstrate in the rest of the work), ample grounds for preferring the Copernican system. This argument, designed to preempt an objection to Copernicanism, is elaborated in the *Apologia* into a general defence of astronomical hypotheses against the scepticism-inducing argument from insufficiency of evidence.

In the first chapter of the *Apologia*, after treating the genus and species of hypotheses Kepler turns to his third 'topic', their *proprium* – that is, their *raison d'être* or distinguishing characteristic. His immediate target is evidently Ursus' assertion that the *proprium* of hypotheses is to be false whilst yielding true conclusions.[20] Kepler's first move is close to his opening gambit in the *Mysterium*. Though false hypotheses may yield a true conclusion by chance, they will always eventually betray their falsity by yielding a false one.[21] As in the *Mysterium* Kepler goes on to raise the crucial difficulty, repeatedly emphasised by Ursus, the existence of disparate but observationally equivalent hypotheses. Again as in the *Mysterium* Kepler's reply has two parts. In the first part Kepler considers putative examples of such observational equivalence – Copernicus' hypotheses and Magini's emendations of the Ptolemaic planetary models designed to yield results in accordance with the Prutenic Tables; the Copernican hypotheses and Christoph Rothmann's inverted Copernican models – and suggests that in each case there are 'physical considerations' on which a choice can be based. He concludes by induction on these examples that in every such case there will be physical grounds which warrant selection of one of the rivals.[22]

The second part of the reply is considerably more elaborate than its

[20] *Tractatus*, sig. Biv, v, 23.   [21] *Apologia*, p. 140.
[22] *Apologia*, p. 141.

counterpart in the *Mysterium*.[23] In introducing it Kepler makes it clear that it is addressed *ad hominem* to those like Ursus who consider predictive accuracy to be the sole criterion of adequacy for hypotheses. He then takes up the case of the observational equivalence – *aequipollentia* is his term – between the hypothesis of diurnal rotation of the earth and that of diurnal rotation of the firmament. As before, he claims that the difference between the two hypotheses is irrelevant to the business of saving the phenomena since what is common to them, postulation of a *motuum separatio* between the earth and the firmament, suffices for predictive purposes. The second example Kepler considers is the one to which Osiander had appealed, the observational equivalence for a suitable choice of parameters between eccentric and epicyclic planetary models. Here again, Kepler claims, the predictive power of the observationally equivalent hypotheses is attributable to a common postulate entailed by both. Further, in this case there is, he maintains, no real difference between the hypotheses. But his description of the common ground between the hypotheses is a cryptic one: 'though the motion of the planet remains steady it spends longer in one half of the circle'. And the reason he gives for denying that the hypotheses really differ is yet more cryptic: 'And we certainly do not ascribe eyes and human reasoning to the planets, so that they can mark a point here or there with compasses; and authors introduce these specific views I have mentioned as conceits of their own...' Fortunately these obscurities are illuminated later in the *Apologia* when he reverts to Osiander's example.[24] Here he makes it clear that the observationally equivalent eccentric and epicyclic models are to be considered as different geometrical devices for representing one and the same underlying astronomical hypothesis, a hypothesis about the planet's real orbit. From this we may infer that Kepler's baffling remark about ascription of intelligence to the planets is to be taken as a gesture in the direction of the following argument. Since there are no solid orbs, the two hypotheses can be supposed to differ in substance only if we take them as representations of different mechanisms for the impulsion of the planets by 'navigating' intelligences. But since a planetary intelligence cannot be supposed to perceive unoccupied places this is absurd.[25] So this alleged example of inconsistent but observationally equivalent hypotheses is a sham. The two models are merely different representations of one and the same astronomical hypothesis.

Let us turn from exegesis to assessment of Kepler's rebuttal of the argument

[23] *Apologia*, pp. 142–3.   [24] *Apologia*, p. 153.
[25] Cf. *Astronomia nova*, I, 2 (*K.g.W.*, III, 61–71), where Kepler argues explicitly for the equivalence of the two hypotheses from the impossibility of planetary intelligences which navigate with respect to an unoccupied point in space.

from insufficiency of evidence. It has two components. The first may be reconstructed as follows. The sceptic's examples of observationally equivalent hypotheses are supposed to show that there are irresoluble theoretical conflicts in astronomy. But these examples seem convincing only because astronomers often ignore sources of evidence other than the capacity to predict apparent celestial coordinates – because they 'consider only the numbers'. In the *Apologia* Kepler repeatedly lumps together these other sources of evidence as 'physical considerations'. He offers only a small number of explicit examples. Thus Ptolemy's use of the equant is held to constitute physical evidence against his system; and the capacity of the Copernican and Tychonic systems to explain why the superior planets are nearest to the earth when in opposition to the sun provides physical evidence in their favour.[26] As noted earlier, in his choice of explicit examples Kepler is punctilious in avoiding those which militate against the Tychonic system. But he alludes in passing both to the physical arguments he had offered in support of the Copernican system in his *Mysterium* and to further arguments for it that he has derived from his reading of Gilbert's *De magnete*.[27] From this we may infer that the physical considerations to be deployed in resolving theoretical conflict in astronomy include not only the criteria of simplicity and coherence, but also arguments based on his harmonic, architectonic and dynamic speculations.

The second component of Kepler's rebuttal proceeds by showing that for each pair of observationally equivalent hypotheses adduced by the sceptic there exists a third hypothesis, entailed by both, that is adequate for predictive purposes. This second argument does not provide a general answer to the sceptic. For it is not generally true that when inconsistent hypotheses predict the same phenomena in some domain there exists a non-trivial third hypothesis, entailed by both, that predicts those phenomena. It is, however, true of the specific examples adduced by Osiander and Ursus, and Kepler makes it clear that this part of his rebuttal is addressed *ad hominem* to those like Osiander and Ursus who deny the relevance of 'related sciences' to astronomy. The first component of Kepler's rebuttal provides, however, a quite general strategy for rebuttal of the sceptic. Crucial for this strategy is the distinction between the substantive and the conventional components of a theory, the distinction Kepler makes in distinguishing 'true astronomical hypotheses' from the various 'geometrical hypotheses' which may be used to express them. For this distinction makes it plausible to claim that every case in which there appear to be no adequate grounds for choice between rival hypotheses is one of the following types: *either* the hypotheses are not really inconsistent,

[26] *Apologia*, p. 145.   [27] *Apologia*, p. 146.

having the same content but employing different conventions to express it; *or* the hypotheses are not really evidentially equivalent, since their implications in other disciplines provide grounds for a rational choice.[28]

Kepler's reply to Ursus' third argument, the argument by analogy with other disciplines, is brief and effective.[29] The alleged examples of solution of equations by approximation he demolishes with exemplary clarity, pointing out that no derivation of a true conclusion from a false premiss is involved. His refutation of the other example, representation of the unknown in an equation by the symbol 'I', requires a word of explanation. According to Kepler 'I' here is to be regarded as an 'unknown name' which enables us 'by the rules of the art' to discover the true name. Odd though this may sound, it is perfectly coherent. We would say here that 'I' designates a variable whose value is to be determined. But the concept of a variable ranging unrestrictedly over quantities, though adumbrated by Viète, was not generally current in 1600. Instead the symbol 'I' is regarded as designating a name whose identity is to be revealed; it is a name of the name of the solution. This treatment is a halfway house between the 'naive' view of Ramus and Ursus that algebraic symbols are peculiar names of numbers and the 'sophisticated' view that they designate variables ranging over quantities.

The last of the arguments for scepticism that Kepler detects in Ursus' *Tractatus*, an induction on the 'crass absurdities' revealed by the history of astronomy, is countered in the fourth and fifth of the refutations of the first chapter of the *Apologia*.[30] Kepler first notes that Ursus' contention would have to be conceded if whatever has seemed to some to be absurd were in fact absurd. But in fact, he insists, when faced with inconsistent hypotheses philosophers do have the capacity to resolve the issue. In particular, he intimates, he has to hand arguments which show that Copernicus' hypotheses are not absurd. Kepler goes on to suggest that though there remain flaws and open questions in current astronomy, there is much in astronomy that is no longer in doubt. This point he illustrates by adducing, in historical order, some of the discoveries of astronomers about the form of the world, a

---

[28] The literature on modern analogues of this thesis is immense: see, e.g., C. Glymour, 'Theoretical realism and theoretical equivalence', in R. Buck and R. S. Cohen, eds., *Boston Studies in the Philosophy of Science*, 8 (1970), 275–88; W. V. O. Quine, 'On empirically equivalent systems of the world', *Erkenntnis*, 9 (1975), 313–28; and W. Newton-Smith, 'The underdetermination of theory by data', *Aristotelian Society Supplementary Volume*, 52 (1978), 71–91. If evidential warrant is interpreted narrowly in terms of predictive power the thesis is clearly untenable, as Kepler realised. The more generous the notion of evidential warrant employed the more plausible the thesis becomes, but at the cost of rendering the criteria of evidence open to sceptical or relativist challenge.   [29] *Apologia*, p. 149.

[30] *Apologia*, pp. 146–7.

catalogue which culminates in the claim, tactful to Tycho, but historically inaccurate even if applied to the narrowest of professional elites of the period, that almost no one doubts 'what is common to the Copernican and Tychonic hypotheses, namely that the sun is the centre of motion of the five planets'. Here Kepler offers an induction on the genuine past discoveries of astronomers about the nature of the cosmos as a direct counter to Ursus' induction on the absurdity of past hypotheses.

Kepler's claim that progress has occurred in astronomy is backed up by accounts of the means whereby such progress has been achieved. Both in maintaining that natural philosophers can resolve conflict of hypotheses and in alluding to his own reasons for espousing the Copernican hypotheses, he makes it clear that it is above all through deployment of 'physical' criteria that progress has come about. But elsewhere he sketches another method through which astronomy has been improved. This method, which he presents as analogous to the solution of equations by successive approximation, is one in which predictive accuracy is improved by comparing existing hypotheses and eliminating sources of predictive error.[31] Kepler clearly has in mind the elaboration of existing planetary models by modification of their parameters and introduction of further technical devices, equants, prosneuses, Tusi couples, etc. This procedure he attributes to 'practitioners' (*artifices*) and describes as 'artless' (*atechnos*). In thus characterising the method Kepler does not, I think, intend to deny its importance in astronomy. Rather, he implicitly contrasts it with the use of physical criteria to resolve theoretical disputes, a method that is 'artful' (*entechnos*) and practised by astronomers who are prepared to become natural philosophers, the method that he promotes as the primary instrument of progress in astronomy.

Kepler's rebuttals of Ursus' argument from insufficiency of evidence and of his induction on the absurdity of past hypotheses can, with a modicum of reconstruction, be made to appear quite striking in their anticipation of lines of argument deployed in defence of scientific realism in this century. Such reading back of modern arguments and concerns into Kepler is, however, potentially misleading on several scores. Kepler is indeed a realist in astronomy, holding that truth is the aim of the discipline and that it consists in the accurate portrayal of the cosmos; but it should go without saying that Kepler's cosmos is not the mind-independent 'disenchanted' external world of modern scientific realism. Further, Kepler's primary target, the attitude to astronomical hypotheses promoted by Osiander and Ursus, is not, so I shall argue in the next chapter, anti-realist, but rather sceptical. And this type of

[31] *Apologia*, p. 150.

scepticism about astronomy represents a response to specific problems internal to sixteenth-century astronomy rather than a general epistemological position. Kepler is not defending realism in science, but rather the claims of one particular science, astronomy, to have attained some truths about the world and to be capable of attaining more such truths. Further, the terms of the conflict between Ursus and Kepler about the nature of astronomical hypotheses are in crucial respects far removed from the terms of modern debates about the status of scientific theories. With hindsight we can identify as a central issue between Ursus and Kepler the question of the means whereby conflict between theories may be resolved, and in particular the question of the legitimacy of criteria other than empirical evidence. But in thus interpreting the conflict we impose categories that were not in the conceptual repertoire of the period. In the *Apologia* Kepler does, on occasion, use the term *hypothesis* in a sense close to that of 'scientific theory', a definite and consistent body of propositions with an associated domain of observation, measurement and experiment. In Chapter 8 I shall argue that this is an unprecedented usage, and that the emergence in the *Apologia* of a notion close to that of a scientific theory represents one of the most striking innovations of the work. Caution is equally in order in using the modern category of empirical evidence in interpreting Ursus' and Kepler's positions. Hacking has argued cogently that the modern concept of inductive or 'internal' evidence does not become fully articulated until well into the seventeenth century; and he contends that its primary context of emergence in the sixteenth century is in discussion of manifest signs and symptoms of latent processes and structures in the 'low sciences' of alchemy, astrology and medicine.[32] (A morsel of confirmation for Hacking's thesis is to be found in the *Apologia*. For in one of the passages in which Kepler considers the cognitive process by which a hypothesis about 'the form of the world' is derived from observation of celestial phenomena, and thus unambiguously invokes the celestial phenomena as inductive evidence, he draws an illustrative analogy with the process whereby a doctor infers the causes of disease from manifest symptoms.[33]) The distinction Kepler draws throughout the *Apologia* between 'mathematical' and 'physical' considerations relevant to astronomical hypotheses happens to coincide with the modern distinction between empirical and non-empirical evidence; for Kepler's mathematical considerations concern the capacity of astronomical hypotheses to predict apparent celestial coordinates, and his physical considerations have to do with simplicity, coherence and the grounding of astronomy in speculative physics and

[32] I. Hacking, *The Emergence of Probability* (Cambridge, 1975). Chs. 4 and 5.
[33] *Apologia*, p. 151.

metaphysics. But this is not the basic rationale behind Kepler's distinction: rather, the distinction is between the domain of quantitative evidence universally acknowledged to be proper to astronomy and domains of largely qualitative considerations whose relevance to astronomy is debatable.

The distortion imposed by reading back the notions of scientific theory and empirical evidence into Kepler's *Apologia* is not a gross one: approximations to these notions are, I think, genuinely to be found in the work. With a measure of reconstruction we can, as I have done in this exegesis, discern in Kepler's insistence on the relevance of 'physical considerations' to astronomy an attempt to rebut a sceptical argument from observational equivalence of inconsistent theories by appeal to non-empirical evidence. And we can likewise discern in his insistence on the occurrence of theoretical progress in the history of astronomy an attempt to rebut a sceptical argument by induction on the falsity of past theories. But interesting though these anti-sceptical arguments may appear from the standpoint of modern epistemological concerns, they should not lead us to overlook the less evident but perhaps more important themes and concerns which stand out only when the *Apologia* is set against the general background of sixteenth-century views on the status and history of astronomy and the mathematical arts. It is to this more ambitious task of interpretation that the remaining chapters are addressed.

7

The status of astronomy

To identify the attitudes of Ursus, Osiander and others against which Kepler reacts in the *Apologia* it is necessary to make a tentative foray into a virtually uncharted territory, that of sixteenth-century views on the status of mathematical astronomy.

Following Duhem's influential essay *To Save the Phenomena*, the idea that sixteenth-century attitudes to astronomy fall into two main categories, 'realist' and 'fictionalist', has gained wide currency. To the realist camp are assigned Copernicus and Tycho, who promoted their world-systems as representations of the true form of the universe, along with those like Junctinus and Clavius who defended a Ptolemaic astronomy realised in an apparatus of real celestial orbs, and those like Achillini and Fracastoro who pleaded for a revived Aristotelian cosmology of concentric spheres. To the fictionalist camp are assigned a yet motlier crew: those who denied outright the possibility of knowledge of the heavens, Pontano and Frischlin, for example; doubters of the truth of Ptolemaic astronomy who nevertheless conceded its predictive power, such as Agostino Nifo; theologically motivated apologists for Copernicus like Osiander; and the many professional astronomers, Reinhold, Peucer and Praetorius, for example, who insisted on the right of the astronomer to perform his role, that of prediction of the celestial phenomena, without involvement in the recondite uncertainties of natural philosophy. Duhem's tidy partition involves, I shall argue, both an underestimation of the diversity of positions adopted and a failure to appreciate the extent to which they represent responses to specific problems in the astronomy of the period rather than the promotion of general metaphysical or epistemological theses. To arrive at a better understanding of sixteenth-century pronouncements on the credentials of mathematical astronomy we should, I suggest, first consider the place of astronomy in renaissance schemes of the arts and sciences.

The account of the scope of astronomy with which Francesco Giuntini (Junctinus) opens his commentary on the first two books of Sacrobosco's

De sphaera provides a useful, if arbitrary, starting point for our survey.[1] Astronomy Giuntini defines simply as the study of the heavenly bodies and their motion. It has five parts. The first considers the nature of the heavens in general and is typified by Aristotle's *De caelo*. Strictly speaking it is to be considered as a branch not of astronomy but of physics. The second part likewise deals with the heavens in general, but from a mathematical rather than a physical standpoint. It is typified by the first two books of Sacrobosco's *De sphaera*. The third part provides detailed mathematical accounts of the motions of the individual planets and is typified by Ptolemy's *Almagest*. The fourth part is concerned with prediction and tabulation of apparent celestial motions and construction of calendars and almanacs. The fifth part is judicial astrology. This account can be paralleled in the prefaces or opening chapters of any number of Epitomes and Institutes of astronomy, and commentaries on Sacrobosco's *De sphaera* and Peurbach's *Theoricae novae*. Sometimes judicial astrology is more sharply separated from astronomy *sensu stricto*; sometimes Giuntini's second, third and fourth parts are explicitly characterised as 'mathematical astronomy'; and sometimes his fourth part is separated from the others as 'practical' as opposed to 'theoretical' astronomy. But however labelled the parts Giuntini picks out are generally recognized. They represented accurately the main genres of astronomical writing in the period and they reflect the disposition of astronomical studies within the curriculum. For our purposes the barest sketch of these genres and their institutional basis will suffice.

The first of Giuntini's branches of astronomy, which, for want of a better term, I shall call 'metaphysical astronomy', is represented in the period in commentaries on *De caelo* and *Metaphysics*, XII, and by specialised treatises on the substance and nature of the heavens.[2] The gulf between such treatises and the various genres of the remaining 'mathematical' branches of astronomy is a wide one. The central topics addressed – are the heavens composed of form and matter, or of form alone? do the heavens have a potential for material existence or only a potential for location? etc. – are

---

[1] Junctinus, *Commentaria in sphaeram Ioannis de Sacro Bosco accuratissima* (first publn, 1564; Lyons, 1568), 10. Giuntini's partition of astronomy is closely based on the partition of astronomy to be found in the first of Pierre d'Ailly's fourteen questions on Sacrobosco's *De sphaera*: see *Nota eorum quae in hoc libro continentur...Reverendissimi domini Petri de Aliaco...quaestiones subtilissimae numero XIIII* (Venice, 1508), f. 72. As Duhem notes, much else in Giuntini's commentary derives from Ciruelo's commentary on *De sphaera* first published with d'Ailly's *quaestiones* in 1498: P. Duhem, *ΣΩZEIN TA ΦAINOMENA* [*To Save the Phenomena*] (Paris, 1908); transl. by E. Doland and C. Maschler (Chicago, 1969), 84–5.

[2] The genre to which belong the works of Niphus, Achillini, Mercenarius and Zabarella, cited in fn. 19 and fn. 63.

of a purely metaphysical type. As Donahue has shrewdly observed, many of these discussions appear to be concerned not so much with the nature of the heavenly bodies but rather with the prior question whether and in what sense the heavenly entities constitute bodies.[3] References to the views of 'mathematical' astronomers on the disposition and motions of the heavenly bodies are rare and, when present, generally offhand. In the various genres of mathematical astronomy we find the antithesis of this position. Whilst the tension between Ptolemaic and Aristotelian cosmology is generally touched on, at least obliquely, the specialised concerns of the natural philosophers are rarely mentioned, and when mentioned are often dismissed as 'curiosities' or 'subtleties'.

This polarisation of interest reflects both a gulf between the disciplines of mathematics and natural philosophy in the conventional classification of knowledge, the 'ideal' curriculum inherited from the mediaeval period, and a gulf between the disciplines in the real curricula of renaissance universities. In the ideal curriculum mathematical astronomy formed one of the four mathematical arts of the *quadrivium*, which together with the *trivium*, grammar, rhetoric and dialectic, constituted the liberal arts.[4] In theory the liberal arts were propaedeutic to the three philosophies, sacred, moral and natural, and it was of the last of these that metaphysical astronomy formed a small and specialised part. The current state of research into sixteenth-century university curricula does not justify any very substantial generalisations about the teaching of mathematical and metaphysical astronomy in the period. But the few generalisations that can be ventured serve only to emphasise the gulf.[5]

[3] W. H. Donahue, 'The dissolution of the celestial spheres, 1595–1650' (Ph.D. thesis, Cambridge, 1972; New York, 1981), 31: 'What is clearly shown is that the traditional philosophy of the time did not include the idea that the heavens must necessarily be either solid or fluid. They were still much involved in an earlier stage of the argument: whether, and to what extent, the heavenly bodies are bodies.'

[4] On mediaeval classifications of the arts and sciences see L. Baur's introduction to his edition of Gundalissinus' *De divisione philosophiae, Beiträge zur Geschichte der Philosophie des Mittelalters*, IV (Münster, 1903), 2–3; J. A. Weisheipl, 'Classification of sciences in Mediaeval thought', *Mediaeval Studies*, 17 (1965), 54–90; and N. H. Steneck, 'A late Mediaeval *arbor scientiarum*', *Speculum*, 50 (1975), 255–69. It should be noted that in mediaeval schemes of knowledge all arts and sciences were in theory subservient to philosophy. It may therefore be misleading to read back into the mediaeval period a view of mathematical astronomy as an autonomous discipline potentially in conflict with other such disciplines: see, e.g., J. E. Murdoch, 'From social into intellectual factors: an aspect of the unitary character of late mediaeval learning', in J. E. Murdoch and E. D. Sylla, eds., *The Cultural Context of Mediaeval Learning* (Dordrecht, 1975), 271–348.

[5] The following account is largely derived from C. B. Schmitt, 'Philosophy and science in sixteenth century universities: some preliminary comments', in Murdoch and Sylla, *The Cultural Context of Mediaeval Learning*, 485–530; 'Science and philosophy

Where astronomy was taught at all it was taught in the arts course, and such instruction generally covered astrology together with mathematical astronomy. The connection between medicine and astronomy in the curriculum, and in the careers of university teachers, was a close one; understandably so, given the importance of astrology in medical practice of the period.[6] In advising Kepler to abandon the teaching of astronomy in order to acquire a medical qualification the councillors of Graz urged him to pursue a conventional career.[7] The 'metaphysical' astronomy of *De caelo* and its commentators figured as a minor part of the philosophy course. Humanist programmes of educational reform called for emphasis on the liberal arts and advocated substantial changes in the arts curriculum, notably the displacement of scholastic logic by the new humanist dialectic inspired by Cicero and Quintilian, and the expansion of mathematical instruction.[8] But as Schmitt has recently argued, the humanist goals were not consistently realised; in most universities a strictly Aristotelian natural philosophy remained entrenched.[9] Further, the expansion of the role of mathematics in the curriculum was by no means the prerogative of the reformed Protestant universities, but rather ran in parallel with the revival of mathematical teaching in the Jesuit colleges.[10] In those universities, a minority, in which mathematical instruction beyond elementary arithmetic and geometry was provided, Sacrobosco's *De sphaera*, a mediaeval treatment of the rudiments of spherical astronomy,

in the Italian universities in the sixteenth and seventeenth centuries', in M. Crosland, ed., *The Emergence of Science in Western Europe* (London, 1975), 35–56; and 'Astronomy in universities, 1550–1650', in M. A. Hoskin, ed., *General History of Astronomy*, forthcoming (I thank Dr Schmitt for allowing me to see a draft of his contribution).

[6] Giuntini's declaration: 'Without astrology a doctor is blind and no one would trust him' (*Commentaria in sphaeram*, 9) represents a prevalent view.

[7] *K.g.W.*, XIV, 128–9. On the connections between astronomy and medicine see Schmitt, 'Astronomy in universities'; and R. S. Westman, 'The astronomer's role in the sixteenth century: a preliminary study', *History of Science*, 18 (1980), 105–47.

[8] On humanist dialectic see C. Vasoli, *La dialettica e la retorica dell'Umanesimo* (Milan, 1968); and L. A. Jardine, 'Lorenzo Valla and the intellectual origins of humanist dialectic', *Journal of the History of Philosophy*, 15 (1977), 143–64. On humanist promotion of mathematics in university curricula see, e.g., J.-C. Margolin, 'L'Enseignement des mathématiques en France (1540–70): Charles de Bovelles, Fine, Peletier, Ramus', in P. Sharratt, ed., *French Renaissance Studies, 1540–70* (Edinburgh, 1976), 109–55.

[9] Schmitt, 'Philosophy and science in sixteenth century universities'.

[10] On mathematics in the Jesuit *ratio studiorum* see B. Cosentino, 'L'insegnamento delle matematiche nei collegi gesuitici nell'Italia settentrionale. Nota introduttiva', *Physis*, 13 (1971), 205–17; and A. C. Crombie, 'Mathematics and Platonism in the sixteenth century Italian universities and in Jesuit university policy', in Y. Maeyama and W. G. Saltzer, eds., *PRISMATA: naturwissenschaftsgeschichtliche Studien* (Wiesbaden, 1977), 63–94.

provided the basic astronomical text. In the yet smaller minority of courses in which 'higher mathematics' figured, Peurbach and Ptolemy were often cited as texts, from which we may infer that a minority of students studied the third and fourth of Giuntini's parts of astronomy, planetary models and their practical applications.

These features of the curriculum are reflected in the careers of the astronomers whose views we are about to examine. Though many held at some stages in their careers university posts in other arts disciplines, rhetoric, Greek, history, etc., and several either were or went on to become medics, it seems that none ever held a university post in philosophy or theology. When we turn from bare enumerations of the branches of astronomy to accounts of its status, its methodology and its relation to other disciplines we no longer find a consensus, but rather heated controversy over a wide range of positions. Why?

Astronomy fitted awkwardly into the traditional Aristotelian scheme of knowledge. Like medicine it was not readily partitioned into theoretical or contemplative and practical or operative branches as the scheme required. And its relations to mathematics and physics were unclear, and had been unclear it appeared even to Aristotle, who in his *Metaphysics* had declared it to be a branch of mathematics, but in his *Physics* and *De caelo* had treated it as a branch of physics, albeit the one closest to mathematics.[11] In other disciplines which dealt with motions in the broad Aristotelian sense, a sense which includes the full range of qualitative changes as well as locomotion, the division of labour between physics and mathematics was straightforward. Study of the kinds and causes of motion was assigned to physics; study of the extents, durations and degrees of motion, the 'mathematical accidents', to mathematics. But in astronomy this partition of labour was problematic both in theory and in practice: in theory because since the heavenly bodies were held not to be subject to changes of state their motions could not strictly speaking be considered amenable to physics, but must be assigned rather to metaphysics;[12] in practice because the metaphysical account of the nature of the celestial motions that Aristotle had offered was in *prima facie* conflict with the best available account of the 'mathematical accidents' of the celestial motions, the Ptolemaic account. It was this last difficulty that lent these disputes their polemical edge. For all the intellectual

[11] *Physics*, II, 2; *Metaphysics*, I, 8; *Metaphysics*, XII, 8.
[12] Cf. Nabod, *Astronomicarum institutionum libri III* (Venice, 1580), f. 3–5, where astronomy is related to *philosophia prima* because of its concern with the immutable heavenly bodies and to physics because of its concern with the changes in the sublunary realm which the celestial motions bring about.

and institutional gulf between the mathematical astronomy of the liberal artist and the metaphysical astronomy of the natural philosopher, few authors can in fact have been unaware of the tension between Aristotelian cosmology and Ptolemaic astronomy.

As a first step in categorising sixteenth-century attitudes to the postulates of mathematical astronomy it is as well to seek out the grain of truth in the old hagiographic myth according to which mediaeval and renaissance astronomers and natural philosophers believed in a celestial machinery of solid crystalline spheres until Tycho's observations of the comets, in Kepler's words, 'dissipated the clouds of solid orbs'.[13] Setting aside the minor point that Jean Pena and not Tycho appears to have been the first to challenge solid spheres on empirical grounds,[14] we may note that in mediaeval astronomy an ingenious compromise was forged between the Aristotelian cosmology of real concentric celestial spheres and the Ptolemaic planetary models. To provide a physical realisation of a Ptolemaic eccentric a planet was fixed in a spherical shell eccentric to the earth in both surfaces, sandwiched between outer and inner spherical shells of uneven thickness, the outer surface of the outer shell and the inner surface of the inner shell being concentric with the earth. To realise a Ptolemaic epicycle a planet was fixed in a small sphere embedded in a spherical shell concentric in both surfaces with the earth. By combining the two devices each Ptolemaic planetary model could be realised by a system of contiguous spherical shells and spheres which together formed a 'complete sphere' (*sphaera tota*) concentric with the earth in both surfaces. These complete spheres were identified with the Aristotelian concentric spheres. This compromise proved remarkably resilient in mathematical astronomy, providing as it did the framework for the planetary models of Peurbach's *Theoricae novae*, the major textbook of technical astronomy throughout the sixteenth century.[15] (It has even been argued controversially that Copernicus conceived his planetary models as realised by systems of real concentric spheres patterned on those of the *Theorica* tradition.[16]) It is, I suggest, against this background that we should interpret the stance of the

---

[13] *Apologia*, p. 207; cf. *Astronomia nova* (*K.g.W.*, III, 37).

[14] *Euclidis optica et catoptrica* (Paris, 1557), *praefatio*. Solid epicycles would, Pena suggested, produce anomalous refractions of the light from the heavenly bodies.

[15] For a clear account of the mediaeval compromise see E. Grant, 'Cosmology', in D. C. Lindberg, ed., *Science in the Middle Ages* (Chicago, 1978), 265–302. On the *Theorica* tradition see O. Pedersen, 'The decline and fall of the *Theorica planetarum*', in E. Hilfstein *et al.*, eds., *Science and History: Studies in Honor of Edward Rosen*, Studia Copernicana, XVI (1978), 157–85.

[16] See N. M. Swerdlow, '*Pseudodoxia Copernicana*: or, enquiries into very many received tenents and commonly presumed truths, mostly concerning spheres', *Archives internationales d'histoire des sciences*, 26 (1976), 108–58; and E. Rosen's

many sixteenth-century mathematical astronomers who appear sanguine about the apparent conflict between Aristotle and Ptolemy. Thus it is possible to trace a series of commentators on Sacroboscos's *De sphaera* from Cardinal Pierre d'Ailly (1350–1420) to Christoph Clavius (1537–1612), who happily accept the reality both of Aristotle's concentric spheres and of epicycles and eccentrics.[17] Likewise amongst natural philosophers, whilst the nature of the celestial spheres is the subject of elaborate speculation, explicit rejections of the reality of epicycles and eccentrics are relatively rare. Since Aristotelian natural philosophers almost without exception endorse Aristotle's concentric spheres, it seems reasonable to infer that here too adherence to the *Theorica* compromise remained widespread.

In his search for sixteenth-century 'fictionalists', admirable persons in his view, Duhem was remarkably thorough. My own survey, guided by Riccioli, Houzeau and Lancaster, and by Donahue's extensive bibliography, has added only a handful of names to his list of those who deny or doubt the physical reality of the planetary models of the traditional astronomy.[18] But the grounds for such scepticism turn out to be extraordinarily varied, far more varied than Duhem realised.

In surveying this medley of opinions, let us consider first a relatively straightforward attitude, that of those whom Clavius, defending traditional orthodoxy, calls 'Averroists'. They include the natural philosophers Alessandro Achillini (1463–1512) and Agostino Nifo (c. 1469–1538), as well as Girolamo Fracastoro (1478–1553) and Giovanni Amici (1511–38).[19] Following Averroes they deny outright the truth of the Ptolemaic astronomy simply on the grounds that a close reading of Aristotle's pronouncements in his *Metaphysics* and *De caelo* shows that to be consistent with the principles of Aristotelian cosmology planetary models must employ only concentric spheres. Nifo and Achillini are content to follow Averroes in arguing that

spirited 'Reply to N. Swerdlow', *Archives internationales d'histoire des sciences*, 26 (1976), 301–4.

[17] D'Ailly, *De sphaera*; Francesco Capuano, *Expositio sphaerae* (Venice, 1508); Wenceslas Fabri, *Opusculum Ioannis de Sacro Busto cum notabili commento a magnifico viro domino Wenceslao Fabri de Budrveysz* (Leipzig, 1510); Elia Vinetus, *Sphaera Ioannis de Sacro Bosco emendata. Eliae Vineti Santonis scholia...* (Cologne, 1564); Junctinus, *Commentaria in sphaeram*; Clavius, *In Sphaeram Ioannis de Sacro Bosco commentarius* (first publn, 1571; Rome, 1585).

[18] G. B. Riccioli, *Almagestum novum astronomiam veterum novamque complectens* (Bologna, 1651); J. C. Houzeau and A. Lancaster, *Bibliographie générale de l'astronomie* (Brussels, 1880–9); Donahue, 'The dissolution of the celestial spheres'.

[19] Achillini, *Quatuor libri de orbibus* (Bologna, 1494); Niphus, *In Aristotelis libros de coelo et mundo...* (first publn, Naples, 1517; Venice, 1549); Amici, *De motibus corporum coelestium juxta principia peripatetica sine eccentricis et epicyclis* (Venice, 1536); Fracastoro, *Homocentrica sive de stellis* (Venice, 1538).

the truth of Ptolemaic models is not demonstrated by their capacity to save the phenomena, since true conclusions can follow from false premises, and that it is therefore possible that concentric models adequate to save the phenomena can be constructed. Thus Nifo writes:

Averroes argued logically against epicycles and eccentrics in the following way. One should understand that a sound demonstration is one in which the *causa* is necessary for the effect. Now it is granted that when eccentrics and epicycles are posited the appearances follow and can be saved. But the converse is not true. When the appearances are posited, epicycles and eccentrics do not have to be [posited], except for the time being until another and better *causa* is discovered which is necessary [for the appearances]. The proponents of epicycles and eccentrics are therefore in error, because they argue from a posit having several *causas* for the truth of one of them. But these appearances can be saved both in this way and in others which have not yet been discovered.[20]

And he goes on to cite with approval Averroes' supposed attempt to construct an adequate concentric astronomy.[21] Amici and Fracastoro go further and try to provide such an astronomy. Such revivals of concentrism are not surprising. The *Theorica* compromise could hardly survive meticulous reading of *Metaphysica* and *De caelo* of the kind to which Nifo and Fracastoro subjected the works or serious attempts to reconstruct the systems of Eudoxus and Calippus as retailed in the *Metaphysica*. So the revival can reasonably be regarded as a byproduct of the widespread early-sixteenth-century Italian attempt to recover, with the help of Averroes and the classical commentators, a pure Aristotelianism purged of mediaeval distortions and accretions.[22] That concentrism remained a minority stance is surely attributable to the difficulties it faces in accounting for the variation in apparent magnitude most noticeable in the cases of the moon and Mars, difficulties evident, as Kepler notes, to the most astronomically unsophisticated.

A second minority stance involves outright denial of the reality of the traditional planetary models without promotion of an alternative type of planetary model. An early exponent is the humanist poet and courtier Giovanni

---

[20] Niphus, *In Aristotelis libros de coelo*, f. 82. Note that *causa* here, as often in the period, means 'sufficient condition'.

[21] In commentary 45 on *Metaphysics*, XII, 8, Averroes says that he had tried and failed to construct such an astronomy: see F. J. Carmody, 'The planetary theory of Ibn-Rushd', *Osiris*, 10 (1952), 556–86, p. 572.

[22] On the increasing scholarly and philological sophistication of sixteenth-century Italian interpretations of Aristotle see C. B. Schmitt, 'Towards a reassessment of renaissance Aristotelianism', *History of Science*, 11 (1973), 159–93. Nifo and Fracastoro were highly competent Greek scholars; Achillini never learned Greek (Schmitt, 'Towards a reassessment of renaissance Aristotelianism', 170).

Gioviano Pontano (1426–1503). At the beginning of Book 3 of his *De rebus coelestibus libri XIIII*, he maintains that the heavens are entirely spiritual and utterly alien in nature.

If we seek in the heaven things which relate to our eyes and ears, why should we not then seek what relates to our noses? Note too that the sun is not in the middle of the planets in the sense that it is distant by a certain amount from the moon and Saturn. But let those who would determine the middle in accordance with mathematical measures read Aristotle as he attacks numbers and the power and rule of numbers over natural causes.[23]

Pontano goes on to deny that the motions of the planets are ruled by the sun's rays, or by a magnetic virtue in the sun. Rather the planets move by their own choice and volition in the ether like fish in the sea, so that it is no wonder that they move sometimes forwards and at other times backwards in their courses. He concludes that though astronomers are to be praised for fabricating epicycles to save the phenomena, it is absurd to suggest that such things exist.[24] Pontano's markedly sceptical position had few later adherents. A possible case is that of Christian Wursteisen (Vorstitius or Urstisius, 1544–88). In his *Quaestiones in theoricas planetarum Purbachii* he cites with approval Pontano's claim that epicycles and eccentrics are fabrications designed solely for the purpose of saving the phenomena, and he emphasises the limits to our knowledge that arise from the remoteness and inaccessibility of the heavens.[25] Another likely adherent is Nicodemus Frischlin (1547–90), the humanist polymath notorious for his scurrilous lampoons of his colleagues at Tübingen. His curious *De astronomicae artis cum doctrina coelesti et naturali philosophia congruentia libri quinque* is for the most part an orthodox, though inept, exposition of elementary spherical astronomy based on Sacrobosco's *De sphaera*. But in the first book he expatiates on the utter incapacity of our imperfect intellects and senses to apprehend the perfect, simple and rapid motions of the heavens and he castigates as manifestly false both the Ptolemaic and the Copernican systems. He promises to declare later the true nature of hypotheses, and in the fourth book he does so, declaring that epicycles and eccentrics are fabrications at odds with the nature of the heavens.[26]

---

[23] *De rebus coelestibus libri XIIII*, (first publn, Naples, 1512; Basel, 1556), 2113.

[24] Plato's *Timaeus* is one of Pontano's sources of inspiration; another may well be Cicero's *De natura deorum*, Bk 2, in which the view that the harmonious celestial motions are the voluntary motions of divine intelligences is aired at length.

[25] *Theoricae novae planetarum Georgii Purbachii Germani. Quibus accesserunt...Christiani Urstisii quaestiones in theoricas planetarum Purbachii* (Basel, 1568–9), 4–15.

[26] *De astronomicae artis cum doctrina coelesti et naturali philosophia congruentia libri quinque* (Frankfurt, 1586), 40–3 and 258–60.

Pierre de la Ramée (1515–72) and Francesco Patrizi (1529–97) present views on the status of astronomical hypotheses that have similar classical sources and echo in certain respects the scepticism of Pontano. In a famous passage in his *Prooemium mathematicum*, quoted in the next chapter, Ramus pleads for an astronomy without hypotheses, of the kind once possessed by the Babylonians. Such an astronomy was lost, he claims, through the misguided postulation of causes for the phenomena, first the revolving orbs of Eudoxus and then the absurd epicycles and eccentrics introduced by the later Pythagoreans. To restore astronomy to its pristine state 'precepts' should be derived from tables of observations, just as the precepts of other arts are derived by logic from our experiences. It seems that Ramus envisaged a pure calculus for predicting apparent celestial coordinates without appeal to planetary models. This reading is in line with his radical epistemology of the mathematical arts. For Ramus each branch of mathematics is defined by its practical goal, arithmetic as the art of counting, geometry as the art of measuring and astronomy as the art of predicting the celestial phenomena.[27] Methodical presentation of an art proceeds by the famous Ramist dichotomies by which its scope and the types of problems to which its precepts are applicable are successively divided up. Within each subdivision the 'precepts' of an art are to be taught as devices for solving specific practical problems. Thus for Ramus the proofs of geometrical theorems are of secondary importance: understanding of theorems is conveyed by showing how they contribute to the solution of practical problems of mensuration. Underlying this didactic programme is a radically empiricist view of the status of mathematical truths. A geometrical theorem is a generalisation about the results obtained in measurement of real bodies, not a truth about abstract geometrical entities. And since Ramus draws no distinction between pure and applied mathematics, it is reasonable to suppose that he envisaged astronomical precepts too as generalisations about apparent celestial coordinates, not as regularities governing real celestial motions. Ramus' epistemology, unlike his influential didactic theories,[28] has received little attention. But the terms in which he attacks Aristotelian logic and methodology in his *Animadversionum Aristotelicarum libri XX*[29] suggest that he is committed to a sceptical denial of the possibility of access to the causes which underlie phenomena. Thus he denies the validity of the Aristotelian distinction between things 'better

---

[27] On Ramus' theory of mathematical knowledge see J. J. Verdonk, *Petrus Ramus en de Wiskunde* (Assen, 1966).

[28] The standard work on Ramist didactic method is W. J. Ong, *Ramus: Method and the Decay of Dialogue: from Art of Discourse to the Art of Reason* (Cambridge, Mass., 1958).   [29] Paris, 1548.

known to us' and things 'better known in nature', insisting that the sole order to be found is that of evidence to our minds. Given that Ramus proposed his new didactic methodology for both natural philosophy and the mathematical arts, it may well be that his refusal to countenance real celestial motions as causes of apparent celestial motions is symptomatic of a general refusal to countenance hidden causes that are prior to the phenomena.

On the strength of the views which Kepler attributes to him, refusal to distinguish apparent from real celestial motions, rejection of the planetary models of astronomy as fabrications and insistence on the divinity of the celestial bodies and the volition which they exercise in their motions, Patrizi's stance is allied to that of Ramus on the one hand and to that of Pontano on the other. Kepler in fact seriously misinterprets Patrizi, but the misinterpretation is understandable given such passages as the following.

They are called planets and wanderers because they meander with various motions to this side and that of the ecliptic, the sun's path. And sometimes they turn from the sun's circuit to the south and sometimes to the north...But it is through our will and our error that they go astray, and through no error of their own. Nor can they go astray of their own nature. For [their nature] is a sort of soul, a rational soul I declare, and it is endowed with intellect, and neither of these can go astray in the celestial region. Nor is blame attributable to it or to the body which it animates, [a body that is] celestial and as Aristotle and Plato said, divine. And they are not, as most of the mass of astronomers and philosophers think, inanimate bodies. But they are, as Zoroaster rightly called them, and after him Aristotle and Plato, animals.[30]

A close reading of Patrizi's work, an elaborate reconstruction of the *prisca cosmologia* of the Zoroastrians derived from the hermetic *corpus* and a wide variety of other neo-Platonist sources, does not confirm Kepler's impressions.[31] Patrizi does not in fact refuse to distinguish between real and apparent celestial motions; indeed he attributes the apparent rotation of the heavens to the earth's diurnal motion.[32] The planets and fixed stars he holds to be located in a single sphere, and far from denying the regularity of their motions, he insists that the astral intelligences follow divine precepts which ensure a universal harmony and regularity of motions.[33]

---

[30] Patrizi, *Nova de universis philosophia* (first publn, Ferrara, 1591; Venice, 1593), f. 105v, col. 2.

[31] On Patrizi's cosmology see B. Brickman, *An Introduction to Francesco Patrizi's Nova de universis philosophia* (New York, 1941); and J. Henry, 'Francesco Patrizi da Cherso's concept of space and its later influence', *Annals of Science*, 36 (1979), 549–75. On his classical scholarship and sources see G. Saitta, *Il pensiero italiano nell'Umanesimo e nel Rinascimento*, 2nd edn (Florence, 1961), Ch. 9.

[32] Patrizi, *Nova de universis philosophia*, f. 102r–103v.

[33] Patrizi, *Nova de universis philosophia*, f. 106v, col. 2.

Despite variation in detail there is substantial common ground between the positions adopted by Pontano, Frischlin, Wursteisen and Ramus and that misattributed by Kepler to Patrizi. In each case the planetary models of astronomers are dismissed as fabrications valid solely for the purpose of saving the phenomena. In each case our capacity to discover simple and regular celestial motions which underlie the appearances is denied outright. Indeed, Pontano and Ramus go further, apparently doubting the intelligibility of claims about real celestial motions; so their attitude to mathematical astronomy can, by a stretch of the historical imagination, be regarded as 'fictionalist' or 'instrumentalist'. And finally in each case the views are at least partially inspired by Plato and Proclus. It is, I suggest, exponents of this cluster of views whom Kepler castigates as 'deluded seekers after abstract forms who despise matter (the one and only thing after God)'. The phrase 'matter the one and only thing after God' has a Stoic ring. In the Stoic cosmology 'spirit' or 'God' and matter are the two primary principles. The necessity or providence which rules the behaviour of material bodies is the product or emanation of 'spirit'. Those who recognise the dignity of matter, Kepler implies, see it as capable of obedience to the providence of the creator.[34] In the case of the heavenly bodies such obedience is manifested in their conformity to the regular motions prescribed at the creation, precisely the state of the cosmos that Kepler has earlier urged against Patrizi as being most pleasing to God.[35] I conclude that those 'who despise matter' are those who doubt or deny the existence of such an underlying order, and that in describing them as 'deluded seekers after abstract forms' Kepler attributes the denial to the Platonic doctrine that true order is to be found only in the unknowable, eternal and immaterial world of the forms, not in the sensible material world. Though Kepler was wrong to attribute this stance to Patrizi, he was, if my identification of his target is correct, right to regard the most

---

[34] *Apologia*, p. 156. Kepler's allusion to such Stoic commonplaces does not imply any specific affiliation to Stoic cosmology, or indeed any great conversancy with it. A likely source for his characterisation of matter as 'the one and only thing after God' is Diogenes Laërtius' *De clarorum philosophorum vitis*, VII, 134, where the Stoic 'active principle' is identified with God. Kepler's association of Stoic commonplaces with the regularity of the celestial motions has a classical antecedent in Cicero's *De natura deorum*, Bk 2, in which the harmony and regularity of the heavenly motions is adduced as evidence for providential design: A. Birkenmajer, 'Copernic philosophe', *Etudes d'histoire des sciences en Pologne* (Wrocław, 1972), 622, notes the importance of this source for Stoic doctrines in renaissance astronomy. Unfortunately the only general treatment of the sixteenth-century interest in Stoicism, L. Zanta, *La renaissance du Stoicisme au XVIe siècle* (Paris, 1914), concentrates mainly on Stoic moral philosophy. [35] *Apologia*, p. 155.

radically sceptical position in the astronomy of his day as having an ultimately Platonic inspiration.[36]

When we set aside protagonists of these two extreme positions, out-right scepticism and Averroist concentricism, we are left with a substantial number of sixteenth-century authors who, without openly committing themselves to radical scepticism, doubt or deny the capacity of astronomers' planetary models to represent the disposition and motions of the heavenly bodies and insist on a strict distinction between the proper concerns of the mathematical astronomer and those of the natural philosopher. Here, for example, is an account of the distinction, given by Benito Pereira (c. 1460–1553).[37]

Even though the physicist and the astrologer deal with the same heaven, they deal with it in different ways. Geminus, whose long discussion of this topic Simplicius cites in his commentary on text seventeen of the second book of the *Physics*, dealt elegantly with the difference between physics and astrology.[38] Averroes too wrote many things [about this] both in his nineteenth comment on the first book of the *Metaphysics* and in his fifty-seventh comment on the second book *De caelo*.[39] (I refrain from citing his words here, so as not to spend longer on a matter that is not at all important.) For my part I think that six differences between physics and astrology can be established. The first difference is this. Physics considers the substance of the heavens and the stars, whether it is ingenerable and incorruptible, whether it is simple or composite, whether it is elementary or is rather a certain fifth essence. Astrology does not consider these matters at all. Secondly, with respect to the heavens the physicist inquires into all the genera of causes, asking for example, whether the

[36] The role of Platonic and neo-Platonic writings as sources of sceptical doctrine in the period is largely ignored in the classical treatments of the revival of scepticism in the sixteenth century: L. Febvre, *Le Problème de l'incroyance au XVIe siècle: la religion de Rabelais* (Paris, 1942); and R. H. Popkin, *The History of Scepticism from Erasmus to Descartes* (Assen, 1960; expanded edn, *The History of Scepticism from Erasmus to Spinoza*, Berkeley, 1979). However, C. B. Schmitt, *Cicero Scepticus* (The Hague, 1972), touches on the topic; and E. N. Tigerstedt, 'The decline and fall of the neoplatonic interpretation of Plato', *Commentationes Humanarum Litterarum*, 52 (Helsinki, 1974), stresses the extent to which Platonic doctrines were assimilated to the scepticism of the New Academy in the sixteenth century. But the post-classical fortunes of *Plato scepticus* await detailed study.

[37] *De communibus omnium rerum naturalium principiis & affectionibus* (first publn, 1562; Rome, 1576), 47D–48B.

[38] Geminus in Simplicius, *In Aristotelis physicorum libros*, ed. by H. Diels (Berlin, 1882), commentary on *Physics*, II, 2, 193b, 23; transl. by T. L. Heath, *Aristarchus of Samos* (Oxford, 1913), 275–6.

[39] Averroes, *Aristotelis omnia quae extant opera...Averrois Cordubensis...commentarii* (Venice, 1550–2), VIII, f. 8r, col. 2 (commentary on *Metaphysics*, I, 8, 982b, 29–990a, 5), and V, f. 64r, col. 1 (commentary on *De caelo*, II, 9, 291b, 1–23).

heavens have an efficient cause or not, whether they have a material cause, whether the intelligence which moves them is their soul and form, what the final cause of the heaven is, and how it operates. These matters fall outside the concern of the astrologer. Thirdly, of the accidents of the heavens the astrologer chiefly considers magnitude, shape and motion, insofar as certain mathematical measures are to be found in these accidents, for example, the relations of greater, lesser and equal in distance, nearness and size or proportion. The physicist, however, examining all the accidents of the heavens, even those which are dealt with by the astrologer, considers them indeed in a very different way, as they are derived from the nature of the heavens, as they are in conformity with its substance, as they are necessary to it and for the carrying out of its physical operations, and finally as they are linked with sensible matter. Fourthly, the astrologer is not concerned to seek and posit causes that are true and agree with the nature of things, but only causes of such a kind that he can universally, conveniently and constantly give an account of all those things which appear in the heavens. That is why it happens that very often he establishes principles which appear to contradict nature and sound reason: eccentrics, epicycles, non-uniform motions and motions of trepidation, and other things like these, are thought to be of this kind... Fifthly, even though the physicist deals with the same matters as the astrologer, the one demonstrates them *a priori*, whereas the other very often demonstrates them *a posteriori*. For example, the physicist will say that the earth is round because, since all its parts are equally heavy, and seek to approach the centre to the same extent, they press and gather themselves together in a round shape. The astrologer, however, will say that the earth is round because the eclipse of the moon, which occurs because of the interposed earth, is caused to occur in a round and circular form. Sixthly, wherever it happens that both explain the same thing, the physicist gives the specific and natural causes, whereas the astrologer gives the general and mathematical causes. Thus suppose the question is: why is the heaven round? The physicist will say: because it is neither heavy nor light but designed to be moved in an orb. The astrologer, however, will reply: because every part of the heaven is at the same distance from the earth, which is its centre.

Such a combination of doubt or denial of the reality of planetary models with insistence on the strict demarcation of a celestial physics concerned with the nature of the cosmos from a mathematical astronomy concerned only with saving the phenomena, without regard to the truth of the hypotheses employed, becomes increasingly prevalent in the course of the sixteenth century, until by the 1580s it rivals the *Theorica* compromise, at least in the context of didactic works, Epitomes and Institutes of astronomy and commentaries on Sacrobosco and Peurbach.[40]

[40] Duhem, *To Save the Phenomena*, attributes such an attitude to a considerable number of ancient and mediaeval astronomers and natural philosophers. G. E. R. Lloyd, 'Saving the appearances', *Classical Quarterly*, 28 (1978), 202–22, argues for substantial qualification of Duhem's claims about Greek astronomy. Sixteenth-

At one level it is easy to understand this stance. Like the hybrid Aristotelian–Ptolemaic cosmology of the *Theorica* tradition it is a compromising stance, an attempt to pour oil on troubled waters. A number of factors can be appealed to in explanation of its growth in popularity. The rise of Aristotelian purism within natural philosophy may well, as suggested earlier, have undermined the plausibility of the *Theorica* compromise. And with the advent of the Copernican system the area of potential conflict was all too evidently extended from recondite questions about the nature of the heavens to the theologically sensitive and far from recondite question of the motion of the earth. Finally, the expansion of the teaching of astronomy within the universities may well have created pressure for the development of a workable compromise that would forestall the dissemination amongst students of dangerous novelties, novelties apt as Osiander feared 'to throw the liberal arts into confusion'.

Let us call this compromise 'the pragmatic compromise'. What epistemological positions lie behind it? Sixteenth-century pronouncements on the licence of astronomy to use imaginary constructions in the quest for predictive adequacy have often been assimilated, following Duhem, to modern instrumentalist accounts of the status of scientific theories. This is seriously misleading. No protagonist of the pragmatic compromise expounded the strict instrumentalist view that truth and falsity are not predicable of astronomical hypotheses. And even the more relaxed instrumentalism which claims only that predictive success rather than truth is the goal of astronomy can rarely be attributed without qualification.[41] For the scepticism or agnosticism of these authors generally applies only to the postulation of epicycles and eccentrics, not to such basic cosmological issues as the immobility of the earth or the existence and ordering of the planetary spheres.

century attitudes are, indeed, prefigured by a number of mediaeval Averroist natural philosophers who concede the right of mathematical astronomers to adopt Ptolemaic planetary models in the absence of an alternative that is both predictively and physically adequate. But Duhem is, I think, wrong to conflate the Averroist view with the attitude that becomes prevalent in the course of the sixteenth century according to which use of false or doubtful hypotheses by mathematical astronomers is not merely excusable *faute de mieux*, but both legitimate and inevitable.

[41] In 'The forging of modern realism: Clavius and Kepler against the sceptics', *Studies in the History and Philosophy of Science*, 10 (1979), 141–73, I argue against the attribution of strict instrumentalism to Osiander and other protagonists of the pragmatic compromise. In *The Rationality of Science*, (London, 1981), 29, W. H. Newton-Smith distinguishes what I have called 'strict' and 'relaxed' instrumentalism as semantic and epistemological instrumentalism, respectively, attributing only epistemological instrumentalism to Osiander. It is a moot point which type of instrumentalism corresponds to Duhem's 'fictionalism'.

And a number of protagonists, Reinhold and Erasmus Schreckenfuchs, for example, explicitly endorse the view that the heavenly bodies execute their motions harmoniously in accordance with *leges motuum* prescribed by God at the creation.[42] If we insist on placing the pragmatic compromise in a modern category, it must, I think, be that of a sceptical, but not radically sceptical, realism.

Is it possible, rather than attempting to foist a modern metaphysical position on these authors, to relate the pragmatic compromise to one of the various sceptical epistemologies that were revived in the course of the sixteenth century?[43] Elsewhere I have speculated that Melanchthon's and Osiander's adoption of this position may be related to the moderately sceptical Augustinian epistemology of Melanchthon's *De anima*.[44] And it is tempting to relate Ursus' abrasive but unsystematic attack on the pretensions of astronomy to the polemical scepticism of Agrippa von Nettesheim's *De incertitudine et vanitate scientiarum declamatio invectoria*.[45] But a number of

[42] Reinhold, *Theoricae novae planetarum Georgii Purbachii...ab Erasmo Reinholdo Salveldensi pluribus figuris auctae, et illustratae scholiis* (first publn, Wittenberg, 1542; revised edn, Wittenberg, 1553), f. 18r–19v; Schreckenfuchs, *Commentaria in sphaeram Ioannis de Sacrobusto* (first publn, 1556; Basel, 1569), sig. a2, v. On the theological and jurisprudential contexts from which the notion of *lex natura* was transposed into natural philosophy see F. Oakley in D. O'Connor and F. Oakley, eds., *Creation: the Impact of an Idea* (New York, 1969), 54–83. On the Stoic background to the notion of law as immanent in nature see E. Zilsel, 'The genesis of the concept of physical law', *The Philosophical Review*, 51 (1942), 245–79. Both authors treat the introduction of the concept into natural philosophy and natural science as a seventeenth-century phenomenon and overlook the ubiquity of the notion in sixteenth-century astronomical works. Further, Oakley, in sharply contrasting the immanent natural law of the Stoic tradition with the imposed law of the theological tradition, overlooks the extent to which the Stoic tradition had been Christianised in late antiquity. When sixteenth-century astronomers appeal to the *lex motuum* of the heavenly bodies, it is invoked both as imposed by God and as immanent in created nature.

[43] On the sixteenth-century revivals of classical scepticism see the references cited in fn. 36.

[44] N. Jardine, 'The forging of modern realism', 147–53. On Melanchthon's epistemology see also J. Kamp, *Melanchthons Psychologie (seine Schrift de anima) in ihrer Abhängigkeit von Aristoteles und Galenos* (Kiel, 1897); and B. Sartorius, 'Melanchthon und das spekulative Denken', *Deutsche Vierteljahrsschrift für Literatur, Wissenschaft und Geistesgeschichte*, 5 (1927), 644–78. Melanchthon's reaction to Copernicus is discussed in detail in W. Maurer, 'Melanchthon und die Naturwissenschaft seiner Zeit', *Archiv für Kulturgeschichte*, 44 (1962), 199–226; and K. Müller, 'Ph. Melanchthon und das kopernicanische Weltsystem', *Centaurus*, 9 (1963–4), 16–28.

[45] Antwerp, 1530. As with Ursus' invective it is a moot point whether Agrippa's diatribe represents a genuinely sceptical epistemology or an attempt to discredit orthodox doctrine in order to pave the way for unorthodox doctrine: this issue is discussed by C. G. Nauert, Jr., 'Magic and scepticism in Agrippa's thought', *Journal of the History of Ideas*, 18 (1957), 161–82; P. Zambelli, 'Magic and radical

considerations conspire to make a general link between sceptical epistemology and the pragmatic compromise appear unlikely. Were the pragmatic compromise linked to a general sceptical epistemology one would expect to find its exponents adopting similarly sceptical approaches to other disciplines. In fact, however, some exponents of the compromise are orthodox in natural philosophy, Benito Pereira and Alessandro Piccolomini (1525–86), for example, whereas others, such as Reinhold, take a dismissive attitude.[46] And though throughout the period the utility of the mathematical arts is conventionally emphasised,[47] there is no evidence for the elevation of predictive success over truth in disciplines other than astronomy by exponents of this position. Further, the pragmatic compromise cannot be considered as a byproduct of the sixteenth-century revival of sceptical epistemologies since it antedates that revival.[48] Add to this the observation that there are precedents in both Plato and Aristotle for the view that knowledge is especially difficult to attain in the study of the heavens, and the case for a general link between sceptical epistemology and the pragmatic compromise appears weak indeed.

One at first sight baffling aspect of the way the pragmatic compromise is commonly presented deserves notice. In author after author, Valentin Nabod, Piccolomini, Melanchthon, the non-existence of real epicycles and eccentrics is offered as a sufficient warrant for the assertion that astronomical hypotheses are fictions or fabrications.[49] As the modern reader is bound to note, and as Tycho and Kepler showed themselves well aware, this is a *non sequitur*. Even if epicycles and eccentrics do not exist, planetary models which employ them may nonetheless represent the true orbits of the planets. Since

---

reformism in Agrippa of Nettesheim', *Journal of the Warburg and Courtauld Institutes*, 39 (1976), 69–103; and Popkin, *The History of Scepticism*, 23–5.

[46] Pereira, *De communibus principiis*; Piccolomini, *La prima parte delle theoriche overo speculationi dei pianeti* (Venice, 1558); Reinhold, *Theoricae novae planetarum*, sig. Oviii, r.

[47] On humanist-inspired encomia to the utility of the mathematical arts see P. L. Rose, *The Italian Renaissance of Mathematics: Studies on Humanists and Mathematicians from Petrarch to Galileo* (Geneva, 1975).

[48] Although, as noted in fn. 40, Duhem's conflation of this position with classical and mediaeval attitudes to mathematical astronomy is questionable, there is at least one fifteenth-century exposition of the pragmatic compromise, Wojciech of Brudzewo's commentary on Peurbach: *Commentariolum super Theoricas novas planetarum Georgii Purbachii in studio generale Cracoviensis per Magistrum Albertum de Brudzewo diligenter corrogatum, A.D. 1482*, ed. by L. A. Birkenmajer (Cracow, 1900). On Wojciech's position see M. Markowski, *Burydanism w Polsce w okresie przed-kopernikańskim* (Wrocław, 1971), 239–40; and N. Jardine, 'The significance of the Copernican orbs', *Journal of the History of Astronomy*, 13 (1982), 167–94.

[49] Nabod, *Astronomicarum institutionum libri*, f. 20r–v; Piccolomini, *La prima parte delle theoriche*; Melanchthon, *Initia doctrinae physicae* (Wittenberg, 1549), f. 52v–53r.

the use of such a planetary model to make predictions involves calculation of coordinates of the orbit of the planet one may doubt whether these authors could have failed to distinguish the machinery of motion invoked by postulating real epicycles and eccentrics from the real motion which is constructed even when these devices are regarded as 'geometrical fictions'. Further, it might be thought that the subscription by many of these authors to the view that the heavenly bodies obey *leges motuum* prescribed by the creator implies their belief that they move regularly in real orbits. These inferences are, I suggest, unwarranted. The fact that in making predictions astronomers had to calculate what we interpret as coordinates of a real orbit does not imply that they considered the real orbits of the planets as playing an explanatory role in astronomy. And talk of the *leges motuum* of the planets may imply only the unspecific view that there is a divinely ordained harmony and regularity underlying the apparent irregularity of the planetary motions. It does not necessarily imply a conception of real orbits described in accordance with 'laws of motion'. The question remains, however, whether Tycho and Kepler were unprecedented in distinguishing the function of planetary models as putative mechanisms of motion from their function as devices for representing real orbits, and in denying or doubting the validity of the former function whilst conceding the validity of the latter. The following passage from the preface by Erasmus Reinhold (1511–53) to his annotated edition of Peurbach's *Theoricae novae* strongly suggests that they were not.

Now the complete periods or revolutions of the orbs, which are agreed to be constant, clearly attest and confirm that the great diversity of the motions of the planets does not arise from some irregular motion of the celestial orbs as the inexperienced imagine. For even though there occurs in parts of the period an inequality that is not to be ignored, as we have just said, nevertheless it is impossible that the entire period of each planet should be constant unless the motion of the individual orbs were regular. So astronomers offer for this great irregularity that is perceived in the parts of the periodic motions an ingenious and clear cause, namely that equable and in their nature uniform motions appear to us otherwise, because they occur in eccentric orbs or because many simple motions being, so to speak, variously compounded, one irregular motion is produced from all of them...But, given this, the fact that a great number of celestial orbs is assembled is to be attributed to art, or rather to the imbecility of our intellects. For though these seven luminous and most beautiful bodies perhaps have the innate power through divine providence to preserve their law and perpetual harmony through all the variety and irregularity of their motions, for us it would be of the utmost difficulty to grasp and hold in thought, at least *rationabiliter*, that harmony of irregularity, as I call it.[50]

[50] *Theoricae novae planetarum*, f. 18r–19v.

Reinhold's stance is confirmed by a remarkable innovation in his representation of two of Peurbach's planetary models. In the case of the moon and Mercury he shows the composite oval trajectory of the centre of the epicycle as well as the 'elementary motions' from which it is compounded.[51] And Reinhold is not the only exponent of the pragmatic compromise to distinguish the function of planetary models as putative mechanisms from their function as means of constructing complex orbits. In his *Hypotyposes* his pupil and successor at Wittenberg, Caspar Peucer (1525–1602), asserts, more unambiguously than Reinhold, that the function of planetary models is to represent the true motions of the planets, motions that are 'appropriate to the heavens' because compounded out of uniform circular motions; and he denies, categorically, that the epicycles and eccentrics used in planetary models exist in the heavens.[52]

I have argued that the pragmatic compromise in astronomy represents neither a general epistemological stance nor a general endorsement of sceptical epistemology. But though Duhem's search for intellectual ancestors led him to misinterpret their stance, his grouping together of these authors was, I suggest, a shrewd one.[53] For despite the substantial differences between their epistemological positions they share a common concern to present mathematical astronomy and natural philosophy as incommensurable disciplines, and thus to evade the potential conflict between them. As a didactic stance the pragmatic compromise is a natural successor to the increasingly unworkable embedding of Ptolemaic models into Aristotelian orbs, the mediaeval compromise perpetuated in the *Theorica* literature.

The pragmatic compromise is, I suggest, the position which Kepler rightly attributes to Ursus, Osiander and 'those astronomers who confine their thinking within the bounds of astronomy and geometry'. And part, at least, of his plea for the unification of astronomy and physics in the study of the heavens may be seen as a reaction against it.

It would, however, be seriously misleading to treat Kepler's insistence

---

[51] *Theoricae novae planetarum*, diagrams interpolated between f. 38 and 39 and f. 68 and 69. Wojciech of Brudzewo's commentary on Peurbach, a work of which Reinhold appears to make substantial use, alludes to the oval path of the centre of the moon's epicycle.

[52] *Hypotyposes orbium coelestium*... (Strasbourg, 1568), 7–10. The contents of this anonymous work, sometimes misattributed to Reinhold, were later published under the name of its author, Caspar Peucer, in *Hypotyposes astronomicae, seu theoriae planetarum* (Wittenberg, 1571): see O. Gingerich, 'Reinhold, Erasmus', in C. C. Gillispie, ed., *Dictionary of Scientific Biography*, XI (New York, 1975), 365–7.

[53] Hübner offers interesting reflections on Duhem's polemical use of history: K. Hübner, 'Duhems historische Wissenschaftstheorie und ihre gegenwärtige Weiterentwicklung', *Philosophia naturalis*, 13 (1971), 81–97.

on the relevance of 'physical considerations' to mathematical astronomy as without precedent. To start with it should be noted that though the pragmatic compromise may have been effective in papering over the conflict between astronomers' planetary models and Aristotelian cosmology, and hence in facilitating uncontroversial textbook presentations of the core of mathematical astronomy, it was powerless to enforce a rigid separation of the disciplines of natural philosophy and astronomy as then understood. For on a whole range of general questions about the constitution of the world – the *theses* dealt with in *De sphaera* as opposed to the *hypotheses* of a *Theorica* – the disciplines manifestly overlapped. Such questions included the shape of the earth, its position and its mobility. And on many more specialised issues the partition of the two disciplines could not be effectively enforced. Thus, as Kepler notes in Chapter 2 of the *Apologia*, the question of the ordering of the spheres had long been considered amenable to both mathematical and natural philosophical arguments. Whilst the timing and duration of eclipses could be considered as strictly the business of mathematical astronomy, the capacity of the heavenly bodies to cast shadows raised physical questions about the nature of their substance. And the questions of the nature, location and type of motion of comets and new stars openly confounded the attempt to keep astronomy and natural philosophy apart. Finally it should be remembered that judicial astrology, an integral part of astronomy throughout the period, was an area which mixed the predictive techniques of mathematical astronomy with natural philosophical speculation about the powers, virtues and qualities of the heavenly bodies.

Towards the end of the sixteenth century the demarcation between astronomical and natural philosophical study of the heavens is steadily eroded. Tycho and Rothmann, Maestlin, and even Ursus (when he is presenting his own system rather than denouncing the pretensions of astronomy) openly deploy a wide range of physical arguments in debating the issue between the rival world-systems. And amongst natural philosophers Bruno, Cardano, Julius Caesar Scaliger, Andrea Cesalpino and Patrizi, to name but a few, venture into territory traditionally reserved for mathematical astronomers. Indeed, in the last decades of the century the distinction in genre between astronomical and natural philosophical writings becomes hard to maintain. Consider, for example, the voluminous writings on the comets of 1572 and 1577 reviewed in masterly detail by Hellman.[54] Whilst the majority can clearly be assigned to one of the two categories, a substantial minority, including those of Cornelius Gemma, Tycho, Roeslin and Hagecius (Hájek),

[54] C. D. Hellman, *The Comet of 1577: its Place in the History of Astronomy* (New York, 1944).

combine natural philosophical and astronomical considerations in such a way as to defy assignment to either category.[55] Other late-sixteenth-century works which manifest the 'confusion of the liberal arts' so feared by Osiander include Aslaksen's *De natura caeli triplicis libelli tres* of 1597 and Roeslin's *De opere Dei creationis*, both somewhat uncritical mélanges of unorthodox natural philosophy and unorthodox astronomy.

With hindsight it is tempting to attribute this erosion of the boundaries between mathematical astronomy and natural philosophy to the impact of Copernicus. For Copernican astronomy, interpreted realistically, can be seen to challenge the entire system of traditional natural philosophy, and to create the need for a radical reassessment of celestial physics. But great caution is in order here. It is now generally accepted that though Copernicus essays a variety of defensive *ad hoc* emendations of traditional tenets, his natural philosophy remains orthodox in many of its essentials.[56] Recently it has been argued by Krafft, Westman and others that for all the conservatism of his physics, Copernicus breaks new ground by shifting the onus of justification for cosmological claims from natural philosophy to mathematical astronomy.[57] The issue is a complicated one, for much hinges on assessment of the cognitive division of labour between the disciplines at the beginning of the sixteenth century. It is surely misleading to present Copernicus' realist stance, as Krafft does, as a reaction against treatment of astronomical hypotheses as fictions designed to save the phenomena, for such 'fictionalism' only comes to the fore with the pragmatic compromise in the mid-sixteenth century.[58] And Westman's view that before Copernicus astronomy ceded to natural philosophers the right to dispute fundamental cosmological assertions requires careful qualification.[59] I have argued elsewhere that Copernicus' ventures into natural philosophy have substantial precedents in earlier attempts to justify postulation of epicyclic and eccentric orbs in the face of salient Aristotelian natural philosophical objections, and that the physical arguments that Copernicus presents in Book 1 of *De revolutionibus*

[55] Gemma, *De prodigiosa specie, naturaque cometae...* (Antwerp, 1578); Brahe, *De mundi aetherei recentioribus phaenomenis* (Uraniborg, 1588); Roeslin, *Theoria nova coelestium ΜΕΤΕΩΡΩΝ...* (Frankfurt, 1588); Hagecius, *Descriptio cometae qui apparuit anno 1577...* (Prague, 1578).

[56] See, e.g., H. Guerlac, 'Copernicus and Aristotle's cosmos', *Journal of the History of Ideas*, 29 (1968), 109–13; and Birkenmajer, 'Copernic philosophe', 612–46.

[57] F. Krafft, 'Physikalische Realität oder mathematische Hypothese? Andreas Osiander und die physikalische Erneuerung der antiken Astronomie durch Nicolaus Copernicus', *Philosophia naturalis*, 14 (1973), 243–75; Westman, 'The astronomer's role in the sixteenth century', 105–47.

[58] Krafft, 'Physikalische Realität oder mathematische Hypothese?', 266–7.

[59] Westman, 'The astronomer's role in the sixteenth century', 111.

play a substantial role in the justification of his system.[60] Whatever the long-term implications of Copernicus' work, he himself does not offer a radically new division of cognitive labour between mathematical astronomy and natural philosophy. Further, dissemination of the Copernican system, as measured by well-informed published notice, is known to have been extraordinarily slow.[61] And as Westman has shown in detail, many of those who assimilated it interpreted it in conformity with the pragmatic compromise, making use of Copernican planetary models for predictive purposes, but remaining sceptical or uncommitted on the question of its validity as a world-system.[62] Though his preparedness to venture into natural philosophy inspired Kepler, it does not appear to have inspired Copernicus' immediate successors.

If a single internal factor is to be cited in explanation of the erosion of the disciplinary boundaries at the end of the sixteenth century it seems plausible to appeal to the impact of refinement in observational technique on the study of planetary distances and the locations of comets and new stars. For here the methods of mathematical astronomy were seen by a significant number of authors as providing the means to answer questions that had traditionally belonged in the province of natural philosophy. It is in treatises on comets that the breakdown of the disciplinary distinction first becomes clearly evident.

Two more general factors may be tentatively adduced in explanation of the increased involvement of mathematical astronomy with natural philosophy. One is the weakening of the exclusive hold of Aristotelian natural philosophy in the latter part of the century. This weakening is manifest in the proliferation of cosmologies that either are openly non-Aristotelian, such as those of Bruno, Paracelsus, Telesio and Patrizi, or have substantial non-Aristotelian elements, such as those of Cardano, Julius Caesar Scaliger, Cornelius Gemma and Cesalpino. Even within the most strict of Aristotelian traditions, that of Padua, treatment of celestial topics becomes, as Donahue has emphasised, extraordinarily speculative and diverse in the latter part of the century. The image of the sixteenth-century Aristotelian philosopher as a conserver of orthodoxy seems wildly inappropriate here; the opinions of Arcangelo Mercenari (Professor of Natural Philosophy at Padua in 1574)

---

[60] N. Jardine, 'The significance of the Copernican orbs'.

[61] See E. Zinner, *Entstehung und Ausbreitung der coppernicanischen Lehre* (Erlangen, 1943); and J. Dobrzycki, ed., *The Reception of Copernicus' Heliocentric Theory* (Dordrecht, 1973).

[62] R. S. Westman, 'The Melanchthon Circle, Rheticus and the Wittenberg interpretation of the Copernican theory', *Isis*, 66 (1975), 165–93.

and Jacopo Zabarella (Professor of Natural Philosophy at Padua from 1568 of his death in 1589) on the nature of the heavens are rather apt to inspire sympathy with the widespread attacks by humanists and mathematicians of the period on the subtlety and uncertainty of scholastic speculation.[63] Such weakening of the authority of traditional natural philosophy may well have contributed to the evident preparedness of both professional mathematicians like Tycho and Ursus and amateurs like Aslaksen and Roeslin to tackle natural philosophical questions. A second plausible factor is a shift in social context of the practice of astronomy. As Rose has shown in detail, the humanist programme for reform of the mathematical arts through recovery, publication and assimilation of classical mathematics, the programme pursued by Regiomontanus, Maurolyco and Commandino, was pursued in the context of courtly patronage, not that of university teaching.[64] In the latter part of the sixteenth century at the courts of Wilhelm of Hesse, Albert of Saxony, the Emperor Rudolph II and, above all, the surrogate court of Tycho at Uraniborg, a major non-academic context for the practice of astronomy was established.[65] This was a context relatively free of the constraints of the university curriculum, and in particular free of the subservience of the mathematical arts to natural philosophy embodied in that curriculum. Though much further substantiation is needed, it is tempting to follow Westman in regarding the development of mathematical astronomy outside the university environment as an important factor in the breakdown of the barriers between mathematical astronomy and natural philosophy.[66] Establishment of a degree of professional autonomy for mathematical astronomy may well have stimulated the ventures of professional astronomers, Tycho and Rothmann, for example, into natural philosophy. And the relative informality of courtly scholarship may well have encouraged interdisciplinary ventures like those of Hagecius and Roeslin.

In the *Apologia* Kepler cites the examples set by Copernicus and Tycho in support of his insistence on the propriety of involvement by astronomers in natural philosophy.[67] But this appeal to distinguished precedent, whilst

---

[63] Zabarella, *De rebus naturalibus*... (Cologne, 1590); Mercenarius, *Dilucidationes...in plurima Aristotelis perobscura et nonnulla Averrois loca* (Venice, 1574).

[64] Rose, *The Italian Renaissance of Mathematics*.

[65] On Wilhelm of Hesse's cultivation of astronomy see B. Sticker, 'Landgraf Wilhelm IV und die Anfänge der modernen astronomischen Messkunst', *Sudhoffs Archiv*, 40 (1956), 15–25; and B. T. Moran, 'Wilhelm IV of Hesse-Kassel: informal communication and the aristocratic context of discovery', in T. Nickles, ed., *Scientific Discovery: Case Studies* (Dordrecht, 1980), 67–96.

[66] Westman, 'The astronomer's role in the sixteenth century'.

[67] *Apologia*, p. 145.

effective in the context of his attack on the pragmatic compromise, is potentially misleading. For the view of the proper relations between astronomy and natural philosophy that Kepler promotes in theory in the *Apologia* and in practice in his other works is, I shall argue, as far removed from the cautious stances of Copernicus, Tycho and Maestlin as it is from the views of those he openly attacks, the 'deluded strivers after abstract forms' and those 'who confine their attention within the bounds of astronomy and geometry'.

In the *Apologia* Kepler insists on the relevance of 'physical considerations' to astronomy. In particular he argues that such considerations can be used to settle the issue between rival but observationally equivalent hypotheses. There is, as we have seen, ample precedent for such claims. Copernicus, Tycho, Rothmann, Roeslin, and even Ursus in his *Fundamentum astronomicum* had each assumed the propriety of deploying arguments drawn from natural philosophy to promote his preferred world-system and to discredit its rivals. Recognition of the validity of natural philosophy as a court of appeal in astronomical disputes is perfectly consistent with recognition of a distinction between the disciplines, and were this all Kepler had claimed his stance in the *Apologia* could not be regarded as a radical one. However, several passages suggest that his stance is extraordinarily radical and that he is pleading not for a modest rapprochement between the two disciplines, but rather for their conflation. In response to Ursus' aspersions on past astronomical hypotheses, Kepler catalogues the contributions to 'physical knowledge' made by astronomers.[68] In challenging Patrizi's unsound philosophy he contrasts it with the sober philosophising of astronomers.[69] In countering Osiander's scepticism about astronomy he draws an analogy between astronomical hypotheses as causes of apparent motions and the hidden causes of medical symptoms, an analogy which foreshadows the 'aetiological' astronomy of the *Astronomia nova*.[70] And in answer to Ursus' claim that astronomers should concern themselves only with saving the phenomena, Kepler insists that astronomers 'are not to be excluded from the company of philosophers' and should rejoice in portraying 'the inmost form of things'.[71] How are we to interpret such radical but unspecific pronouncements?

As a first step towards an interpretation, let us follow up the clue Kepler offers in citing his *Mysterium cosmographicum* as an example of the way in which the natural philosopher can resolve disputes in astronomy. In particular it proves instructive to contrast the 'demonstration' of the Copernican hypotheses which Kepler offers in that work with the defence of the

[68] *Apologia*, p. 147.
[69] *Apologia*, p. 155.
[70] *Apologia*, p. 151.
[71] *Apologia*, p. 144.

Copernican hypotheses offered by Maestlin in the preface to Rheticus'
Narratio prima, which he appended to Kepler's treatise. Though Maestlin
heaps fulsome praise on Kepler's 'speculations' his preface is essentially
defensive. He feels it necessary to preempt whatever prejudices the reader
may have about the absurdity of the Copernican system, and to this end he
trots out a miscellaneous collection of 'physical considerations' drawn from
Copernicus and Rheticus.[72] Kepler's approach could scarcely be more
different. As we have seen, he opens the first chapter with some preemptive
remarks, first to forestall theological objections, then to forestall the standard
sceptical argument that saving the phenomena is no warrant of truth since
true conclusions can follow from false premises. After these somewhat
cursory preliminaries he announces: 'Now to turn from astronomy to physics
or cosmography, these hypotheses of Copernicus not only do not sin against
the nature of things, but rather greatly benefit it.'[73] Kepler proceeds for the
rest of the chapter to retail the 'benefits to nature' that ensue from the
Copernican hypotheses. And it is as the culmination of this list of
contributions to 'physics or cosmography' that he introduces, in the second
chapter, the revelation by means of the Copernican hypotheses of the
harmonious architecture of the cosmos through which the creator has
expressed his nature. Where for Maestlin the credentials of the Copernican
system are to be established by showing its conformity to the *placita* of natural
philosophy, for Kepler its credentials are attested by its capacity to yield new
and remarkable insights in natural philosophy. For Kepler astronomy can,
as he insists in the *Apologia*, 'establish things in the realm of physical
knowledge'.[74]

From the *Mysterium* alone it is clear that the 'physics or cosmography'
into which he hoped to integrate astronomy is far removed in both form
and content from traditional Aristotelian physics and metaphysics. And by
the time he wrote the *Apologia* Kepler had already revealed in his
correspondence his plans for the integration of mathematical astronomy, the
sound core of judicial astrology and music into a profoundly unorthodox
master science of harmonics, plans that come to partial fruition in his
Harmonice.[75] How did Kepler envisage the place of astronomy in such a
scheme?

Kepler's only extended treatment of the relations between astronomy and
allied disciplines is to be found at the start of the first book of his *Epitome*

---

[72] *K.g.W.*, I, 83–4.   [73] *K.g.W.*, I, 16.
[74] *Apologia*, p. 147.
[75] See especially *K.g.W.*, XIV, 7–16 (July 1599 to Edmund Bruce), 21–41 (August 1599
     to Herwart von Hohenburg), 43–59 (August 1599 to Maestlin).

astronomiae Copernicanae, published in 1618. Here are extracts from his account.[76]

What is astronomy? It is a science setting out the causes of those things which appear to us on earth as we attend to the heavens and the stars, and which the vicissitudes of time bring forth: and when we have perceived these causes, we are able to predict the future face of the heavens, that is, the celestial appearances, and to assign particular times to things in the past... *What is the relation between this science and others?* 1. It is a part of physics, because it seeks the causes of things and natural occurrences, because the motion of the heavenly bodies is amongst its subjects, and because one of its purposes is to inquire into the form of the structure of the universe and its parts... *Concerning the causes of hypotheses.* What, then, is the third part of the task of an astronomer? The third part, physics, is popularly deemed unnecessary for the astronomer, but truly it is in the highest degree relevant to the purpose of this branch of philosophy, and cannot, indeed, be dispensed with by the astronomer. For astronomers should not have absolute freedom to think up anything they please without reason; on the contrary, you should be able to give *causas probabiles* for your hypotheses which you propose as the true causes of the appearances, and thus establish in advance the principles of your astronomy in a higher science, namely physics or metaphysics – yet you are not prevented from using those geometrical, physical or metaphysical considerations about matters pertaining to these higher disciplines that are supplied to you by the very exposition of the specific discipline,[77] provided you do not introduce any begging of the question.[78] This being granted, it comes about that the astronomer (master of what he has set out to do insofar as he has devised causes of the motions which are in accord with reason and fit to give rise to everything that the history of observations contains) now draws together in a single form those things which he had previously determined one at a time. And having set aside what was, up till then, his intended purpose (which was the demonstration of the phenomena, and the things useful for everyday life which flow from this), himself aspires, with the greatest joy in philosophising, to a higher end. To this end he directs all his opinions, both by geometrical and by physical arguments, so that truly he places before the eyes an authentic form and disposition or furnishing of the whole universe. Truly this is the very Book of Nature, in which God the Creator manifested and represented in part and by a kind of writing without words his essence and his will towards mankind.

There are many parallels here with central tenets of the *Apologia*. The claim

[76] *K.g.W.*, VII, 23–5.
[77] *Quae tibi suppeditantur ab ipsa diexodo disciplinae propriae.* Presumably Kepler's thought here is that 'physical or metaphysical' principles may, on occasion, be generated from within the discipline of astronomy itself.
[78] The significance of the warning about *petitio principii* is obscure. Perhaps Kepler has in mind the circularity that would arise were physical or metaphysical consequences of astronomical hypotheses invoked as principles without independent justification.

that astronomy is a part of physics confirms the impression conveyed in the *Apologia* that Kepler envisages not merely a mutually profitable interaction but rather a conflation of the disciplines; and the insistence that the astronomer should aspire not only to save the phenomena but to offer causes of the phenomena and to portray the form of the universe echoes the earlier claims about the role of the astronomer. In addition this account provides important clues to the way in which Kepler envisages the integration of astronomy into his new 'physics or cosmography'. The astronomer is required to ground his astronomical hypotheses in 'a higher science' by providing physical or metaphysical *causas probabiles*. And 'geometrical arguments' play a crucial role in such grounding; so the physics or metaphysics which Kepler has in mind is evidently a quantitative and not, like Aristotelian metaphysics, a qualitative discipline. Let us consider these points in turn.

Probabilis has in Kepler's usage, as generally in the period, a broad spectrum of meanings.[79] Sometimes it has the connotation of mere plausibility.[80] Thus in the *Astronomia nova* Kepler claims that the Ptolemaic hypotheses are *probabiles* whereas the Tychonic and Copernican hypotheses can be confirmed by appeal to 'physical causes'.[81] On occasion it bears the implication of incomplete warrant as opposed to certainty, as when he explains how in the *Astronomia nova* he will mix *probabilia* with *necessaria* as is done in physics, medicine and other sciences in which it is necessary to use *conjecturae* as well as what is attested with certainty by observation.[82] But most often it has the connotation of worthiness, acceptability and adequate warrant, as when Kepler claims that his postulation of a magnetic power in the sun is *probabilissime*.[83] When Kepler urges the astronomer to seek *causas probabiles* for his hypotheses, he means, I suggest, that he is to seek causes that are acceptable because adequately warranted. A clear indication of the strategy Kepler has in mind is to be found in the *Mysterium cosmographicum*.

[79] There is an extensive recent literature on mediaeval and renaissance usage: see, e.g., E. F. Byrne, *Probability and Opinion* (The Hague, 1968); I. Hacking, *The Emergence of Probability* (Cambridge, 1975); and D. Patey, *Probability and Literary Form: Philosophical Theory and Literary Practice in the Augustan Age* (Cambridge, 1984), Ch. 1. All these authors emphasise the gulf between renaissance and modern usages. Kepler's usage is surveyed by I. Schneider in an important paper, 'Wahrscheinlichkeit und Zufall bei Kepler', *Philosophia Naturalis*, 16 (1976), 40–63. As Schneider notes, in his *Hyperaspistes* Kepler explicitly distinguishes the *probabilitas* which depends on popular opinion and authority from that which depends on substantial reasons (*K.g.W.*, VIII, 274). Schneider assimilates Kepler's notion of *probabilitas* arising from reasons rather more closely to the modern notion of weight of evidence than seems warranted by the texts he cites.

[80] Especially when contrasted with *verus* or *verisimilis* (cf. the contrast between Greek *pithanos* and *eikos*).   [81] *K.g.W.*, III, 22.

[82] *K.g.W.*, III, 19.   [83] *K.g.W.*, III, 23 and 35.

The preface and the first twelve chapters of the *Mysterium* are apt to appear to the modern reader somewhat disorganised. Kepler does, indeed, introduce his main cosmological postulate, the interpolation of the planetary orbits into the nested Platonic solids, but instead of proceeding at once to show its conformity with the Copernican hypotheses he starts by embellishing it with an extraordinary variety of theological, historical, astrological and numerological speculations. Only in Chapter 13 does he settle down to the task of empirical confirmation of the postulate. Kepler himself marks the transition as follows: 'So far nothing has been declared except certain signs and tokens (*eikota*) of the theorem I have proposed. Let us now turn to the distances between the orbs of astronomy and to geometrical demonstrations.'[84] It seems reasonable to interpret the opening chapters of the work in which these signs and tokens are offered as Kepler's attempt to render his postulate *probabilis*, adequately warranted and hence worthy of approval and acceptance. Some of the speculations Kepler presents in support of his postulate, such as the far-fetched analogies he uses to associate the traditional qualities of the planets with the regular solids, may seem playful and lighthearted.[85] And the account of God's plan for the cosmos offered in the preface is unmistakably edifying in intent.[86] Since the creator manifests his nature in the cosmos, the edifying power of a postulate is surely for Kepler a token of its truth. And the playfulness too may have a role in the hunt for causes. Thus in his *Tertius interveniens* of 1610, when defending the doctrine of signatures against Feselius' mockery, Kepler characterises the cosmos as the product of God's play and concludes that when reason emulates the divine game it does so not childishly but in accordance with a natural tendency implanted by God.[87] D. P. Walker is surely right to infer from this that 'even when Kepler seems to be "playing" with analogies, he may be imitating the divine game of the creator, and should perhaps be taken seriously'.[88] With the *probabilitas* of his architectonic postulate established, Kepler turns in Chapter 13 to the task of establishing the conformity between the maximum and minimum planetary distances from the sun predicted by his postulate and those calculated from observations using the Copernican hypothesis. This part of the work conforms to the final instructions of the account in the *Epitome*, 'to draw together in a single form those things that had previously been considered one at a time', in order 'partly by geometrical and partly by physical arguments' to 'place before the eyes an authentic form and disposition or furnishing of the whole universe'.

[84] *K.g.W.*, I, 43. [85] In Ch. 9; *K.g.W.*, I, 34–6.
[86] *K.g.W.*, I, 13–14. [87] *K.g.W.*, IV, 245–6.
[88] *Studies in Musical Science in the Late Renaissance* (Leiden, 1978), 57.

In the *Mysterium* Kepler nowhere explicitly reflects on the quantitative nature of his new 'physics or cosmography'. But in another early work, *De fundamentis astrologiae certioribus* of 1601, he explicitly contrasts his position with that of Aristotle.

We derive both the powers of the planets and their number in a manner which is different from that in which Aristotle derives his four elements from the combinations of four qualities, and is perhaps no more inelegant. All variety arises from contrarieties, and the first [variety] arises from the first [contrariety]. Aristotle, wishing to philosophise at a higher level than geometry and more generally, admits as the first contrariety in metaphysics that which is between sameness and difference. It seems to me that diversity in created things arises only from matter or by reason of matter. But where matter is, there is geometry. So that which Aristotle declares to be the primary contrariety, namely between sameness and difference without a mean, I find likewise in the realm of geometry, considered philosophically, to be the primary contrariety, but with a mean, in such a way that what for Aristotle was one term, difference, we divide into two terms, more and less. So since geometry represented the exemplar for the creation of the whole world, this geometrical contrariety accords not improperly with the furnishing of the world which consists of the diverse powers of the planets.[89]

The context here is that of Kepler's attempt to provide physical and metaphysical 'causes' for the efficacy of conjunctions and aspects in judicial astrology.[90] Kepler's contrast between the geometrical character of his metaphysics and the qualitative character of Aristotle's is highly significant for an understanding of the relation he envisages between metaphysics and astronomy, astrology and other mathematical arts. By insisting on the capacity of geometry to comprehend all 'diversity in created things', Kepler undermines the traditional distinction between the subject matter of natural philosophy and that of the mathematical arts, the former being concerned with the intrinsic qualities and qualitative changes of things and the latter with their quantitative characteristics, their 'mathematical accidents'. The Aristotelian injunction against violating the 'homogeneity' of a discipline by using principles relating to different subject matters, an injunction that could be used to maintain the impropriety of mixing natural philosophy with mathematics, is deprived of its force.[91] As a result the traditional sharp distinction between mathematical astronomy and the 'physical' study of the

---

[89] *K.o.o.*, I, 423.

[90] This account, which differs markedly from that of the *Mysterium*, is adumbrated in his letters of May and August 1599 to Herwart.

[91] For a discussion of the principle of homogeneity and its Aristotelian sources see S. Gaukroger, *Explanatory Structures: a Study of Concepts of Explanation in Early Physics and Philosophy* (Hassocks, Sussex, 1978), 102–5.

heavens, variously appealed to both by those like Achillini who called for a mathematical astronomy subservient to Aristotelian physics and by those like Reinhold and Pereira who opted for the pragmatic compromise, can no longer be sustained. Further, it becomes possible to regard geometrical demonstration, as Kepler does, as the method which underlies all levels of study of the cosmos, serving both to relate astronomical hypotheses to observations and to relate them to their postulated physical or metaphysical causes.

The quantitative character of Kepler's metaphysics, and in particular its concentration on ratios and proportions, is central to the elaborate epistemology of his *Harmonice*. It is because the human mind is formed in the image of the creator's mind that we are able to understand part at least of the plan according to which he created the world. In particular we are able to share in the delight in harmonies in proportion and disposition which led him to create the world as he did. Kepler distinguishes sensible harmonies, harmonies of sounds, shapes and dispositions of bodies evident to the senses, from the hidden or intellectual harmonies of the world. The harmonies of music, architecture and the planetary aspects are sensible harmonies. But the harmonies which reveal God's plan in the creation of the cosmos are intellectual harmonies, and to these we do not have direct access.[92] The warrant for theses about intellectual harmonies in the dispositions and motions of the heavenly bodies lies only partly in their inherent appeal to the intellect. In addition they must be shown to be in conformity with astronomical hypotheses that are adequate to save the phenomena with the minutest precision. The harmonies that God has embodied in the world Kepler classifies in accordance with an elaborate, and on occasion *ad hoc*, theory of 'concrete numbers' which form the alphabet of the book of nature. These concrete numbers, ratios of dimensions of geometrical figures constructable by a particular recipe, play a crucial role in the epistemology of the *Harmonice*.[93] Exploiting a theme prevalent in the period, that of understanding as a capacity to create or recreate, Kepler grounds our ability

---

[92] For the distinction between sensible and intellectual harmonies see *K.g.W.*, VI, 211–19.

[93] On musical aspects of Kepler's harmonic theory see Walker, *Studies in Musical Science*. The role of harmonic theory in Kepler's astronomy and astrology is explored in detail by G. Simon, *Kepler: astronome, astrologue* (Paris, 1979); and by J. B. Brackenridge, 'Kepler, elliptical orbits, and celestial circularity: a study in the perversity of metaphysical commitment', *Annals of Science*, 39 (1982), 117–43 and 265–95. Both Simon and Brackenridge emphasise the continuity of Keplerian metaphysics from the *Mysterium* to the *Harmonice*.

to appreciate the archetypal harmonies of the world on the constructability, and hence intelligibility, of the concrete numbers.[94]

Read in the context of others of Kepler's works, the *Apologia* can be seen to promote a radical rejection of conventional views on the relations between mathematical astronomy and natural philosophy. Kepler is not merely opposing the widespread view that mathematical astronomy should be content to save the phenomena and keep clear of natural philosophical questions; he is equally opposing the attitude, typified by his teacher Maestlin, which sees in natural philosophy merely a source of qualitative arguments to be deployed piecemeal to resolve the issues between conflicting hypotheses. Rather, he envisages the grounding of astronomical hypotheses within a quantitative natural philosophy by geometrical demonstration from theses about the harmonious disposition of the cosmos. The gulf between Maestlin and Kepler on the role of physics in astronomy is revealed by Maestlin's reaction to Kepler's announcement in a letter of 1616 that he intends to deal with the physical causes of the moon's anomalies in his forthcoming Epitome of Copernican astronomy. Maestlin protests that Kepler should deal with astronomy 'through astronomical causes and hypotheses, not physical ones'.[95] At first sight this is a puzzling reaction from one who had approved and promoted the *Mysterium*, and had appended to it his own defence of the Copernican hypotheses on physical grounds. But there is surely no inconsistency here. By 1616 Maestlin had become aware of the extent to which Kepler's attempt to integrate astronomy into physics diverged from his own stance as a professional mathematical astronomer prepared merely to draw cautiously on natural philosophy for defence of his hypotheses against the charge of absurdity.

In the *Apologia* Kepler understandably never explicitly contrasts his project for a celestial physics practised by philosopher–astronomers with the cautious

---

[94] A. C. Crombie has explored renaissance developments of this theme in the context of the mechanical arts in *Styles of Scientific Thinking*, forthcoming, Ch. 4; and he has emphasised its role in Mersenne's epistemology in 'Marin Mersenne and the seventeenth century problem of scientific acceptability', *Physis*, XVII (1975), 186–204. On classical sources of the theme see H. Blumenberg, '"Nachahmung der Natur": zur Vorgeschichte der Idee des schöpferischen Menschen', *Studium generale*, 10 (1957), 266–83. I am indebted to Dr A. Pérez-Ramos for enlightenment on this issue.

[95] *K.g.W.*, XVII, 187. E. J. Aiton, 'Johannes Kepler and the *Mysterium cosmographicum*', *Sudhoffs Archiv*, 61 (1977), 180, suggests very plausibly that whereas Maestlin felt able to approve of the appeal to final causes, an appeal with ample precedent, he objected to the introduction of efficient causes, merely adumbrated in the *Mysterium*, but vastly elaborated in Kepler's later works.

ventures into natural philosophy of Copernicus, Tycho and Maestlin. But the contrast is implicit in his remarks in the second chapter on the physical arguments that have been used since antiquity to resolve the question of the ordering of the inner planets.[96] Here Kepler appears to use the term 'physical' as opposed to 'astronomical' pejoratively, and comments adversely on the inconclusiveness of such arguments. Superficially this seems odd in a treatise which so consistently promotes the importance of physics for astronomy. But the oddity is dispelled once it is realised that for Kepler resolution of theoretical dispute in astronomy does not come about through piecemeal citation of miscellaneous physical testimony for and against hypotheses, but through systematic derivation of a system of hypotheses from the architectonic, harmonic and dynamic causes to be established in his new 'physics or cosmography'.

It is a moot point how far at the time Kepler wrote the *Apologia* he had elaborated the complex epistemology and ontology of the *Harmonice*. Kepler mentions his projected work on harmonies once in the *Apologia*,[97] and from his correspondence it is clear that he had planned it in considerable detail. By the time he wrote the *Apologia* he had read Proclus' *Commentary on the First Book of Euclid's Elements*, a major source for the epistemology of the *Harmonice*,[98] and from his letters of July 1599 to Edmund Bruce and of August 1599 to Herwart it is clear that he had already developed the crucial theory of concrete numbers, a theory which he was soon to apply in his account of astrological aspects and their efficacies in *De fundamentis astrologiae certioribus* of 1601.[99] The cryptic remark at the beginning of the first chapter of the *Apologia*, in which Kepler relates our capacity to comprehend nature to an innate light of the mind which thrives on geometrical figures,[100] may therefore be regarded as a hint at the epistemology of the *Harmonice*, an epistemology in which our capacity to apprehend the underlying order in the cosmos is grounded in a specific account of the geometrical alphabet God used in writing the book of nature. And behind the extended analogy of the third chapter of the *Apologia* between the architect and the astronomer as constructor of hypotheses, one may glimpse the equation of knowledge with imaginative construction in emulation of the creation on which the epistemology of the *Harmonice* is premissed.[101]

[96] *Apologia*, p. 179.   [97] *Apologia*, p. 157.

[98] He first refers to the work in a letter of September 1599 to Herwart (*K.g.W.*, XIV, 63).

[99] *K.g.W.*, XIV, 7–16 (to Edmund Bruce) and 21–41 (to Herwart). On Kepler's new account of the aspects see Simon, *Kepler: astronome, astrologue*, Chs. 1 and 2; and Brackenridge, 'Kepler, elliptical orbits, and celestial circularity', 130–42.

[100] *Apologia*, p. 138.   [101] *Apologia*, pp. 185–6.

At the cost of rather drastic historical simplification Kepler's programme for the integration of mathematical astronomy into a new cosmography and his plea for the astronomer to become a natural philosopher can be seen as the culmination of a progression of sixteenth-century attitudes. Until the latter decades of the century the field is held by compromises designed to stave off a potential conflict between traditional natural philosophy and mathematical astronomy, first the 'naive' *Theorica* compromise and then the more sophisticated pragmatic compromise. The pragmatic compromise is in turn eroded as professional astronomers and natural philosophers make ever bolder forays into each other's territory. And this increasing boldness on the part of astronomers can plausibly be related both to internal factors in astronomy and to a shift in the social context of the profession of astronomy. But there are substantial grounds for circumspection here. The simplification is a drastic one. It imposes a linear chronological order on very complex data. Further, the explanations that have been offered for the increased preparedness of astronomers to venture into natural philosophy are highly speculative. Much further research into the institutional history of astronomy is needed if they are to be confirmed, and their historical impact would be greatly enhanced if parallel developments in the profession of other mathematical arts – mechanics, optics, music, etc. – could be established. Even if such confirmation were to be had, it would be unwise to exaggerate the scope and force of such explanations. For they do little to explain the content of Kepler's remarkable vision of a mathematical astronomy integrated into a new cosmography. At best they help us to understand the possibility of presentation of such a vision in 1600 and to explain some of the ways in which Kepler felt called on to defend it in the *Apologia*.

8

Historiography and validation

What is the subject matter of the *Apologia*? History and philosophy of astronomy seems a natural answer. Judged as an essay in that genre it is, of course, bound to appear to the modern reader seriously incomplete, if not defective. Thus the historical part, whilst on occasion revealing a commendable historical sense in its interpretation of crucial classical texts, is elsewhere startlingly naive, shows conversancy with only a narrow range of the available primary sources,[1] and leaps directly from ancient to contemporary astronomy without appreciation of the contributions of the Arabs. And the philosophical part, whilst brilliant in certain of the strategies it deploys to rebut the sceptic, is apt to appear woefully inconclusive in the absence of any sustained attempt to justify the use of physical criteria in the resolution of theoretical disputes. It is not hard to find charitable explanations of such *lacunae*. Thus Kepler's apparently offhand treatment of the role of physical criteria is (as was argued in Chapter 4) partly dictated by his tactful decision to refrain from active promotion of Copernicanism. And the form of his historical narrative is heavily constrained by his commitment to a point by point refutation of Ursus' history of hypotheses. But it is, I suggest, wrongheaded to regard such 'defects' as striking or in urgent need of explanation. What is genuinely striking and in need of explanation is rather the very fact that it is possible to recognise in the *Apologia* a cluster of concerns that are central to the modern discipline of history and philosophy of science.

Kepler is not without sixteenth-century precedent in composing a history of astronomy. But, as I shall show, his concern with theoretical progress is almost entirely without precedent. Likewise there are, as we have seen, ample

---

[1] Crucial sources not used by Kepler include Proclus' *Hypotyposis* (which he wished to consult: see Ch. 1, p. 28) and Simplicius' *Commentary on De caelo*. Simon Grynaeus' edition of the *Hypotyposis* was published in 1540 and Giorgio Valla's Latin translation in 1541, both at Basel; part of Valla's translation had been included in his *De expetendis et fugiendis rebus* (Venice, 1501). Simplicius' commentary was first published in 1526, and in Latin translation in 1540; Kepler cites Simplicius' commentary once in the *Mysterium* (*K.g.W.*, 1, 31), but this citation may well be derived from a secondary source.

sixteenth-century precedents for Kepler's concern with the status of astronomy and the methods proper to its practice; nor is he entirely original in attempting to defend astronomy against sceptical attack.[2] But I have found no substantial precedents in any field for his concern with the criteria for resolution of theoretical disputes, or for his linking of claims about progress with a specific methodology for the achievement of progress. These new concerns of Kepler's are all distinctive and central concerns of the history of science and philosophy of science as they later came to be understood. The *Apologia* marks, I suggest, the birth of history and philosophy of science as a form of reflection on the status and credentials of the natural sciences.

There are obvious dangers in formulating such general claims about originality in terms of the modern categories of natural science, its epistemology and its history. For nothing corresponding to these categories is to be found in classifications of the fields of human inquiry used in the period – they are not, to use a fashionable jargon, actors' categories. Let us therefore start by considering the genre of the *Apologia* in terms of renaissance classifications of the arts and sciences.[3]

Unsurprisingly, given that natural science is not itself a renaissance category, nothing corresponding directly to the modern categories of history and philosophy of science is to be found. But in the case of philosophy of science the sixteenth-century disciplines which prefigure the modern discipline are easily located – they are 'psychology' and 'logic'. Nothing approaching the modern distinction between psychology and epistemology is to be found in the period. Sixteenth-century treatises *De anima*, *De intellectu*, *De cognitione*, etc., a vast field virtually untouched by recent scholarship, freely combine

---

[2] Kepler had precedents in Clavius' defence of Ptolemaic astronomy against sceptical attack, and in his teacher Maestlin's defence of astronomical hypotheses against the charge that they are 'mere figments which correspond to nothing outside the mind': *De astronomiae hypothesibus...* (Heidelberg, 1582), f. A2, v; and *Epitome astronomiae* (first publn, Heidelberg, 1582; Tübingen, 1624), 28–9. Maestlin had earlier defended the reality of the constructions of astronomy against Frischlin's sceptical position: *Judicium de opere astronomico D. Frischlini*, MS, cited in D. F. Strauss, *Leben und Schriften des Dichters und Philologen Nicodemus Frischlin* (Frankfurt, 1856), 330.

[3] Mediaeval classifications of knowledge have been studied in some detail, but apart from the relevant sections of P. O. Kristeller's important paper, 'The modern system of the arts: a study in the history of aesthetics', *Journal of the History of Ideas*, 12 (1951), 496–527, there is little literature on the sixteenth century, in which there was a proliferation of new schemes. Some information, not always reliable, is to be found in R. Flint, *Philosophy as Scientia Scientiarum and a History of Classifications of the Sciences* (Edinburgh, 1904); and F. Watson's *The Beginnings of the Teaching of Modern Subjects in England* (London, 1909) admirably conveys the gulf between modern curricula and their sixteenth-century counterparts.

speculative faculty psychology, 'metaphysical' treatments of the principles of individuation of minds and persons, and 'epistemological' accounts of the means whereby knowledge is attained in the arts and sciences.[4] Logic, 'the art of arts and science of sciences, opening the way to principles of all methods', as Peter of Spain's famous definition has it, ranged far beyond formal logic, covering also the methods whereby knowledge is acquired and handed on in the various arts and sciences. As Gilbert and others have shown, the latter part of the sixteenth century saw a proliferation of often polemical treatises on method in all disciplines. And it became standard practice to emulate Aristotle's *Physics* and Galen's *Ars parva* by prefacing works with disquisitions on the methods of inquiry and presentation most appropriate to their subject matters.[5] Many of the accounts of the status of astronomical hypotheses reviewed in the last chapter are derived from such methodological prefaces to textbooks of astronomy. And in the *Apologia* Kepler himself, in his account of the etymology of the term *hypothesis*, by treating it as a word of second intention proper to logic, implicitly characterises the first chapter on the nature of hypotheses as an essay in logic.

In the case of history of science no such prefigurement of the modern category by renaissance categories can be made out. In his *Dialogi decem de historia* of 1560 Francesco Patrizi offers an enumeration of the species of history that is unusual only in its pedantic attempt at exhaustiveness.[6] He starts out with the conventional distinction between natural and human history. Human history he proceeds to partition by means of a series of dichotomies. It may concern thoughts and sayings or it may concern deeds;

---

[4] On renaissance faculty psychology see E. R. Harvey, *The Inward Wits: Psychological Theory in the Middle Ages and the Renaissance* (London, 1975). Responses to Aristotle's *De anima* are surveyed by F. E. Cranz, 'The renaissance reception of the *De anima*', in *Platon et Aristote à la Renaissance: XVIe colloque international de Tours* (Paris, 1976), 359–76; extensive references to the secondary literature on this are to be found in C. H. Lohr, 'Renaissance Latin Aristotle commentaries', *Studies in the Renaissance*, 21 (1974), and then in successive instalments in *Renaissance Quarterly*. Secondary literature has tended to centre around the notorious disputes on the immortality of the soul, but a fascinating indication of the philosophical richness of renaissance reactions to the *De anima* is given by H. Skulsky, 'Paduan epistemology and the doctrine of the one mind', *Journal of the History of Philosophy*, 6 (1968), 341–61.

[5] On renaissance treatises on method see N. W. Gilbert, *Renaissance Concepts of Method* (New York, 1960); W. Risse, *Die Logik der Neuzeit. 1 Band, 1500–1640* (Stuttgart, 1964); and C. Vasoli, *La dialettica e la retorica dell'Umanesimo* (Milan, 1968). I have argued elsewhere that the continuity between these treatises and seventeenth-century discussions of scientific method has been greatly exaggerated: N. Jardine, 'Galileo's road to truth and the demonstrative regress', *Studies in the History and Philosophy of Science*, 7 (1976), 227–318.

[6] Dialogue entitled *De historiae diversitate* in the edition of Basel, 1576, 410–23.

it may concern members of one city, nation or race, or of many; it may concern practitioners of one discipline or of many, and so on. In Patrizi's elaborate classification, as in other less elaborate classifications of the period, we can recognise analogues of many of the modern genres of historical writing: historical biography, legal and constitutional history, ecclesiastical history, and even social history and history of art. But none of Patrizi's categories directly prefigures the history of science. At best we can recognise in certain of his categories the genres from which much of the raw material for later histories of the mathematical arts was derived. Thus under the heading 'sayings and thoughts of men' there is a place for records of the opinions of practitioners of one particular discipline, and under biography there is a place for lives of practitioners of a single discipline. Into these categories fall such classical doxographic works as Pseudo-Plutarch's (Aëtius') *Placita philosophorum* and Diogenes Laërtius' *De clarorum philosophorum vitis*. Further, Patrizi sees a need for a new – that is, non classical – category, that of history of artefacts, clothing, weaponry, etc. Into this category falls the substantial literature on the history of inventions which grew up in the wake of Polydore Vergil's *De inventoribus rerum* of 1499.[7]

No historiographical category even approximately corresponding to history of science was recognised in the sixteenth century. Nor, as far as I have been able to tell from my reading of the literature of the period, are there to be found independent works that can with hindsight be assigned to that category.[8] Ursus and Kepler are, however, not without precedent in surveying the past of astronomy. But in the sixteenth century such surveys never form independent works. Rather, like treatments of the methodology of astronomy, they are to be found embedded in dedicatory letters, prefaces and prolegomena to astronomical treatises, and in orations and introductory

---

[7] J. Ferguson, *Bibliographical Notes on Histories of Inventors and Books of Secrets* (Glasgow, 1882), and F. M. Feldhaus and C. Graf von Klinkowstroem, 'Bibliographie der erfindungsgeschichtlichen Literatur', *Geschichtsblätter für Technik und Industrie*, 10 (1923), 1–26, indicate the considerable extent of this literature in the sixteenth century. The classical and early renaissance sources for this genre are explored in B. P. Copenhaver, 'The historiography of discovery in the Renaissance: the sources and composition of Polydore Vergil's *De inventoribus rerum, I–III*', *Journal of the Warburg and Courtauld Institutes*, 41 (1978), 191–214.

[8] The failure of the standard general histories of historiography to mention such works does not, alas, yield much confirmation, for they scarcely mention historiography of science even for later periods in which it is a well-established genre. Lack of mention of such works in the few major seventeenth-century treatments of the history of mathematical science does, however, provide support for this claim: see, e.g., G. J. Vossius, *De universae mathesios natura et constitutione liber* (Amsterdam, 1650); and G. B. Riccioli, *Almagestum novum astronomiam veterum novamque complectens* (Bologna, 1651).

lectures on the excellence of astronomy. I shall call such historical accounts 'prefatory histories'. They provide the background against which we must consider the historical chapters of the *Apologia* if we are to locate and assess its historiographical originality.

Tycho Brahe's *De disciplinis mathematicis oratio* (delivered in 1574, but not published until 1610) contains a prefatory history of astronomy that can fairly be used to typify the genre.[9] The oration opens with an encomium to the excellence, dignity, utility and edifying power of mathematical studies. Their excellence he attributes to the certainty of their demonstrations and to the range of the 'speculations' they inspire. After a brief survey of the variety of the mathematical arts, first the primary disciplines, arithmetic and geometry, and then their various 'progeny', optics, architecture, astronomy, etc., Tycho turns to the question of their origins. Geometry arose, as Herodotus tells, in Egypt, where it was used after the annual floods to reestablish the field boundaries. Its practical value is illustrated by the marvellous devices Archimedes used to repel the Roman naval attack on Syracuse: Tycho evidently assumes his audience's familiarity with the tale, variously recounted in Livy, Plutarch and Polybius,[10] for he does not bother to say what the devices were. The usefulness and excellence of mathematical studies are further attested by their pedagogic and propaedeutic value – again Tycho is offhand, for this is standard humanist fare. These preliminaries dispensed with, Tycho now turns to the noblest of the progeny of arithmetic and geometry, indeed the noblest of all branches of mathematics, astronomy. As befits so noble a subject it is of the very greatest antiquity, having been invented by Adam, preserved from the flood by Seth's descendants, who inscribed their astronomical lore on pillars of stone, and thence passed on to Abraham.[11] From Abraham it passed to the Egyptians and thence to Pythagoras and the Greeks. Tycho then traces a succession of astronomers, Timocharis, Hipparchus of Rhodes, Ptolemy, al-Battani and, finally, Copernicus, whose system whilst physically absurd is mathematically superior to Ptolemy's. To Ptolemy and Copernicus all current knowledge in astronomy is owed. Tycho concludes with an encomium to astronomy, emphasising the extreme difficulty of the subject and the edification to be derived from contemplation of the everlasting law of motion of the stars implanted by divine providence.

---

[9] *T.B.o.o.*, I, 143–73; published in a truncated form at The Hague in 1610 by Kord Aslaksen.

[10] Livy, *Historiae ab urbe condita*, XXIV, 34; Plutarch, *Vita Marcelli*, 15–17; Polybius, *Historiae*, VIII, 4–6.

[11] This comes from Josephus, *Antiquitates judaicae*, I, 69–70.

Tycho's oration is entirely devoid of originality in its themes and organisation. The historiographical motifs he employs, *successio* whereby a discipline is handed on from master to pupil and *translatio studii* whereby it is transplanted from one race or nation to another, are standard. These motifs are already to be found in Hugh of St Victor's (d. 1140) historical account of the arts,[12] and in what is perhaps the earliest European historical account of astronomy, Plato of Tivoli's preface to his translation of al-Battani's *De motu stellarum* (second or third decade of the twelfth century).[13] The views on the status of astronomy which Tycho's skeletal history promotes are likewise standard. The emphasis on the useful and wonderful devices produced by the mathematical arts is a constant feature of such encomia and prefaces, with the same *exempla*, Archimedes' marvellous weaponry, Archytas' mechanical dove, and so on, trotted out again and again. It reveals the close affinity between these prefatory histories and the histories of invention of Polydore Vergil and his imitators. And at a more general level it reflects both humanist promotion of the 'useful arts' and the prevalent theme of human understanding as manifest through imitation of the creative acts of the divine intelligence.[14] The claims made for the propaedeutic value of mathematics can be paralleled in any number of encomia to the mathematical arts and prefaces to mathematics textbooks of the period. Promotion of mathematics as an alternative to scholastic logic as the ideal introduction to other disciplines formed an important part of humanist programmes for educational reform. Its educational value was generally supposed to lie not in any direct contribution it might make to other disciplines, but rather in its power to train the mind in rigorous reasoning and the discernment of truth from falsity, in its edifying nature and, above all, in its practical utility.[15] Even

---

[12] The 'Didascalicon': a Mediaeval Guide to the Arts, transl. and ed. by J. Taylor (New York, 1961). On Hugh of St Victor's historiography see R. W. Southern, 'Aspects of the European tradition of historical writing', *Transactions of the Royal Historical Society*, Series 5, 21 (1971), 151–9.

[13] *Praefatio Platonis Tiburtini in Albategnium*, in *Rudimenta astronomica Alfragani. Item Albategnius...de motu stellarum...* (Nuremberg, 1537), sig. A, r.

[14] Cf. Ch. 7, fn. 94. On machines and automata as simulacra of the creation see F. Huber, 'The clock as intellectual artefact', in K. Maurice and O. Mayr, eds., *The Clockwork Universe: German Clocks and Automata, 1550–1650* (New York, 1980), 9–18.

[15] These are the main grounds for promotion of the teaching of the mathematical arts in, for example, Juan Luis Vives' *De tradendis disciplinis* (Antwerp, 1531) and Henri de Monantheuil's *Oratio pro mathematicis artibus* (Paris, 1574). Utility is emphasised in promotion of all the liberal arts throughout the period, and, indeed, forms part of the connotation of the terms *ars*: see, e.g., Kristeller, 'The modern system of the arts'.

Tycho's praise of Copernicus exploits a standard theme. For, though Tycho does not actually use the term, it is as a *restaurator* that Copernicus is praised. When a century earlier Regiomontanus had hailed Peurbach as the greatest astronomer of the moderns and had pleaded for further revival of astronomy and the mathematical arts, his theme was brilliantly original; but by the time Tycho wrote, the themes of restoration and renaissance of the mathematical arts were commonplace.[16] And with his elevation of astronomy over other arts, Tycho addresses himself to a question whose treatment is standard practice in such prefaces, the question of the precedence of his discipline.[17]

Despite occasional idiosyncrasies, such as a surprisingly rich list of Arab astronomers in Luca Gaurico[18] and an unusually bombastic and wide-ranging account of the past and present uses of geometry in Cardano,[19] there is a remarkable uniformity in the twenty-five or so published renaissance prefatory histories of the mathematical arts that I have been able to trace. Three departures from the norm deserve mention: Regiomontanus' *Oratio introductoria in omnes scientias mathematicas*; Johannes Stadius' *Astronomiae aetas, usus, peregrinatio, incrementum, utilitas*; and Ramus' brief and highly polemical sketch of the history of astronomy in antiquity which enlivens the second book of his *Prooemium mathematicum*.[20]

Regiomontanus' introduction to his series of lectures at the University of Padua in 1464 on Alfraganus and astronomy is a superb specimen of humanist oratory. In it he plays ingeniously on the standard Ciceronian theme of *translatio studii*, presenting it not merely as the handing on of knowledge from one race or nation to another, but also as the process of translation and criticism of texts which mediates such transmission. Rose has shown in detail how Regiomontanus uses this historical theme to promote his vision of a mathematical renaissance in Europe to be achieved through a systematic programme of translation, publication and emulation of the mathematical

---

[16] Cf., for example, Melanchthon, who in his preface to Sacrobosco's *De sphaera* in *Libellus de sphaera* (Wittenberg, 1545) hails Peurbach and Regiomontanus as *restauratores*.

[17] On the precedence of the arts see E. Garin, *La disputa dell'arti nel Quattrocento* (Florence, 1947).

[18] Gaurico, *Oratio de astronomiae seu astrologiae inventoribus* (1508), and under a different title in *Opera* (Basel, 1575), I, 1–8.

[19] Cardano, *Encomium geometriae recitatum anno 1535 in Academia Platina Mediolani*, in *Opera* (Lyons, 1663), IV, 440–5.

[20] Regiomontanus, *Oratio introductoria in omnes scientias mathematicas*, in *Rudimenta astronomica Alfragani*; Stadius, *Astronomiae aetas, usus, peregrinatio, incrementum, utilitas*, in *Tabulae Bergenses* (Cologne, 1560); Ramus, *Prooemium mathematicum* (Paris, 1567).

masterpieces of antiquity, a programme he attempted to carry out through his press at Nuremberg.[21] But for our purposes the really striking feature of his history of astronomy lies in the significance he attaches to the growth of astronomy. Near the end of the oration he contrasts the certainty and stability of mathematics, including astronomy, with the uncertainty attested by perpetual disputations of peripatetic philosophy. He then apostrophises astronomy as follows.[22]

You ascend to the heights of the heavens. You show that the sun is 160 times the size of the earth, but the moon approximately equal to the 40th part of the earth. You relate all the stars in definite proportions to the size of the earth. You find out the thickness of the heavens. You promise to measure the size of the flaming earthy vapours in the uppermost region of the air that are called comets, and to measure their distance from the earth.

In associating the growth of astronomy with growth of knowledge of the form of the world, rather than merely with increased predictive power and practical utility, Regiomontanus is apparently without precedent, and clearly anticipates Kepler.

Stadius' work, which serves as an introduction to his *Tabulae Bergenses*, is striking as perhaps the first self-contained treatise on the history of astronomy.[23] At the end of the account he accurately summarises its content as follows.

Now we have shown how from the first genesis of things astronomy was cultivated by the Assyrians, Phoenicians, Chaldaeans and Babylonians and the way in which it passed across to Egypt and ascended thence to Greece; and again how its growth surmounted difficulties up to the time of Hipparchus and the birth of Christ; how from then to the age of Ptolemy it flourished in Greece and Italy, then languished again; how it sought again first Italy and thence Greece, but soon migrated to Spain together with the Arabs, whence it came to Italy as if restored with *belles lettres*, and thence through the beneficence of Frederick, Emperor of Austria, it travelled to Pannonia and thereupon acquired such an affinity with Germany that it has never before been so sympathetically received or liberally cultivated by any people (far be it from me to speak invidiously).[24]

[21] P. L. Rose, *The Italian Renaissance of Mathematics: Studies on Humanists and Mathematicians from Petrarch to Galileo* (Geneva, 1975), Ch. 4. On Regiomontanus' programme of publication see E. Zinner, 'Die wissenschaftlichen Bestrebungen Regiomontans', *Beiträge zur Inkunabelkunde*, n.F., 2 (1938), 89–103.

[22] *Oratio introductoria*, f. Biv, r.

[23] Stadius taught history and mathematics at the University of Louvain, then became a professor of mathematics at the Collège de France; the *Biographie nationale de Belgique*, 23 (1924), col. 528, considers the date strikingly early for composition of such a history.

[24] *Astronomiae aetas*, 22–3. The Emperor Frederick III is referred to in his capacity as patron of Peurbach.

Stadius' work is unusually rich in detail, and shows an appreciation of the gradual nature of the growth of astronomy that is lacking in some of the earlier accounts which attribute the invention of astronomy to a few sages of antiquity. And in his claims for the superiority of Copernicus' account of the precession of the equinoxes to that of Thabit ibn Qurra he touches on the theoretical development of astronomy.[25] But his primary concern is with the practical aspects of astronomy, its usefulness in providing calendars, almanacs and tables.

The treatment of the history of mathematics in Ramus' *Prooemium mathematicum* carries to an extreme the humanist concern with practical utility. The work, addressed to Catherine of Medici, is a plea for patronage. And this partly explains Ramus' concern to detail the benefits, both military and commercial, which accrued from the practice of mathematics in the ancient world, and which are now accruing to the princes of Italy and Germany – he positively hectors his own patroness on this point – whose patronage has been farsightedly lavish. In the course of Book 2 of the *Prooemium* Ramus ventures the following historical explanation for the present parlous state of astronomy.[26]

Of all the liberal arts astrology is the one that is most obscure and weighed down by the most onerous burden of hypotheses, from which it could be extricated and

---

[25] *Astronomiae aetas*, 22–3. An earlier account of the history of astronomical tables which shows an appreciation of the links between the theoretical and practical development of astronomy is that given by Rheticus in the preface to the reader of his *Ephemerides novae...* (Leipzig, 1550); French translation in H. Hugonnard-Roche *et al.*, 'Georgii Joachimi Rhetici *Narratio Prima*: édition critique, traduction française et commentaire', *Studia Copernicana*, xx (1982), 221–5. A substantial part of this preface is quoted by Kepler in his *Mysterium cosmographicum* (*K.g.W.*, I, 63–4).

[26] *Prooemium mathematicum*, 212–16. In a letter to Maestlin of October 1597 Kepler jocularly claimed for himself and Copernicus the Regius Professorship offered by Ramus (*K.g.W.*, XIII, 140–1):

If Ramus wants banished those hypotheses which, though they are believed, are postulated not proved, and if he praises that astronomy without hypotheses which is satisfied with the provision of just the nature of the celestial orbs, as indeed he certainly seems to imply earlier and later on, then I, or Copernicus, or both of us, have won and the Ramean Chair is our due. But if, on the other hand, Ramus entirely rejects all hypotheses, whether true and natural or false, that is, as I have said above, stupid, both in my opinion and by your judgement. Truly, as it reflects on the honour of both of us, I prefer to call myself Regius Professor than to call Ramus a fool.

Maestlin, serious-minded as ever, missed the joke, supposing that Kepler had really been offered a Professorship at Paris (*K.g.W.*, XIII, 151). On the significance of Kepler's joke see E. Aiton, 'Johannes Kepler and the astronomy without hypotheses', *Japanese Studies in the History of Science*, 14 (1975), 49–71. In the *Astronomia nova* (title-page verso; *K.g.W.*, III, 6) Kepler quotes from 'So the fabrication of hypotheses is absurd' to the end of the passage, and then heatedly denounces Ramus' views.

relieved by diligent cultivation of the elements first of logic, then of arithmetic and geometry. For even if the books written about all the other disciplines were to be utterly destroyed, these arts would immediately be restored by innumerable grammarians, rhetoricians, logicians and masters of other arts. But were all the books of astrology to be consumed by some fire, as happened to the library at Alexandria, I ask not how many astrologers, but which astrologer from which country, would revive astrology. The responsibility for this difficulty in the study of the heavens lies with the infinite number of hypotheses, a kind of fabrication [*commenti*] unused and unheard of in the other arts. The principles of grammar, rhetoric, logic, arithmetic, geometry, and of any discipline you care to name except astrology are experiences and observations assembled and drawn up [*collectae ac inductae*] by masters of logic into general and universal precepts [*documenta*], and no other causes of these are sought; rather, they are then posited as foundations for the rest. But astrology from Aristotle's time onwards, or rather from [the time of] that mind which was not content with prior observations and experiences, has sought out prior causes to bring about the progressions, retrogressions, stations, slowings down and speedings up. This, to be sure, is the origin of hypotheses. For in those days Eudoxus of Cnidus was the first to discover the hypotheses of revolving orbs, which Aristotle together with Calippus corrected and amended. Nor did [Aristotle] think that these spheres in the heaven were fabrications, but rather judged them to be real and true. Indeed he honoured them as divine bodies, hence the gods, fifty-five in number deified in the twelfth book of his philosophy. But not long afterwards the Pythagoreans made this theology of Aristotle's quite ridiculous when they profaned and turfed out of the temple all these gods of concentric orbs and introduced epicycles and eccentric orbs.[27] And afterwards in the third and even the fourth centuries there was no end to the making up of new hypotheses. Likewise in our age too Copernicus, an astrologer not merely fit to be compared with the ancients but altogether admirable in astrology, having rejected the whole tradition of hypotheses, revived hypotheses, remarkable ones, though not new, which demonstrated astrology not from the motion of the stars, but from that of the earth. Truly astrologers, both ancient and modern, have almost stifled astrology with hundreds of tables combined with hypotheses. For it is sufficiently agreed on the basis of Proclus' commentary on Plato's *Timaeus* and the Greek commentators on Aristotle that the astrology of the ancient Babylonians, Egyptians and Greeks before Eudoxus was without hypotheses, and that the motions of the celestial bodies were reckoned and eclipses were predicted by it.[28] So it cannot be objected that hypotheses were invented and retained of

[27] A source of this claim, reiterated by Ursus and rebutted by Kepler, is Proclus, *Hypotyposis*, I, 18.

[28] Proclus contrasts an inferior modern astronomy based on hypotheses with Babylonian astronomy, derived from observations made over a long period: *Procli Diadochi in Platonis Timaeum commentaria*, ed. by E. Diehl (Leipzig, 1903), 125, 4–126, 5. Ramus apparently takes Simplicius' *Commentary on De caelo*, 488, 20–4, and 493, 4–5, which presents Eudoxus as the first to attempt to meet Plato's challenge to save the phenomena by means of uniform circular motions, to imply that no one

necessity on the grounds that they could not have had a method for calculating the celestial motions without them, since they did so for so many centuries. Nor is the logic of some author to be appealed to as a pretext for objection on this point, since hypotheses were invented in the face of all the logical rules for constructing an art. So the fabrication of hypotheses is absurd. But the fabrication occurs in Eudoxus, Aristotle, Calippus and Simplicius, who thought that hypotheses were true. And even the orbs without stars [*anastros*] were revered as gods.[29] But in later generations the far more absurd story was told that the truth about natural things can be demonstrated from false causes. It follows that first logic, as I have said, then the mathematical elements of geometry and arithemtic, would provide the greatest assistance in establishing the purity and dignity of this most excellent art. And would that Copernicus had rather set his mind on the establishment of such an astrology without hypotheses. For it would have been far easier for him to describe an astrology of his stars corresponding to the truth than to move the earth like some toiling giant, so as to let us look on the stars at rest with respect to the motions of the earth. Why from so many noble German academies does not someone who is both a philosopher and a mathematician come forth to earn the proffered palm amidst eternal praise? And if the fruit of my ephemeral service may be offered as a reward for so great a virtuosity, I promise you a Regius Professorship at Paris as the prize for an astrology without hypotheses. I shall happily fulfil this promise by giving up my own Professorship.

Ramus' condemnation of astronomical hypotheses amounts to a denial of the legitimacy of theoretical speculation in astronomy. But the account preserves from its classical sources, Proclus' *Hypotyposis* and *Commentary on Timaeus* and Simplicius' *Commentary on De caelo*, at least the germs of an appreciation of theoretical development in the history of astronomy. Though Ursus differs sharply from Ramus on the question of the possibility of a perfect astronomy without hypotheses,[30] Ramus' account clearly provides a major inspiration and source for his own account. For all their scholarly shortcomings and extreme dogmatic biases the accounts of Ramus and Ursus are apparently unprecedented in their appreciation of theoretical change in astronomy.

Before we attempt a general assessment of sixteenth-century historiography of the mathematical arts two further works should be mentioned. Sir Henry Savile's (1549–1622) unpublished *Prooemium mathematicum*, dated 1570,

---

before Eudoxus had employed hypotheses in astronomy. Ramus draws extensively on Proclus and Simplicius in the fuller treatment of the issue in his letter of September 1563 to Rheticus published in J. Freigius, *P. Rami Professio Regia*... (Basel, 1576), sig. ☞1r–2r.

[29] Cf. *Scholae metaphysicae* (first publn. 1566, Hanover 1610), 151, where Ramus castigates Aristotle for postulating gods without stars in addition to the seven planetary gods.

[30] *Tractatus*, sig. Ci, v, 11–25.

forms the introduction to a series of mathematics lectures delivered at Oxford.[31]
It echoes Ramus' *Prooemium*, one of its major sources, in its extreme
preoccupation with the practical utility of the mathematical arts, and it rivals
anything to be found in Ramus' works in its denunciations of the sterility
of Aristotelian natural philosophy. Historiographically the work is remarkable
in including an unprecedentedly full biography of ancient mathematicians
from 'Seth's sons' to Ptolemy. Savile's historical biography is, however,
surpassed in range and detail by the massive *Vite dei matematici* composed by
Bernadino Baldi (1553–1617) in the years 1587–96. From Rose's fascinating
account it is clear that Baldi's work shows an unprecedented concern with
the theoretical content of past mathematics.[32] As translator of Hero and
Pappus he elaborates on Regiomontanus' theme of a renaissance of mathematics through the recovery and assimilation of the masterpieces of antiquity.
And as a pupil of Commandino he cleverly deploys his history in an attempt
(in Rose's words) 'to legitimise the Archimedean revival of the sixteenth
century by proving the claim of Archimedes to be the true perfector of
mechanics and geometry'. I have found no comparably sophisticated
legitimatory use of history in any of the prefatory histories of mathematics
and astronomy.

Certain aspects of the sixteenth-century prefatory histories deserve special
attention if we are to locate the historiographical originality of Kepler's
Apologia. One salient feature is the apparent lack of concern with the details
of the views of the authors who are assigned places in the succession of
practitioners. Reference to their contributions is generally in the form of bare
mentions of titles or subjects of works, of machines constructed or of
particularly striking beliefs and achievements: Apollonius of Perga wrote a
book on conic sections; Perdix, nephew of Daedalus, invented compasses;
Thales predicted a solar eclipse; and so on. Given the paucity of detail about
the hypotheses of past authors it is hardly surprising that few (Ramus and
Ursus are the obvious exceptions) manifest any real conception of theoretical
change, let alone a conception of theoretical progress. In the earlier histories
the mathematical arts are assumed either to have been fully known to Adam
or to have been invented by some sage or hero of antiquity: Abraham, Moses
or an unknown Chaldaean. In the later histories there is a gradual appreciation
that the mathematical arts cannot have sprung into being fully fledged, like
Athene from the head of Zeus, but were rather perfected over a long period

---

[31] Bodleian, MS Savile, 29–32. On Savile see R. S. Westman, 'The astronomer's role
in the sixteenth century: a preliminary study', *History of Science*, 18 (1980), 129–30.

[32] Rose, *The Italian Renaissance of Mathematics*, Ch. 11. Parts of the work have been
published: *Cronica de' matematici*...(Urbino, 1707); 'Vite inedite di matematici
italiani', *Bolletino di bibliografia e di storia delle scienze*, 19 (1886), 335–640, and 20
(1887), 197–308.

at the hands of many inventors.[33] And implicit in Tycho's oration, explicit in Stadius' history, is the claim that modern astronomy has surpassed that of the ancients. But it would be unwise to read back into these attitudes anything approaching the enlightenment belief in an open-ended accumulation of positive knowledge or even the Baconian vision of progress towards a finite and complete natural philosophy to be achieved through systematic and collaborative endeavour. As Rossi, Burke and others have argued, the concept of the historical development of civilisation, and with it the arts and sciences, that prevails in the latter part of the sixteenth century, the concept to be found in, for example, Vasari, Bodin and Le Roy, is not that of open-ended progress; rather, the vision is that of successive stages in a cycle of growths and decadences, a cycle which is often supposed to be linked to, if not caused by, long-term cycles in the dispositions of the heavenly bodies.[34] Nor are the growth phases of the cycles considered in terms of growth in theoretical insight into the nature of the world; rather, growth is considered in terms of the invention of useful and wonderful artefacts.[35] When such protagonists of the moderns as Villalón and Bodin vaunt the achievements of their age it is to the technology of navigation and the arts of war that they appeal; and when Stadius and Clavius praise recent advances in astronomy it is improved tables, instruments and calendars they have in mind, not better theories.[36] Until Kepler no one is found to champion modern astronomy for its fulfilment of the promise Regiomontanus had seen in it, that of attainment of new insights into the form of the world.

Partial explanations of the failure of the prefatory histories of the mathematical arts to evince concern with theoretical change are not hard to come by. Dedicatory letters, prefaces and one-hour orations do not, after all,

[33] A. B. Ferguson, '"By little and little": the early Tudor humanists on the development of man', in J. G. Rowe and W. H. Stockdale, eds., *Florilegium historiale: Essays Presented to Wallace K. Ferguson* (Toronto, 1971), 125–50, charts the growth of awareness of the gradual origins of human culture.

[34] P. Rossi, *Philosophy, Technology and the Arts in the Early Modern Era* (New York, 1970), Ch. 2 (transl. by S. Attanasio, *I filosofi e le macchine* [Milan, 1962]); P. Burke, *The Renaissance Sense of the Past* (London, 1969), Ch. 4. On the classical and biblical sources for this cyclic conception of history see G. W. Trompf, *The Idea of Historical Recurrence in Western Thought: from Antiquity to the Reformation* (Berkeley, 1979). On the links between astrology and historical recurrence see, e.g., E. Garin, *Lo Zodiaco della vita* (Bari, 1976), Ch. 1.

[35] Rossi, *Philosophy, Technology and the Arts*; A. Keller, 'Mathematical technologies and the growth of the idea of technical progress in the sixteenth century', in A. G. Debus, ed., *Science, Medicine and Society in the Renaissance: Essays to Honour Walter Pagel*, 1 (London, 1972), 11–27; J. A. Maravall, *Antiguos y Modernos* (Madrid, 1966), Pt 4, Chs. 3 and 4.

[36] Stadius, *Astronomiae aetas*; Clavius, *In Sphaeram Ioannis de Sacro Bosco commentarius* (first publn, 1571; Rome, 1585).

provide much scope for the tasks of reconstruction and textual analysis required for detailed accounts of past planetary models and world-systems. And the scope is further diminished by the formalities of composition with their insistence on definition and partition of the subject matter, an edifying peroration and, in the case of an oration, lavish civilities. For detailed reconstruction we must rather turn to the bodies of the works thus introduced or to commentaries on the ancient texts. Thus, to take a single example, Clavius includes in the introduction to his commentary on Sacrobosco's *De sphaera* the most skeletal and stereotyped of histories of astronomy; but the body of the text retails the views of past astronomers on such theoretical questions as the number and order of the spheres in considerable detail.[37] Further explanation for this aspect of the prefatory histories is to be found in the weight of the classical doxographic tradition, a tradition characterised by Lucien Braun as dedicated to the preservation of knowledge 'in the form of a bare multiplicity of sayings, actions, circumstances, works, successions, etc.'.[38] As Braun has shown, the historiography of philosophy is another field in which the doxographic approach, typified in antiquity by Diogenes Laërtius, remains the dominant historiographic mode throughout the sixteenth century.[39] But these factors leave much unexplained. In the course of the fifteenth and sixteenth centuries legal, constitutional and ecclesiastical history established themselves as independent genres in which detailed reconstruction of changes in men's theoretical beliefs was pursued through increasingly scholarly and historically sensitive analysis of primary sources.[40] If humanist philology could yield so rich a harvest in these areas,

[37] In Sphaeram Ioannis de Sacro Bosco, 4 (De inventoribus astronomiae) and 63 (De ordine sphaerarum).

[38] L. Braun, *Histoire de l'histoire de la philosophie* (Paris, 1973), 87. A *caveat* is in order: in many cases the doctrinal snippets and anecdotes are presented, often with edifying intent, as signs and symptoms of the characters of past philosophers. Only when read as attempts to present past philosophical doctrine do such doxographic histories appear as 'bare multiplicities', just as sixteenth-century natural histories are apt to appear unsystematic if read as attempts to describe the structure of organisms: cf. M. Foucault, *Les Mots et les choses* (Paris, 1966), Ch. v.

[39] *Histoire de l'histoire*, Ch. 2. Braun concentrates on general histories of philosophy. However, in the context of more specialised and polemical treatises a more detailed and scholarly historiography of philosophy emerges: notable cases include Giulio Castellani's *Adversus Marci Tullii Ciceronis academicas quaestiones disputatio* of 1558, discussed in detail by C. B. Schmitt, *Cicero Scepticus* (The Hague, 1972), Ch. v; and Jacopo Mazzoni's *In universam Platonis et Aristotelis philosophiam praeludia* of 1597, treated by F. Purnell, 'Jacopo Mazzoni and Galileo', *Physis*, 14 (1972), 273–94.

[40] On the impact of philology on historiography in these fields see Burke, *The Renaissance Sense of the Past*, Ch. 3; and D. R. Kelley, *Foundations of Modern Historical Scholarship: Language, Law and History in the French Renaissance* (London, 1970), Ch. 3.

why did it yield so little in the historiography of the mathematical arts? The question is one to which I shall return.

The *Apologia* is by no means Kepler's only venture into historiography. His *Rudolphine Tables* are prefaced with a brief history of astronomy.[41] Book 3 of his *Harmonice* offers reflections on the nature of ancient music and on the rise of polyphony.[42] In his *De stella nova* of 1606 he commits himself decisively in the *querelle* of the ancients and moderns with a joyous celebration of the achievements of the modern age.[43] And throughout his career he remained fascinated by problems of chronology, a fascination which led to his treatises on Christ's birthday and his massively scholarly *Eclogae chronicae* of 1615.[44] Kepler's historiography often echoes the themes of the prefatory histories. Thus his historical introduction to the *Rudolphine Tables*, whilst unusual in its silence on the Jewish origins of astronomy and in its concluding section where the relevance of natural philosophy to astronomy is emphasised, is otherwise close in form and content to its sixteenth-century prototypes. The celebration of modern achievements in his *De stella nova*, in certain respects highly original, is conventional in its emphasis on printing, feats of navigation and technological achievement. The long letter to Maestlin of June 1598, in which he defends his plan to have constructed a planetarium to illustrate the cosmography of *Mysterium*, contains a list of 'mathematical inventions' of antiquity typical of the prefatory histories in its emphasis on useful and ingenious mechanical devices.[45] In the *Apologia* there are certain passages, notably at the beginning of the second chapter, in which Kepler makes a fairly uncritical use of the basic motifs of the prefatory histories, *successio* and *translatio studii*. Finally, and significantly, for all his decisive commitment to the reality of progress in astronomy, in the *Apologia* as elsewhere Kepler clearly holds in common with the prefatory histories that ancient doctrines do in certain cases have a special authority. This is most obvious in the case of his belief that the Pythagoreans had anticipated both the Copernican system and the *Mysterium* of his *Mysterium cosmographicum*, but it surfaces also, I think, in the special weight he attaches to Proclus' mathematical epistemology in the *Harmonice*. It should be noted, however,

---

[41] *K.g.W.*, x, 36–44.

[42] K.g.W., VI, 91–185. On Kepler's treatment of the history of music see M. Dickreiter, *Der Musiktheoretiker Johannes Kepler* (Berne, 1973), 148–50; and D. P. Walker, *Studies in Musical Science in the Late Renaissance* (Leiden, 1978), 40–2.

[43] *K.g.W.*, I, 329–32, discussed below.

[44] *K.g.W.*, v, 129–201, 7–126; *K.g.W.*, v, 221–370.

[45] *K.g.W.*, XIII, 218–32. On Kepler's projected planetarium see F. D. Prager, 'Kepler als Erfinder', in F. Krafft *et al.*, eds., *Internationales Kepler-Symposium, Weil der Stadt 1971* (Hildesheim, 1973), 386–92.

that Kepler's belief in the authority of the Pythagoreans is a qualified one: in the *Harmonice*, in which he repeatedly hails the Pythagorean insight into the harmonies of the world, he is unabashed in demolishing and even mocking the Pythagorean theory of harmony. (By way of digression I cannot resist citing a story told by Kepler to Maestlin which is suggestive of the ambivalence of his attitude to ancient authority. Kepler reports that to avoid the charge of odious novelty he told the Rector of the Protestant *Stiftsschule*, Johannes Regius, that the 'discovery' of his *Mysterium* was known to the ancients, only to be told that in that case he was wasting his time.[46]) But for all these echoes of themes of the prefatory histories, if the *Apologia* is read as a whole in conjunction with the celebration of modern achievement in the arts and sciences of the *De stella nova*, substantial contrasts between Kepler's historiography and that of his predecessors emerge. Let us consider some of these.

1) In contrast to the lack of concern in the prefatory histories with the details of past theories, the historical narrative of the *Apologia* involves meticulous reconstruction of the astronomical hypotheses of antiquity based on close reading and comparison of texts.

2) In contrast to the generally uncritical use of sources in the prefatory histories, typified by Tycho's acceptance from Josephus of the story about the preservation of astronomy from the flood, Kepler's use of sources is generally critical and at its best shows a sophisticated historical sense. Thus where the prefatory histories are often content to use Homer, Josephus and Pliny alongside Aristotle, Ptolemy and Proclus as sources for ancient astronomy, Kepler observes that Pliny was not an astronomer, but rather an eclectic compiler who may well not have understood all he derived from others.[47] And in assessing the claims of Martianus Capella, Macrobius and Vitruvius to have partially anticipated the Tychonic system, a question that remains controversial today, Kepler makes a valiant if somewhat far-fetched attempt to establish a pedigree for geoheliocentric cosmology by tracing a textual chain of derivation and emendation back to Plato.[48] But it is in the

---

[46] Letter of December 1598 to Maestlin (*K.g.W.*, XIII, 251).

[47] *Apologia*, pp. 200–1.

[48] B. Eastwood, 'Kepler as historian of science: precursors of Copernican heliocentrism according to *De revolutionibus*, I, 10', *Proceedings of the American Philosophical Society*, 126 (1982), 367–94, provides a thorough analysis of Kepler's attempt in Ch. 4 of the *Apologia* to trace the pedigree of the geoheliocentric idea back to Plato, and shows how Kepler's enthusiasm led him to misinterpret texts from Pliny, Vitruvius and Macrobius. It should, however, be noted that Kepler's particular line of misinterpretation remained very much in evidence until well into this century. Further, the polemical context should be borne in mind: Kepler is out to show that Ursus' denial of Tycho Brahe's originality cannot be sustained even if the most

two detailed reconstructions he essays, that of the systems of Eudoxus and Calippus as retailed by Aristotle and that of Apollonius of Perga's account of the planetary stations as retailed by Ptolemy, that Kepler's combination of common, philological and historical sense shows up to best advantage. Of course, as we now know, and as Kepler would surely have realised had he read the relevant passages in Simplicius' *Commentary on De caelo*, his reconstruction of Eudoxan and Calippic astronomy proceeds along entirely mistaken lines. But for all that his handling of the text is impressive in its combination of minute attention to philological detail with consideration of questions of motive and provenance. Thus in Kepler's reconstruction of the Eudoxan system, as in the currently accepted reconstruction of Schiaparelli, the solar model has a component that is apparently redundant. Why? Kepler speculates very plausibly that Eudoxus wished to attribute a small motion in latitude to the sun to preserve the analogies between the sun and moon.[49] Equally impressive in its use of considerations of motive and context is his line by line demolition of the interpretation of Aristotle on which Fracastoro rests his reconstruction of the systems of Eudoxus and Calippus. But the historiographical high point is surely the crucial refutation of Ursus' claim that the Tychonic system was anticipated by Apollonius of Perga.[50] Here the meticulous analysis of a slender text and the marshalling of evidence to support a carefully qualified negative conclusion constitute a masterpiece of historical reconstruction and assessment.

3) The notion of growth of the mathematical arts that is invoked in the prefatory histories is that of practical and technological progress. Kepler by contrast deploys a full-blooded notion of theoretical progress in astronomy through which ever greater knowledge of 'the inmost form of things' is achieved. Further, as has been shown in Chapter 6, his claims about the fact of theoretical progress are linked to an explicit account of the means whereby it is brought about through the emendation of existing hypotheses and the resolution of conflict between rival sets of hypotheses.

4) Even the more perfunctory of the prefatory histories embody assumptions about the historical causes of the growth of astronomy and the mathematical arts; for the historiographical motifs they exploit, *successio* and

sympathetic efforts are made to extract geoheliocentric ideas from the available texts.

[49] *Apologia*, p. 167. Schiaparelli argues, like Kepler, that one of the three spheres in Eudoxus' system is designed to explain a supposed motion in latitude of the sun rather than precession of the equinoxes: G. Schiaparelli, 'Le sfere omocentriche di Eudosso, di Callippo e di Aristotele', *Memorie di Reale Istituto Lombardo. Classe di sci. mat. e nat.*, 13 (1877); reprinted in *Scritti sulla storia della astronomia antica*, I, 2 (Bologna, 1926), 24–42.   [50] *Apologia*, pp. 185–97.

translatio studii, are not purely descriptive notions. Thus *successio* explains growth as the product of elaboration at the hands of a sequence of masters and pupils; and *translatio studii* associates the growth and decay of a subject with the rise and fall of empires and nations, thus presupposing an explanatory link between political stability and prosperity and the flourishing of the arts and sciences. On occasion more specific explanatory factors are invoked. Thus Regiomontanus sees the renaissance of mathematics as a natural consequence of a programme of translation, edition and publication of the masterpieces of antiquity; Clavius associates recent innovations in astronomy with the quest for better calendars; and both Ramus and Stadius associate the prosperity of the mathematical arts with princely patronage.[51]

More sophisticated sixteenth-century speculation on the historical factors which have fostered the growth of learning is to be found in another context, a context of which Kepler was definitely aware. Around 1560 Philipp Melanchthon and François Baudouin had called for a new type of history, *historia integra*, which would, in Melanchthon's words, provide a complete 'portrait of the human race'.[52] At the hands of Jean Bodin, Louis Le Roy, Lancelot Voisin, Sieur de la Popelinière and others there developed a genre of universal history devoted to the origins of civilisation, the differentiation, rise and fall of societies and nations, the changes in men's customs, laws and forms of government, and, above all, the causes of these phenomena.[53] The growth of learning and its historical conditions is a constant topic in these universal histories. In Bodin's *Methodus* and *De republica* the emphasis is predominantly utilitarian. Like Juan Luis Vives, whose *De tradendis disciplinis* of 1531 prefigures many aspects of these histories of learning, Bodin presents the arts as derived from *expérience* in response to men's need to defend their common society.[54] The particular aptitudes of the various races to different arts and sciences he seeks to explain in terms of environmental and

51 Regiomontanus, *Oratio introductoria*; Clavius, *In Sphaeram Ioannis de Sacro Bosco*; Ramus, *Prooemium mathematicum*; Stadius, *Astronomiae aetas*.

52 Melanchthon, preface to the second part of the *Chronica Carionis* (Wittenberg, 1558) (*M.o.o.*, IX, col. 1976); Baudouin, *De institutione historiae universae* (Paris, 1561).

53 On universal histories see G. Huppert, *The Idea of Perfect History: Historical Erudition and Historical Philosophy in Renaissance France* (Urbana, Ill., 1970); D. R. Kelley, 'Historia integra: François Baudouin and his conception of history', *Journal of the History of Ideas*, 25 (1964), 35–57; and Kelley, *Foundations of Modern Historical Scholarship*, 128–48.

54 Bodin, *Methodus*, transl. by B. Reynolds (New York, 1945), 10 and 50. The opening chapters of Vives' *De tradendis disciplinis* (Antwerp, 1531), described by his translator Foster Watson as 'the first modern descriptive history of civilisation', anticipate many of the themes of the later universal histories: F. Watson, *Vives on Education* (Cambridge, 1913).

astrological factors. The rise and fall of societies, and with them the arts and sciences, he associates with cyclical changes in the disposition of the heavenly bodies and, especially, with planetary conjunctions. Near the end of the *Methodus* he eulogises 'the supreme flowering of the arts in the present age'.[55] The emphasis here is on technological achievements, printing (which 'alone can easily vie with all the discoveries of the ancients'), navigation and war machines, but there are passing references to theoretical discoveries in alchemy and astronomy.[56] In Le Roy *expérience* of nature is again presented as the prerequisite for development of the arts and sciences, but the utilitarian emphasis is less marked.[57] Leisure within a stable society rather than the practical and defensive needs of society is invoked as the crucial factor.[58] He follows Bodin, though with less enthusiastic detail, in considering geography and the stars as important influences in the rise and fall of cultures and their learning. But unlike Bodin, and anticipating Kepler, he speculates more specifically on the local factors which have fostered the modern flowering of learning which so far surpasses that of the ancients. He recognises in passing the importance of printing and universities in the dissemination of learning, whilst, again like Vives, emphasising the importance of the modern rejection of scholasticism and undue dependence on the authority of books.[59] But for Le Roy it is above all the geographical discoveries of the age which have fostered progress through the rich store of new *expériences* that they have yielded.[60]

Kepler's views on the historical factors that have fostered the growth of astronomy are never explicitly aired in the *Apologia*. But there is a revealing passage in which he rejects the view that the growth of astronomy in antiquity had a linear character with pupils uniformly improving on the discoveries of their masters. Ursus, he claims, writes as if there had been astronomical schools and successions in antiquity and as if there had been 'maintenance or public and formal change of astronomical dogmas'. The picture is a false one: no such schools are recorded and communication even between contemporary astronomers was demonstrably poor due, Kepler surmises, to lack of printing.[61] It is not merely the questioning of the prevalent historiographic concept of *successio* that is remarkable here. Implicit

---

[55] Bodin, *Methodus*, 149 *et seq*. There is striking precedent for his views on the influence of celestial and geographical factors on human learning in Plato's *Timaeus*, 22B–23B.   [56] *Methodus*, 145 and 301 *et seq*.

[57] Le Roy, *De la vicissitude ou variété des choses en l'univers* (Paris, 1575).

[58] *De la vicissitude des choses*, f. 113r. Cf. Aristotle (*Metaphysics*, I, I, 981b), who attributes the origin of mathematics in Egypt to the leisure of the priesthood.

[59] *De la vicissitude des choses*, f. 99v–100r.

[60] *De la vicissitude des choses*, f. 114r.   [61] *Apologia*, p. 181.

in this passage is a contrast between the ancient world, with its uneven development of astronomy due to lack of teaching of the subject and poor communication, and the steady growth of astronomy in the modern world, in which astronomy has a place in the curriculum and printing provides a medium for the dissemination of knowledge. The suspicion that this is to read too much into the passage is allayed by Kepler's celebration of modern achievements in the twenty-ninth chapter of his *De stella nova* of 1606.[62]

Kepler's primary theme here is the influence of planetary conjunctions on human affairs. They act, he maintains, through a stimulating effect on that human faculty which makes men social by nature in such a way that 'the minds of many men more easily come together in an undertaking'. Kepler then retails some of the achievements of the modern age attributable to such collaborative enterprises. Public order has been restored after the mediaeval barbarism; military technology and strategy have made unprecedented advances; the Spaniards have achieved remarkable feats in navigation, and their discoveries have led to a great increase in trade and prosperity. Above all the invention and steady improvement of printing testify to the industry of the new age. So far Kepler's account runs closely parallel to those of Bodin and Le Roy. But now comes an extraordinary and moving passage.[63]

After the birth of printing books became widespread. Hence everyone throughout Europe devoted himself to the study of literature. Hence many universities came into existence, and at once many learned men appeared so that the authority of those who clung to barbarism declined. Men's longings were not satisfied until there appeared the founder of a new order which was openly engaged in the study of literature.[64] Since then almost all the authority of the religious orders has passed to the new order, except insofar as those older orders, whilst busy with their own affairs, also apply themselves to the study of literature. Voyages and trade have provided Europeans with the opportunity to spread afar the Christian faith among

---

[62] *K.g.W.*, I, 325–35.

[63] *K.g.W.*, I, 330–2. In translating this rather difficult passage I have found useful E. Rosen's excellent translation of a part of it, 'In defense of Kepler', in A. R. Lewis, ed., *Aspects of the Renaissance* (Austin, Tex., 1967), 141–58.

[64] Apparently a reference to St Ignatius Loyola and the educational activities of the Jesuits. In view of the confessional difference it is perhaps surprising that Kepler alludes to the Jesuit role in the revival of literary studies and passes over that of the Lutheran educational reformers. Kepler had, however, been on good terms with the Jesuits at the University of Graz. Both at Graz and at the Clementinum in Prague Jesuit activity was indeed predominantly oriented towards *litterarum studia*. On the educational activities of the Jesuits at Graz, and Kepler's contacts with them, see J. Andritsch, 'Gelehrtenkreise um Johannes Kepler in Graz', in P. Urban and B. Sutter, eds., *Johannes Kepler 1571–1971: Gedenkschrift der Universität Graz* (Graz, 1975), 159–95; on the early history of the Clementinum see A. Kroess, *Geschichte der Böhmischen Provinz der Gesellschaft Jesu*, Vol. I (Vienna, 1910), 71–157.

barbarous and previously unknown peoples. On the other hand from the universities, and [their] licence to hold disputations,[65] from the abundance of books and from the convenience of printing, as well doubtless as from learning and public unrest, there has in the end sprung that immense and forever memorable secession of very many regions of Europe from the see of Rome...I would ascribe all this to the Holy Spirit were not so many such things evil...Given that opposite parts of the world do not evidently differ in nature, who ever heard of preparing a fleet to go to the antipodes, there to contend about sovereignty [*ut illic de summa rerum decertent*], as if it were not just as expedient to stay at home?[66] Did it ever happen under the Roman Empire that a Sarmatian or German fleet, sailing from the Baltic, circumnavigated Europe, came to Italy, then sent ahead fast runners who took the shortest route through the Mediterranean lands and brought corn, thus bringing down the high price of grain?[67] What did they have in antiquity to approach present-day knowledge in the art of war? What shall I say of today's mechanical arts, countless in number and incomprehensible in subtlety? Do we not today bring to light by the art of printing every one of the extant ancient authors? Does not Cicero himself learn again how to speak Latin from our many critics? Every year, especially since 1563,[68] the number of writings published in every field is greater than all those produced in the past thousand years. Through them there has today been created a new theology and a new jurisprudence; the Paracelsians have created medicine anew and the Copernicans have created astronomy anew. I really believe that at last the world is alive, indeed seething, and that the stimuli of these remarkable conjunctions did not act in vain.

Many of the ingredients for these speculations on the historical factors that have fostered recent growth in the arts and sciences are, as we have seen, to be found in sixteenth-century universal histories. But it is a far cry from

[65] Rosen, 'In defense of Kepler', 143, renders this freely but eloquently: 'the universities with their academic freedom'. The context, however, suggests the more precise meaning. Throughout the sixteenth century the formal disputations of the universities served as occasions for the expression of theological dissent and as stimuli to confessional partisanship; see D. R. Kelley, *The Beginnings of Ideology: Consciousness and Society in the French Reformation* (Cambridge, 1981), Ch. 4.

[66] *Antipodes* often refers to the Americas in the period. The context makes it clear that Kepler refers to intellectual dispute about sovereignty, not to fighting for sovereignty (which his Latin would admit as a reading). He may well have in mind the debate about the nature of man sparked off by the Spanish conquests, a debate which centred on the question whether the American Indians possessed a sovereignty of their own or fell into the Aristotelian category of 'slaves by nature': see, e.g., L. Hanke, *Aristotle and the American Indian: a Study of Race Prejudice in the Modern World* (Chicago, 1959).

[67] In the wake of the bad Italian harvests of 1586–90 the Dutch had established an extensive and profitable trade in Baltic grain throughout the Mediterranean: F. Braudel, *The Mediterranean World in the Age of Philip II*, transl. by S. Reynolds (London, 1975), 599 *et seq.*, 634 *et seq.*

[68] The year of a great conjunction.

mere recognition, as in Bodin and Le Roy, that printing and education have played roles in the revival of learning to Kepler's brilliantly perceptive hints at the connections between printing and educational reform, the creation of a public domain of knowledge, and progress in the arts and sciences. And when we turn to the 'internal' causes of progress it is a yet further cry from the claim of Vives, Ramus and Le Roy that 'experience' is crucial, or from the insistence of Clavius and others that astronomical hypotheses should be designed to save the phenomena whilst containing nothing 'repugnant to natural philosophy', to Kepler's articulation in the *Apologia* of a detailed methodology whereby progress in astronomy is achieved.

5) The central concern of sixteenth-century prefatory histories is, manifestly and explicitly, to promote and validate astronomy as a discipline. The narratives are designed to demonstrate the dignity which the subject derives from its antiquity, the esteem in which sages and potentates have held it and, above all, the practical benefits that have accrued and are likely to accrue from its cultivation. Kepler too is openly concerned with legitimation. He is out to validate the discoveries of astronomers in the face of Ursus' sceptical denigration and to vindicate Tycho's claim to originality against Ursus' allegation that his world-system was derived from Apollonius of Perga and Copernicus. But at a deeper level Kepler's validatory concerns are more ambitious. He is concerned to validate both his own view of astronomy as a discipline that has yielded ever greater insight into the constitution of the world and his own account of the means whereby such progress can come about through integration of mathematical astronomy into natural philosophy. Such concerns are not entirely without precedent; Ramus and Ursus had, indeed, used the history of astronomy to support epistemic stances, in Ramus' case a radically empiricist view of astronomy and the mathematical arts, in Ursus' case a radical scepticism about astronomy. But their exploitations of history fall far short of Kepler's in subtlety and coherence. The epistemology and methodology of astronomy that Kepler is out to defend are not to be seen merely as a set of tenets about the existing status and practice of the art; rather, they provide a model for a new kind of astronomy and a programme for a new kind of practice of astronomy, a practice which he himself pursued. If, following Mannheim, we consider ideology as relating declared policy and actual strategy,[69] and following Shils insist that an ideology demands actual embodiment in persons playing new roles,[70] we may fairly claim that Kepler envisages a new ideology for

[69] K. Mannheim, *Ideology and Utopia*, transl. by L. Wirth and E. Shils (New York, 1952).

[70] E. Shils, *The Intellectuals and the Powers* (Chicago, 1972).

astronomy (and, by implication, for the mathematical arts generally) and that he uses history in the service of that ideology.

It is time to reflect on the historical conditions for Kepler's historiographical and epistemological innovations. The factors I shall consider are philology, technological progress, printing and the genesis of the concept of a theory. Kepler is, unusually for his period and his profession, intensely self-conscious as a writer, and he himself provides hints of the importance of such factors in his articulation of a concept of progress in astronomy. Let us consider these factors in turn.

In his plaintive letter to Maestlin of February 1601 Kepler describes the *Apologia* as largely philological rather than mathematical.[71] There can be no doubt that philological expertise deployed in the reconstruction from the sources of past systems of astronomical hypotheses is a precondition for the composition of a history which retails the theoretical development of astronomy. Kepler's sophistication as an interpreter of sources is perhaps the most readily explicable aspect of his historiographical achievement. At Tübingen he had been a pupil of the historian and philologist Martin Crusius and of the poet and orator Erhart Cellius.[72] He was acquainted with at least one distinguished historian at the predominantly arts-oriented University of Graz, the Chancellor, Johannes Decker.[73] His passion for biblical chronology, shared with his teacher Maestlin, a field in which extensive reconstruction and comparison of texts was needed to excavate the least nugget of hard fact, had involved him in extensive philological studies. And his patron Herwart's interest in the significance of astrological passages in classical authors, an issue on which he bombarded Kepler with queries, had led him into reluctant philological reflections on obscure passages in Lucan, Manilius and others.[74] Whatever his reluctance to enter the field, Kepler was well prepared by his interests and intellectual circle for the philological aspects of his historical enterprise.

That philological expertise alone need not engender a sophisticated historiography is well attested by later histories. Thus Vossius' history of the mathematical arts and Riccioli's monumental history of astronomy show far

---

[71] *K.g.W.*, XIV, 165.

[72] On Kepler's contacts with Crusius see E. Rosen, 'Kepler's mastery of Greek', in E. P. Mahoney, ed., *Philosophy and Humanism: Renaissance Essays in Honour of Paul Oskar Kristeller* (Leiden, 1976), 310–19. On his relations with Cellius see P. Sanders, 'The regular polyhedra in renaissance science and philosophy' (forthcoming Ph.D. thesis, University of London).

[73] Andritsch, 'Gelehrtenkreise um Johannes Kepler in Graz'.

[74] See, e.g., *K.g.W.*, XIII, 132–40 and 351–2. Kepler's reluctance in these enterprises may be inferred from a letter to Maestlin in which he describes Herwart as having 'pestered him for two whole years' on one such question (*K.g.W.*, XIV, 46).

more minute and extensive scholarship than does Kepler's history; but the upshot in each case is a narrative which, whilst of far greater descriptive range and detail, remains largely within the doxographic tradition of the prefatory histories.[75] As Braun has shown in detail, the impact of philology on the history of philosophy yielded in the sixteenth century no substantial break with the classical doxographic genres, and even in the seventeenth century tended rather to increase the scope and detail of description than to usher in a more sophisticated historiography.[76] The impotence of philology alone to inspire a consideration of theoretical development should, I think, occasion little surprise. Whilst growth and decay at the level of technology can be regarded as a factual matter that cannot but be appreciated once a sufficiently detailed study of the documents and monuments of the past is undertaken, theoretical growth is not thus transparent to historical inspection. It is surely true that awareness of technological progress, an awareness which became widespread in the sixteenth century, puts pressure on the historian to seek factors which explain that progress. But the very diversity of the factors to which sixteenth-century historians appeal for this purpose, celestial influence and environment in Bodin, stability of society and geographical discovery in Le Roy, shows how far from self-evident is the link between technological advance and progress at the level of theory. The notions of a theory, of conflict of theories and of elaboration and succession of theories do not emerge automatically from the historical evidence, but are rather constructs used by the historian of science to interpret and explain that evidence.

In considering the role of printing we need to distinguish sharply two very different issues. First there is the question, for recent historians a highly contentious one, of the direct impact of printing on the growth of science and technology.[77] On this question, as we have seen, Kepler took a definite stance, presenting printing as the crucial factor in the intellectual achievements of his age. Apart from this vexed first-order historical question there is the question that concerns us, that of the impact of printing on the historical consciousness of theoretical change in the arts and sciences. Such an impact can, I suggest, be diagnosed at more than one level. First, and perhaps most

---

[75] Vossius, *De universae mathesios natura*; Riccioli, *Almagestum novum*.

[76] Braun, *Histoire de l'histoire*, Ch. 2.

[77] On the diverse assessments of the impact of printing on the growth of science see E. L. Eisenstein, *The Printing Press as an Agent of Change* (Cambridge, 1979), 453 *et seq*. Eisenstein's own assessment rests on the claim that printing rapidly surpassed scribal reproduction in effective preservation of texts in stable form; A. T. Grafton, 'The importance of being printed', *Journal of Interdisciplinary History*, 11 (1980), 265–86, argues that she exaggerates the inaccuracy of *scriptoria* and the accuracy of early print-shops.

obviously, printing facilitated, indeed may have been a precondition for, the extensive comparison of past texts from which an appreciation of theoretical change emerged. Further, printing surely facilitated – and may well have been, in fields in which no strong scribal or oral traditions existed, a historical precondition for[78] – the emergence of the conception of a public domain of knowledge to which individual *auctores* may variously contribute.[79] And this, in turn, is surely linked to the idea that there are common norms which regulate 'the maintenance or public and formal change of dogmas'.

In these two complementary developments, the rise of philology and the invention of printing, we have, I suggest, factors crucial for the genesis of an appreciation of the history of the mathematical arts as evincing theoretical growth. What we still lack, however, is any explanation for the most distinctive and original features of Kepler's epistemology and historiography, the vision of progress through a succession of specific theories mediated by specific norms for the conduct of inquiry and the assessment of novelties.

Kepler himself, in his celebration of the modern age, juxtaposes the new Protestant confession with the new Paracelsian medicine and the new Copernican astronomy. His friend Helisaeus Roeslin had drawn a more explicit parallel between religious and natural philosophical sectarianism.[80]

But in the fullness of time it was found that certain of the tenets and hypotheses [of Galen, Ptolemy and Aristotle] not only were inadequate, but entailed many absurdities. This moved some in this latter age to plan a defection and to rise up against them: Nicolaus Copernicus against Ptolemy, Theophrastus Paracelsus against Galen and Aristotle, and then Petrus Ramus against the latter in the philosophical arts. They did so at the very same time that many contentions arose in Europe in religion and sacred matters, and a certain Thechelles revolted against the Mahometan sect.[81] This age of ours is indeed a marvel and pregnant with great things.

Awareness of theoretical conflict plays a crucial role in the epistemology and historiography of the *Apologia*, as it does in Kepler's oecumenical theology. Thus the crux of his rebuttal of scepticism in the first chapter is his account of the means whereby apparently irresoluble theoretical conflicts in astronomy may be resolved. And the resolution of such conflicts evidently provides at

---

[78] The qualification is important. At least in fields in which a limited range of texts held authority, theology, jurisprudence and Aristotelian physics, for example, the concept of a public domain of knowledge constituted by a *traditio* of interpretation and commentary was surely prevalent before the invention of printing.

[79] Cf. Eisenstein, *The Printing Press as an Agent of Change*.

[80] *De opere Dei Creationis* (Frankfurt, 1597), 4–5.

[81] Thechelles = Shah Kuli, leader of a great rebellion against the Ottomans in 1511, had proclaimed himself as the Mahdi and Isma'il, the first Shah of Persia, as the reincarnation of the Godhead: see *The New Cambridge Modern History*, Vol. 1 (Cambridge, 1957), 406.

least part of the means whereby he supposes progress in astronomy to come about.

The connections between conflict, epistemology and historiography in other sixteenth-century fields have been well documented. Thus Popkin has studied the development of sceptical and anti-sceptical philosophical tactics in theological polemic.[82] Polman and others have explored the uses of ecclesiastical history in the legitimation of rival confessions.[83] And Kelley has brilliantly shown how in France in the latter part of the century political conflict fostered increasingly sophisticated uses of legal and social history to legitimise rival positions on ideologically sensitive issues.[84] At another level too conflict and historiography may be connected. For it is a natural step from awareness of the magnitude of the differences between contemporary societies and systems of belief to an awareness of social and intellectual change.

That Kepler in composing his epistemology and history of astronomy was acutely aware of conflict in astronomy is obvious enough. It was, after all, as a champion of Copernicanism against the Ptolemaic world-system that he had first deployed in his *Mysterium* the main anti-sceptical strategy of the first chapter of the *Apologia*. And for all his care to avoid the active promotion of Copernicanism in the *Apologia*, he makes it quite clear in the fourth chapter that it is a Copernican and not a Tychonic world-system that he sees as the culmination of progress in astronomy. But at this very general level we still find little to explain the most original features of his epistemology and historiography. It is not, I think, awareness of conflict in astronomy *per se* that is crucial in determining the form of Keplerian epistemology and historiography, but rather his awareness of such conflict as embodied in inconsistent theories.

As a rough and ready approximation we may characterise a scientific theory as a systematic body of hypotheses that is related to a systematic practice of prediction, observation and instrumentation in some domain of inquiry. No sixteenth-century term has this connotation. *Theoria* itself is used collectively for contemplative as opposed to practical or operative knowledge.[85] It has also a specialised partitive use in astronomy, where like

---

[82] R. H. Popkin, *The History of Scepticism from Erasmus to Descartes* (Assen, 1960; expanded edn, *The History of Scepticism from Erasmus to Spinoza*, Berkeley, 1979).

[83] P. Polman, *L'Elément historique de la controverse religieuse du XVIᵉ siècle* (Gembloux, 1932). See also E. Scherer, *Geschichte und Kirchengeschichte an der deutschen Universitäten* (Freiberg, 1912); and E. Menke-Glückert, *Geschichtsschreibung der Reformation und Gegenreformation* (Habilitationsschrift, Osterweck, 1912; Leipzig, 1971).

[84] Kelley, *Foundations of Modern Historical Scholarship*; Kelley, *The Beginnings of Ideology*.

[85] Cf. Calepinus, *Dictionarium* (Basel, 1564), where *theoria* is glossed as *speculatio, meditatio, contemplatio*.

theorica it may refer to a planetary model. Other candidates, *speculatio, ratio, ordo, doctrina* and *systema*, whilst sometimes used to describe what with hindsight we would regard as individual theories, have far wider connotations.

It would be a major undertaking to reconstruct in detail the various sixteenth-century anatomies of belief. But for present purposes a few commonplaces suffice. The primary categories are the entire discipline – *ars, doctrina* or *scientia* – and the particular tenet on some *quaestio* within a discipline – *sententia, placet, thesis, hypothesis*. Particular schools of thought are of course recognised – *Copernici hypotheses, placita Stoicorum, medicina Galeni, Aristotelis sententiae*, etc. – but these are conceived indefinitely as the bodies of belief of some author and his followers, not as definite and systematically related bodies of propositions that can be isolated from their historical contexts for purposes of assessment. Even in astronomy, where the terms *systema mundi, hypothesis* (in a generic sense) and *ordo mundi* are widely applied to world-systems, there are good grounds for denying the presence of a concept close to that of a scientific theory. For in all the cases I have come across the reference is to a small number of central theses – the earth has a triple motion, the sun is at the centre, for example – not to an entire astronomical system complete with its planetary models. Whilst in modern usage the term 'theory' is sometimes applied to just such sets of central theses (as when 'Darwinian theory' is used as a synonym for 'the principles of natural selection'), it is surely the usual concept of a theory as a body of principles so elaborated as to make contact with observation and prediction that is crucial for the emergence of accounts of theoretical change and its dynamics.

In the *Apologia* Kepler does quite explicitly introduce a notion close to that of a theory in the usual modern sense. The crucial words are: '...when we speak in the plural of astronomical hypotheses...we designate thereby a certain totality of the views of some famous practitioner'.[86] That Kepler envisages such totalities as systematically related bodies of propositions, not as mere collections of *sententiae*, is emphasised a little later when he writes of the 'demonstrations through diverse syllogisms' which link their component propositions.[87] At the philological level Kepler's usage is not radically innovative. It involves merely the elision of two uses of the term *hypothesis* prevalent in astronomical writings of the period: its use for the central postulates of a world-system and its use for individual planetary models. But conceptually it is profoundly innovative.

[86] *Apologia*, p. 139.   [87] *Apologia*, p. 140.

It may well be that the emergence of the concept of a theory can be understood only in the context of general changes in men's presuppositions about the relation of language to the world. At all events, the following more local reflections are highly speculative.

The nearest sixteenth-century counterpart to the notion of a theory is that of the collection of *sententiae* or *placita* of some author or school. The move from this notion to that of a theory involves a shift from a focus on bodies of belief to a focus on bodies of propositions. It is natural to see this shift as fostered by the shift from teachers to books as primary vehicles for the dissemination of knowledge. For, to put the point crudely, where bodies of belief are instantiated in persons, propositions are instantiated in permanent form in books. To the extent that printing led to a shift from teachers to books as agents of instruction and communication it is surely an important factor.

Emergence of the concept of a scientific theory requires more than the hypostatisation of beliefs as propositions. It requires in addition that bodies of belief be represented as definite bodies of propositions that are systematically related to each other and to some domain of practice. For all the proliferation of rival schools of thought in natural philosophy and the mathematical arts in the latter part of the sixteenth century, fields in which the rival bodies of belief are thus representable are hard to come by. The conflict between the 'new medicine' of Paracelsus and Galenic medicine, for example, fails by both criteria, involving systems of belief that can be neither represented by definite bodies of propositions nor represented by bodies of propositions systematically related to some body of practice. The same is true, I think, of the conflict between Aristotelian natural philosophy and the new cosmologies of Bruno, Telesio and Patrizi. Indeed the single absolutely clearcut case of rivalry between what with hindsight can be regarded as fully articulated scientific theories appears to be that of the rival world-systems. So it should occasion no surprise that it is in astronomy that the concept of a theory first emerges, and with it the representation of rivalry between systems of belief as inconsistency between theories. That this conceptual reaction to the conflict should be so long delayed is perhaps rather more surprising. A partial explanation is perhaps to be found in the prevalence of the pragmatic compromise, discussed in the last chapter. For this attitude undoubtedly led astronomers to concentrate on planetary models rather than on world-systems considered as unified constructions complete with central postulates, planetary models, derived tables and all.

Kepler's *Apologia* is epistemologically original in its focus on the resolution of theoretical conflict, and historiographically original in its conception of a scientific progress which links practical improvement with the accumulation of theoretical knowledge. And, perhaps most remarkably of all, it is original in its combination of epistemology and historiography into a distinctive mode of reflection on the status of a mathematical science. I have made a case for the importance of philology, printing and the emergence of the concept of a theory as historical conditions for these innovations. The case is, let me emphasise again, highly tentative. Its substantiation would involve detailed exploration of the ways in which knowledge was organised and disseminated in the sixteenth century and of the development of historiographical and methodological traditions in disciplines other than astronomy.

For want of publication the *Apologia* remained without direct impact on Kepler's successors. But it is, nevertheless, historically significant in its innovations. To the historian of science it is of interest in revealing more clearly than any other work of Kepler's the central assumptions about the nature of mathematical science and the proper conduct of inquiry which underlie his decisive first-order contributions to astronomy, optics and music theory.[88] Less obviously, but, I think, just as importantly, it is of interest to historians and philosophers of science in revealing the historical conditions which shaped certain of our central assumptions about the nature of science and scientific inquiry. In the final chapter I shall touch lightly on these implications of the work.

[88] As has been noted in the Introduction, no systematic attempt is made to substantiate this claim in the present work. Useful secondary sources for the reader who wishes to consider the methodology of Kepler's mature works in the light of the *Apologia* include: A. Koyré, *The Astronomical Revolution*, transl. by R. E. W. Maddison (Paris, London and Ithaca, N.Y., 1973), Pt II, on the methodology of *Astronomia nova*; G. Buchdahl, 'Methodological aspects of Kepler's theory of refraction', *Studies in the History and Philosophy of Science*, 3 (1972), 265–98, on the methodology of *Ad Vitellionem paralipomena*; Walker, *Studies in Musical Science*, Ch. 4, on the methodology of Kepler's harmonic theory; J. B. Brackenridge, 'Kepler, elliptical orbits, and celestial circularity: a study in the perversity of metaphysical commitment', *Annals of Science*, 39 (1982), 117–43 and 265–95, on the methodology of Kepler's astrology; and J. V. Field, 'Kepler's geometrical cosmology' (unpublished Ph.D. thesis, London, 1981), Chs. v and vi, on the methodology of the *Harmonice*.

9

Final reflections

Kepler's *Apologia* is, in the hackneyed phrase, a seminal work. It is seminal for Kepler's mature works, adumbrating as it does the central assumptions about the nature of astronomy, the methodology of astronomical research, and the role of the astronomer in the institution of the arts and sciences which inform his later works. A monument to this aspect of the *Apologia* appears as the frontispiece to the *Rudolphine Tables* (see illustration, p. 288).[1]

In Kepler's curiously contrived baldachino astronomical progress, both practical as represented by the instruments and tables and theoretical as represented by planetary models, is symbolised by the succession of pillars from the rude trunks of the Chaldean astronomers at the rear to Copernicus' and Tycho's splendid Doric and Corinthian columns.[2] The dome itself undoubtedly represents Kepler's own new astronomy resting on the theoretical insights of Copernicus and the observational achievements of Tycho. For on the dome geometry, as Kepler's muse, displays a Keplerian elliptical orbit;[3] and in the inner left-hand panel of the podium Kepler, exploiting the analogy between architecture and astronomy developed in such detail in the third chapter of the *Apologia*, depicts himself as designer of the dome. (This aspect of the iconography is passed over in silence in Hebenstreit's explanatory poem *Idyllion*, doubtless out of consideration for Tycho's heirs, who were concerned to make sure that Kepler did not in any way vaunt

---

[1] Ulm, 1627 (*K.g.W.*, x, 7).
[2] An engaging account of the frontispiece and its symbolism is given by O. Gingerich, 'Johannes Kepler and the Rudolphine Tables', *Sky and Telescope*, 21 (1971), 328–33. As Gingerich notes, the original sketch for the proposed engraving was modelled on the foyer of Tycho's observatory at Stjerneborg on the island of Hveen. The depiction of Tycho's world-system there with its motto *Quid si sic?* was, as we have seen, one of the targets of Ursus' mockery (*Tractatus*, sig. Dii, r).
[3] In the accompanying poem by Johann Baptist Hebenstreit, Rector of the Ulm Gymnasium, the figure is identified more specifically as 'the doctrine of triangles'. That she and the figure on her left (identified as 'logarithmics' by Hebenstreit) represent geometry and arithmetic, respectively, is perhaps implied by their central placement directly beneath the wings of the Imperial eagle, for geometry and arithmetic were conventionally said to be the two wings of astronomy.

his own contributions over Tycho's.)[4] The integration of mathematical astronomy into natural philosophy is indicated by the presence on the dome of figures emblematic of the study of magnetism and dynamics – note the sun at the fulcrum of her balance – alongside the more conventional handmaidens of astronomy: geometry, arithmetic and optics. Here in Kepler's final masterpiece we find depicted all the major innovative themes of the *Apologia*.

The *Apologia* is seminal in another respect as well. In it there emerge, apparently for the first time in European literature, concepts that have remained central to our understanding of science and scientific inquiry: the concept of a scientific theory as a definite and consistent body of propositions systematically related to each other and to some domain of practice; the concept of a rational methodology for the resolution of theoretical conflict; and the concept of scientific progress mediated by a rational method and evincing an ever more accurate portrayal of the world. The task of explaining the emergence of such concepts at a particular time in a particular intellectual milieu at the hands of a particular author is a formidable one. In this work only the barest speculations on the historical factors at work have been essayed.

Many lines of thought, variously elaborated by sociologists and archaeologists of knowledge, are conducive to the belief that a study of the historical conditions for the emergence of concepts may reflect on their validity, and may indeed constitute a distinctive mode for the critique of concepts. Demarcation of the range of explanations that impugn the concepts or beliefs whose genesis they explain is a hard and contentious matter, especially with respect to the more ambitious types of sociological explanations; and in the realm of history of science there has been a tendency among both protagonists and antagonists of such 'external' explanations greatly to exaggerate their destructive potential. Certain connections between current validity and original conditions are, however, relatively unproblematic. Thus a study of the conditions of emergence of our concepts may expose to criticism hidden commitments and presuppositions that surreptitiously constrain both the range of questions and the types of answers available to us. More specifically, where a study of the emergence of a concept reveals that its original applications are either atypical of or outside its current domain of application, its contemporary validity may be called in question. In a constructive vein,

---

[4] The original sketch sent to Tycho's heirs had none of these potentially offensive features: Gingerich, 'Johannes Kepler and the Rudolphine Tables', 333. On the difficulties with Tycho's heirs see *K.g.W.*, x, 29*–31*.

a study of related concepts that were displaced by a concept that has become problematic, whilst unlikely to suggest fully-fledged viable alternatives, may intimate ways of escape from current quandaries.

Many of the central presuppositions about the nature of science and scientific inquiry that are crucial to the argument of the *Apologia* remain with us, if not unchallenged, at least prevalent in philosophical and historical interpretations of the natural sciences. Salient examples include: the assumption of an explanatory link between theoretical and practical progress; the hypostatisation of scientific theories as definite and consistent sets of propositions; the supposition that there is a rational methodology for the resolution of theoretical dispute; and the representation of theoretical progress as an ever more accurate and complete portrayal of the world. Do our historical speculations on the conditions for the emergence of these tenets in the *Apologia* provide bases for their critique? Great caution is needed here. Thus, while Kepler is original in his concentration on the resolution of theoretical conflict, the concept of a rational methodology has its European roots not in late renaissance astronomy, but in Greek antiquity. And though Kepler is original in the way in which he links theoretical progress with practical advance, he is hardly original, even in the field of astronomy, in postulating such a link. In Proclus' *Hypotyposis*, in Simplicius' *Commentary on De caelo*, and in Theon of Smyrna's *Expositio rerum mathematicarum*, for example, the two modes of advance are clearly associated.[5] But our historical speculations are at least obliquely relevant to the question of the validity of two central Keplerian assumptions, both of which remain widely prevalent today: the assumption that theories are representable by definite and consistent bodies of propositions, and the assumption that the truth of a theory consists in its accurate depiction of the world. These assumptions have great plausibility in the context in which they emerged, that of the rival world-systems of late-sixteenth-century astronomy. Let us consider them in turn.

Each of the three world-systems, Ptolemaic, Copernican and Tychonic, can be represented as a small consistent set of propositions about the disposition and types of motion of the heavenly bodies. A version or articulation of a world-system, obtained when it is fleshed out with detailed

---

[5] Simplicius, *Commentary on De caelo*, 488–506; partly transl. by T. L. Heath, *Greek Astronomy* (London, 1932), 67–70. Theon of Smyrna, *Expositio rerum mathematicarum ad legendum Platonem utilium*, ed. by E. Hiller (Leipzig, 1878). L. Edelstein, *The Idea of Progress in Classical Antiquity* (Baltimore, 1967), shows that, contrary to the suppositions of many earlier scholars, notions of gradual elaboration of the arts and sciences became well established in the later classical period.

planetary models, can be represented by expanding this small set of propositions to a larger set which includes in addition a specification of each planetary model. So the various levels of conflict in astronomy, both the conflict between protagonists of whole world-systems and the conflict between protagonists of rival versions of a given world-system, can be represented in the same way, as inconsistencies between sets of propositions held to be true by the protagonists.

Are commitments to world-systems and their versions typical of the affiliations which constitute the current state of the natural sciences and which give rise to scientific controversy? On at least three counts it seems that they are not.

1) World-systems and their versions had, unlike most of the current theories of the exact sciences, but a single domain of predictive application, the apparent celestial coordinates, and a single domain of reference, the heavenly bodies; further, again atypically, these were domains that could readily be specified without obvious appeal to theoretical postulates. The problems of commensurability which beset attempts to give a general account of theoretical conflict are here hardly pressing (though surely present in principle).

2) An individual astronomer's total commitment, Kepler's 'totality of the opinions of some practitioner', is in this case readily partitioned into a 'core' which expresses his commitment to a world-system and a periphery which bespeaks the version of the world-system he espouses or proposes. (That is why Kepler is able to use one and the same phrase, 'the Copernican hypotheses', to refer sometimes to Copernicus' central assumptions only and sometimes to his entire system, without engendering serious confusion.) But typical bodies of theoretical belief are not readily partitioned in this way. In general it is as absurd to ask for the essential core of a body of scientific belief as it is to ask: 'Which specific portion of the human body is *the* minimum sufficient for the maintenance of life?'

3) As Polanyi, Kuhn and Foucault have so persuasively urged, typical theoretical affiliations involve, in addition to beliefs that can readily be expressed in propositional form, a whole series of commitments and dispositions that are not readily thus specified: tacit acceptance of particular instrumental techniques, tacit attachment of relative weights to particular problems and their solutions, tacit observance of certain norms in the conduct of argument, and so on. In the history of science, conflict typically involves such inarticulate levels of commitment as well as the level of explicit theory.

However apposite the concept of a scientific theory as a definite body of propositions with a definite domain of application may have been to the astronomy of 1600, it becomes hard to allay the suspicion that it is inapposite over much of the range of the current natural sciences.

It may well be that in seeking alternative ways of representing affiliation and conflict in science there are lessons to be learned from other sixteenth-century taxonomies of belief fated to lose their hold in the face of the emergent law/theory/field hierarchy. No one could seriously hope for a revival of scholastic or Ramist classifications of the arts and sciences. But in the quest for a new taxonomy the scholastic principle of classification is suggestive in its conception of a 'science' as constituted by a distinctive range of *quaestiones* and a distinctive repertoire of methods of inquiry and presentation; and the Ramist principle of classification is suggestive in its conception of an 'art' as constituted by *praecepta* elicited from the quest for solutions to particular practical problems.

Let us now turn to the second of the central Keplerian assumptions, the assumption that the truth of a 'body of hypotheses' consists in its accurate portrayal of 'the form of the world'. In the course of their deconstructive endeavours, opponents of accounts of truth as correspondence have variously assigned historical blame for their entrenchment in the Western metaphysical tradition. Alleged culprits range from Plato's visual vocabulary of cognition through to Copernicus' decentralisation of the earth, Galileo's dehumanising mathematisation of the world and Descartes' veil of self-intimating ideas.[6] Whatever its remote metaphysical ancestry, Kepler's particular brand of realism becomes more readily comprehensible when we consider its immediate context, astronomy.

Given the uncontroversial premiss that the heavenly bodies have a definite disposition and definite motions, the view that the truth or falsity of a body of astronomical hypotheses consists in its success or failure in accurately describing that disposition and those motions is apt to seem inescapable. Further, a fully articulated world-system provides a recipe for the construction of a scale-model of the universe, a model of the type that Kepler was so keen to have constructed to illustrate the system of his *Mysterium cosmographicum*. Together these considerations conspire to make 'portrayal of the form of the world' appear not as a metaphor, but as a literal description of the relation that must hold between true hypotheses and the world. Except in restricted areas of the applied sciences it is rare to find theories which prescribe recipes for the construction of scale-models of some part of the universe. And when

---

[6] Cf., for example: M. Heidegger, *Introduction to Metaphysics* (1953), transl. by R. Manheim (New Haven, 1959), 115–96; J. Dewey, *Experience and Nature* (1925), 2nd edn (New York, 1929), 248–97; R. Rorty, *Philosophy and the Mirror of Nature* (Princeton, 1980), Ch. 1.

the entities and properties introduced by hypotheses include fields, gravitational waves, electromagnetic potentials and baryon numbers, the view that there are definite states of the world that determine the truth or falsity of theories loses much of the cogency that it has for hypotheses that deal only with material bodies and primary qualities. However compelling the metaphor of truth as portrayal may once have been in astronomy and in fields of inquiry that were in the course of the seventeenth century prosecuted under the *aegis* of the corpuscular–mechanical philosophy,[7] it seems far from inevitable over much of the range of contemporary science.

Truth as portrayal is not the only visual metaphor that sustains Kepler's robust realism. In his response to Patrizi, Kepler presents the task of the astronomer as that of distinguishing 'the true motions of the planets from those which are accidental and derived from the phantasms of the sense of sight' and of discovering 'the reasons for our deluded sight perceiving those regular motions otherwise than they have in reality been ordained'.[8] In astronomy the distinction between true and apparent motions is indeed explained by appeal to the standpoint of the observer. And in the context of Copernican astronomy the image of the attainment of truth through the discounting of the observer's point of view becomes peculiarly telling. Here we have a crucial anticipation of the prevalent modern conception of scientific truth as a representation of the world undistorted by the peculiarities of the observer and his vantage point. Yet here again the underlying metaphor is scarcely compelling outside the field of astronomy.

Of course, such reflections on one context for the historical emergence of realist imagery merely reinforce suspicions about the validity of realist interpretations of science that are already lively in twentieth-century philosophy. But the way in which Kepler's realist construal of astronomy provides him not only with a philosophical account of the status of astronomical hypotheses, but also with a mode of interpretation of the history of his subject, offers special food for thought. Current philosophy of science offers many variously compelling alternatives to the realist account of the status of scientific theories, but it remains unclear how any such alternative can serve the other primary function of Keplerian realism, that of rendering the history of science intelligible.

The conception of science as organised into a hierarchy of hypotheses, theories and disciplines may well be doomed. And there are many who regard

---

[7] On the connections between corpuscular mechanism and a conception of the activity of the mathematical scientist as one of modelling an impersonal and objective nature see R. Lenoble, *Esquisse d'une histoire de l'idée de nature* (Paris, 1969), Ch. IV. [8] *Apologia*, p. 155.

the realist construal of science, with its vision of progress as an ever more complete and undistorted representation of the world, as equally ill-fated. But threatened though they are, these themes continue to exert a pervasive influence on the philosophy and historiography of science and on the popular understanding of science. It was in Kepler's *Apologia* that such images of science, hardy if not perennial, began to take shape.

Index of names

Index of names

Buck, R., 221
Bulloch, P., viii
Bürgi, J., 10, 11, 32–3, 36, 64
Burke, P., viii, 270
Burmeister, K. H., 44, 153
Byrne, E. F., 251

Caesalpinus, *see* Cesalpino
Calcidius, *see* Chalcidius
Calepinus, 283
Calippus (Callippus, Kalippus), 46, 93, 104,
    109–15, 145, 163, 170, 172–9, 232, 267,
    268, 274
Callippus, *see* Calippus
Camerarius the Elder, Joachim, 78
Camp, Tengnagel von, *see* Tengnagel
Capella, *see* Martianus Capella
Capuano, F., 231
Cardano, G., 115, 179, 244, 246, 264
Carmody, F. J., 232
Caspar, M., ix, 9, 20
Cassirer, E., 2
Castellani, G., 271
Catherine of Medici, 266
Cellius, E., 280
Cesalpino (Caesalpinus), A., 244, 246
Chalcidius (Calcidius), 186, 201
Chevalley, 150
Chiaramonti, Scipio, 78
Cicero, 3, 4, 74–6, 130, 152, 201–3, 228,
    233, 236, 278
Ciruelo, P., 226
Clavius, C., 64, 180, 225, 231, 259, 270–1,
    275, 279
Clement of Alexandria, 46
Cohen, R. S., 221
Collimitius (Tannstetter), G., 128, 199–200
Commandino, F., 182, 184, 247, 269
Copenhaver, B. P., 261
Copernicus, 1, 3, 5, 9, 15, 18–20, 28, 33,
    36, 41, 43, 46–8, 53–8, 60–73, 89, 90,
    93–4, 97–8, 103, 116, 118–22, 124,
    127–9, 133, 140–1, 145–7, 150–5, 160,
    167, 180, 183–4, 186, 188–9, 192, 198,
    200, 206–7, 215–16, 218, 221, 225, 230,
    240, 245–9, 256, 262, 264, 266–8, 279,
    282, 287, 291–2
Cornarius, J., 201
Corraduc, R., 26
Cosentino, B., 228
Craig, J., 20
Cramer, D., 27, 30, 32

Cranz, F. E., 161, 164, 176, 260
Crates, 116, 180
Crombie, A. C., viii, 228, 255
Crosland, M., 228
Crusius, M., 280
Cunningham, A., viii
Curtius, J., 23

Daedalus, 269
D'Ailly, *see* Pierre d'Ailly
De Bèze, *see* Bèze
Debus, A. G., 152, 270
Decker, P. J., 280
De la Popelinière, *see* Popelinière
De la Ramée, *see* Ramus
Demay, P., 35
De Monantheuil, *see* Monantheuil
Demosthenes, 137
Descartes, 292
De Villalón, *see* Villalón
Dewey, J., 292
Dickreiter, M., 157, 272
Diehl, E., 267
Diels, H., 237
Diogenes Laërtius, 46, 101, 103, 106–7,
    111, 115–16, 158, 160, 163, 167–8, 173,
    178–81, 236, 261, 271
Dobrzycki, J., 28, 65, 68, 168, 180, 198,
    246
Doland, E., 2, 212, 226
Donahue, W. H., 227, 231, 246
Dreyer, J. L. E., ix, 11, 31–2, 36, 44, 187,
    193
Duhem, P., 2, 212, 225–6, 231, 238–9, 241,
    243
Duncan, A. M., 53, 84, 139
Dyck, W. von, ix, 9

Eastwood, B., 3, 5, 129, 200, 202, 204, 273
Edelstein, L., 290
Eisenstein, E. L., 281–2
Epicurus, 115, 178
Erasmus, 79, 140
Euclid, 17, 52, 59, 85, 118, 122, 134, 183,
    186, 191
Eudoxus, 46, 93, 102, 104–16, 145, 159,
    163–81, 232, 267, 268, 274

Fabri, W., 231
Fabricius, D., 27–8
Febvre, L., 237
Feldhaus, F. M., 261

Index of names

Index of names

Index of names

Index of names

Index of names